Practical Advances in Petroleum Processing
Volume 2

Errata

Practical Advances in Petroleum Processing
Edited by Hsu and Robinson

Mr. Naomasa Kondo, whose critique of the volumes appears on the back cover, is the **Managing Director** of the Cosmo Oil Company, Ltd.

Practical Advances in Petroleum Processing
Volume 2

Edited by

Chang S. Hsu

ExxonMobil Research and Engineering Company
Baton Rouge, Louisiana, USA

and

Paul R. Robinson

PQ Optimization Services
Katy, Texas, USA

 Springer

Chang S. Hsu
ExxonMobil Research and Engineering Co.
10822 N. Shoreline Avenue
Baton Rouge, Louisiana 70809
USA
chang.samuel.hsu@exxonmobil.com

Paul R. Robinson
PQ Optimization Services
3418 Clear Water Park Drive
Katy, Texas 77450
USA
paul-robinson@houston.rr.com

Cover design by Suzanne Van Duyne (Trade Design Group)

Front cover photo and back cover photo insert: Two views of the OMV plant in Schwechat, Austria, one of the most environmentally friendly refineries in the world, courtesy of OMV. Front cover insert photo: The Neste Oil plant in Porvoo, Finland includes process units for fluid catalytic cracking, hydrocracking, and oxygenate production. The plant focuses on producing high-quality, low-emission transportation fuels. Courtesy of Neste Oil.

Library of Congress Control Number: 2005925505

ISBN-10: 0-387-25811-6
ISBN-13: 978-0387-25811-9

9 8 7 6 5 4 3 2 1

springeronline.com

CONTENTS

15. Conventional Lube Basestock Manufacturing
B. E. Beasley
1. Lube Basestock Manufacturing 1
2. Key Base Stock Properties 3
   2.1 Lube Oil Feedstocks 4
3. Base Stock Composition 5
4. Typical Conventional Solvent Lube Processes 5
   4.1 Lube Vacuum Distillation Unit (VDU) or Vacuum
       Pipestill (VPS) - Viscosity and Volatility Control 6
   4.2 Solvent Extraction - Viscosity Index Control 6
   4.3 Solvent Dewaxing - Pour Point Control 6
   4.4 Hydrofinishing - Stabilization 6
   4.5 Solvent Deasphalting.. 7
   4.6 Refined Wax Production 7
5. Key Points in Typical Conventional Solvent Lube Plants 8
6. Base Stock End Uses 8
7. Lube Business Outlook 9
8. Feedstock Selection 9
   8.1 Lube Crude Selection 9
9. Lube Crude Assays 11
10. Vacuum Distillation 12
    10.1 Feed Preheat Exchangers 15
    10.2 Pipestill Furnace 15
    10.3 Tower Flash Zone 15
    10.4 Tower Wash Section 15
    10.5 Wash Oil 16
    10.6 Purpose of Pumparounds 16
    10.7 Tower Fractionation 16
    10.8 Fractionation Packing 16
    10.9 Bottoms Stripping Section 18
    10.10 Side Stream Strippers 18
    10.11 Overhead Pressure 18

        10.12 Tower Overhead Pressure with Precondensers 19
        10.12a Tower Overhead Pressure without Precondensers.......... 19
        10.13 Tower Pressure - Ejectors 19
        10.14 Factors Affecting Lube Distillate Feed 20
11. Pipestill Troubleshooting 20
        11.1   Material Balance and Viscosity Measurements 20
        11.2   Tower Pressure Survey 21
12. Solvent Extraction 22
        12.1   The Characteristics of a Good Extraction Solvent 24
        12.2   Extraction Process 25
        12.3   Extraction Process Variables 28
        12.4   Solvent Contaminants 28
        12.5   Solvent Recovery 28
                12.5.1 Raffinate Recovery 29
                12.5.2 Extract Recovery 29
        12.6   Minimizing Solvent Losses 29
                12.6.1 Recovery Section 29
                12.6.2 Other Contributors to Solvent Losses 29
13. Corrosion in NMP Plants 30
14. Extraction Analytical Tests 30
15. Dewaxing 31
16. The Role of Solvent in Dewaxing 32
17. Ketone Dewaxing Processes 34
        17.1   Incremental Ketone Dewaxing Plant 34
        17.2   DILCHILL Dewaxing 35
        17.3   Dewaxing Process Variables 37
18. Process Variable Effects 37
        18.1   Crude Source Affects Dewaxed Oil Yield 37
19. Solvent Composition 38
        19.1   Miscible and Immiscible Operations 38
        19.2   Effect of Viscosity on Filtration Rate 40
        19.3   Effect of Chilling Rate on Filtration Rate and Dewaxed
               Oil Yield 40
        19.4   Effect of Temperature Profile 41
        19.5   Effect of Solvent Dilution Ratio 41
                19.5.1 Filtration Rate 41
                19.5.2 DWO Yield 42
        19.6   Effect of Water 42
        19.7   Effect of Increased Raffinate VI 43
        19.8   Effect of Pour Point Giveaway on Product Quality and
               Dewaxed Oil Yield 43
20. Scraped Surface Equipment 43
21. Filters 45

   21.1 Filter Operation/Description 45
   21.2 Filter Media 47
  22. Cold Wash Distribution 50
  23. Wash Acceptance 52
  24. Wash Efficiency 54
  25. Filter Hot Washing 55
  26. Dewaxed Oil/Wax-Solvent Recovery 57
  27. Solvent Dehydration 58
  28. Solvent Splitter 58
  29. 2-Stage Dewaxing 59
  30. Deoiling 59
  31. Propane Dewaxing 63
   31.1 Effect of Water 66
  32. 2-Stage Propane Dewaxing 66
   32.1 Propane Deoiling 66
   32.3 Propane Filter Washing with Hot Kerosene 66
  33. Dewaxing Aids 67
  34. DWA Mechanism 68
  35. Asphalene Contamination 69
  36. Regulatory Requirements 69
  37. Glossary 70
  38. Acknowledgements 77
  39. References and Additional Readings 77

16. Selective Hydroprocessing for New Lubricant Standards
I. A. Cody
  1. Introduction 79
  2. Hydroprocessing Approaches 83
  3. Chemical Transformations 85
   3.1 Ring Conversion 85
   3.2 Paraffin Conversion 88
   3.3 Saturation 91
  4. Process Combinations 96
   4.1 Ring Conversion-Hydroisomerization-Hydrofinishing 96
   4.2 Extraction-Hydroconversion 99
  5. Next Generation Technology 101
  6. References 103

17. Synthetic Lubricant Base Stock Processes and Products
Margaret M. Wu, Suzzy C. Ho, and T. Rig Forbus
  1. Introduction 105
   1.1 Why Use Synthetic Lubricants? 106
   1.2 What is a Synthetic Base Stock? 106

1.3 A Brief Overview of Synthetic Lubricant History 107
2. Overview of Synthetic Base Stocks 108
3. Synthetic Base Stock - Chemistry, Production Process,
    Properties and Use 109
    3.1 PAO 109
        3.1.1 Chemistry for PAO Synthesis 110
        3.1.2 Manufacturing Process for PAO 112
        3.1.3 Product Properties 112
        3.1.4 Comparison of PAO with Petroleum-based Mineral
               Base Stocks 113
        3.1.5 Recent Developments - SpectraSyn Ultra as Next
               Generation PAO 116
        3.1.6 Applications 116
    3.2 Dibasic, Phthalate and Polyol Esters - Preparation,
         Properties and Applications.. 118
        3.2.1 General Chemistry and Process 118
        3.2.2 Dibasic Esters 119
        3.2.3 Polyol Esters 120
        3.2.4 Aromatic Esters 121
        3.2.5 General Properties and Applications of Ester Fluids 121
    3.3 Polyaklylene Glycols (PAG) 123
        3.3.1 Chemistry and Process 123
        3.3.2 Product Properties 124
        3.3.3 Application 125
    3.4 Other Synthetic Base Stocks 125
4. Conclusion 126
5. Acknowledgement 127
6. References 127

18. Challenges in Detergents and Dispersants for Engine Oils
James D. Burrington, John K. Pudelski, and James P. Roski
1. Introduction 131
2. Engine Oil Additive and Formulation 131
    2.1 Detergents 132
    2.2 Dispersants 134
3. Performance Chemistry 137
4. Current Dispersant and Detergent Polymer Backbones 138
5. Future Polymer Backbones 140
6. Future Trends 142
    6.1 Advanced Fluids Technology 143
    6.2 Technologies for New Product Introduction 144
    6.3 Performance Systems 146
7. Summary and Conclusions 146

    8. Acknowledgements 147
    9. References 147

19. The Chemistry of Bitumen and Heavy Oil Processing
Parviz M. Rahimi and Thomas Gentzis
    1. Introduction 149
    2. Fractional Composition of Bitumen/Heavy Oil 150
    3. Heteroatom-Containing Compounds 154
    4. Properties of Asphaltenes (Solubility, Molecular Weight,
       Aggregation) 157
       4.1 Chemical Structure of Asphaltenes 159
       4.2 Thermal Chemistry of Asphaltenes 160
    5. Chemistry of Upgrading 163
       5.1 Reaction of Feedstock Components - Simplification of
           Upgrading Chemistry 168
    6. Application of Hot Stage Microscopy in the Investigation of
       the Thermal Chemistry of Heavy Oil and Bitumen 171
       6.1 Effect of Feedstock Composition 171
       6.2 Effect of Boiling Point 172
       6.3 Effect of Additives 174
       6.4 Effect of Deasphalteming 174
    7. Stability and Compatibility 175
       7.1 Physical Treatment 175
           7.1.1 Effect of Distillation 175
           7.1.2 Effect of Addition of Diluent 177
           7.1.3 Thermal/Chemical Treatment 177
    8. References 179

20. Mechanistic Kinetic Modeling of Heavy Paraffin Hydrocracking
Michael T. Klein and Gang Hou
    1. Introduction 187
    2. Approach and Overview 188
    3. Model Development 191
       3.1 Reaction Mechanism 191
       3.2 Reaction Families 192
           3.2.1 Dehydrogenation/Hydrogenation 192
           3.2.2 Protonation/Deprotonation 192
           3.2.3 Hydride and Methyl Shift 194
           3.2.4 PCP Isomerization 194
           3.2.5 β-Scission 194
           3.2.6 Inhibition Reaction 195
       3.3 Automated Model Building 196
       3.4 Kinetics: Quantitative Structure Reactivity Correlations 198

3.5 The C_{16} Paraffin Hydrocracking Model Dignostics 198
4. Model Results and Validation 199
5. Extension to C_{80} Model 201
6. Summary and Conclusion 202
7. References 203

21. Modeling of Reaction Kinetics for Petroleum Fractions
Teh C. Ho
1. Introduction 205
2. Overview 206
   2.1 Partition-Based Lumping 206
   2.2 Total Lumping 207
   2.3 Reaction Network/Mechanism Reduction 207
   2.4 Mathematical Approaches to Dimension Reduction 208
3. Partition Based Lumping 209
   3.1 Top-down Approach 209
   3.2 Bottom-up Approach 211
       3.2.1 Mechanistic Modeling 212
       3.2.2 Pathways Modeling 215
       3.2.3 Quantitative Correlations 217
       3.2.4 Carbon Center Approach 218
       3.2.5 Lumping via Stochastic Assembly 218
4. Mathematical Reduction of System Dimension 220
   4.1 Sensitivity Analysis 220
   4.2 Time Scale Separation 221
   4.3 Projective Transformation 221
       4.3.1 First Order Reactions 221
       4.3.2 Non-Linear Systems 223
       4.3.3 Chemometrics 224
   4.4 Other Methods 224
5. Total Lumping: Overall Kinetics 224
   5.1 Continuum Approximation 225
       5.1.1 Fully Characterized First Order Reaction Mixtures 226
       5.1.2 Practical Implications 227
       5.1.3 Partially Characterized First Order Reaction
              Mixtures 228
              5.1.3.1 Plug Flow Reactor 229
              5.1.3.2 CSTR 230
              5.1.3.3 Diffusional Falsification of Overall
                       Kinetics 231
              5.1.3.4 Validity and Limitations of Continuum
                       Approach 232
              5.1.3.5 First Order Reversible Reactions 232
              5.1.3.6 Independent nth Order Kinetics 233

5.1.3.7 Uniformly Coupled Kinetics 233
    5.1.4 Upper and Lower Bounds 234
    5.1.5 One Parameter Model 235
    5.1.6 Intraparticle Diffusion 236
    5.1.7 Temperature Effects 237
    5.1.8 Selectivity of Cracking Reactions 237
    5.1.9 Reaction Networks 238
  5.2 Discrete Approach: Nonuniformly Coupled Kinetics 238
    5.2.1 Homologous Systems 239
    5.2.2 Long-Time Behavior 239
  6. Concluding Remarks 241
  7. References 242

22. Advanced Process Control
Paul R. Robinson and Dennis Cima
  1. Introduction 247
  2. Useful Definitions 247
  3. Overview of Economics 249
  4. Source of Benefits 250
  5. Implementation 253
  6. Costs 254
  7. References 255

23. Refinery-Wide Optimization with Rigorous Models
Dale R. Mudt, Clifford C. Pedersen, Maurice D. Jett, Sriganesh Karur,
Blaine McIntyre, and Paul R. Robinson
  1. Introduction 257
  2. Overview of Suncor 257
  3. Refinery-Wide Optimization (RWO) 259
  4. Rigorous Models for Clean Fuels 261
    4.1 Feedstock and Product Characterization 262
    4.2 Aspen FCC Overview 262
    4.3 Aspen Hydrocracker 266
      4.3.1 Reaction Pathways 269
      4.3.2 Catalyst Deactivation Model 271
      4.3.3 AHYC Model Fidelity 272
    4.4 Clean Fuels Planning 272
      4.4.1 Hydrogen Requirements for Deep Desulfurization ... 272
      4.4.2 Effects of Hydrotreating on FCC Performance 274
  5. Conclusions 278
  6. Acknowledgements 278
  7. References 278

24. Modeling Hydrogen Synthesis with Rigorous Kinetics as Part of Plant-Wide Optimization
Milo D. Meixell, Jr.
    1. Introduction 281
    2. Steam Reforming Kinetics 283
        2.1 Methane Steam Reforming Kinetic Relationships 283
        2.2 Naphtha Steam Reforming Kinetic Relationships 286
        2.3 Coking 292
        2.4 Catalyst Poisoning 294
    3. Heat Transfer Rates and Heat Balances 295
        3.1 Firebox to Catalyst Tube 297
        3.2 Conduction Across Tube Wall 299
        3.3 Fouling Resistance 299
        3.4 Inside Tube to Bulk Fluid 300
        3.5 Bulk Fluid to Catalyst Pellet 300
        3.6 Within the Catalyst Pellet 301
        3.7 Convection Section 301
        3.8 Fuel and Combustion Air System 302
        3.9 Heat Losses 302
    4. Pressure Drop 302
        4.1 Secondary Reformer Reactions and Heat Effects 303
        4.2 Model Validation 304
            4.2.1 Validation Case 1 (Naphtha Feed Parameter Case)...305
            4.2.2 Validation Case 1a (Naphtha Feed Simulate Case)...307
            4.2.3 Validation Case 2 (Butane Feed Parameter Case).....307
            4.2.4 Validation Case 3 (Primary and Secondary Reformer Butane Feed Reconcile Case) 309
    5. References 311
    Appendix A Simulation Results 313
    Primary Reformer 313
    Adiabatic Pre-Reformer 317
    Oxo-Alcohol Synthesis Gas Steam Reformer 317
    Appendix B Case Study of Effects of Catalyst Activity in a Primary Reformer 318

25. Hydrogen Production and Supply: Meeting Refiners' Growing Needs
M. Andrew Crews and B. Gregory Shumake
    1. Introduction 323
    2. Thermodynamics of Hydrogen 324
    3. Technologies for Producing Hydrogen 326
        3.1 Steam Methane Reforming (SMR) Technologies 326
            3.1.1 Maximum Steam Export 326

3.1.2 Limited Steam Export 327
3.1.3 Steam vs. Fuel 328
3.1.4 Minimum Export Steam 329
3.2 Oxygen Based Technologies 330
3.2.1 SMR/O_2R 330
3.2.2 ATR 331
3.2.3 POX 332
3.2.4 Products 332
3.2.5 H_2/CO Ratio 332
3.2.6 Natural Ratio Range 333
3.2.7 CO_2 Recycle 333
3.2.8 Import CO_2 335
3.2.9 Membrane 335
3.2.10 Cold Box 335
3.2.11 Steam 335
3.2.12 Shift Converter 335
3.2.13 Other Considerations 335
3.3 Technology Comparison 336
3.3.1 Process Parameters 337
3.3.2 Export Steam 339
3.3.3 Economic Considerations 340
3.3.4 Oxygen Availability 340
3.3.5 Hydrocarbon Feedstock 340
3.3.6 H_2/CO Ratio 340
3.3.7 Natural Gas Price 340
3.3.8 Capital Cost 340
3.3.9 Conclusions 341
3.4 Hydrogen Purification 341
3.4.1 Old Style 341
3.4.2 Modern 342
4. Design Parameters for SMR's 343
4.1 Function 343
4.2 Feedstocks 344
4.3 Fuels 344
4.4 Design 344
4.5 Pressure 345
4.6 Exit Temperature 346
4.7 Inlet Temperature 346
4.8 Steam/Carbon Ratio 347
4.9 Heat Flux 347
4.10 Pressure Drop 348
4.11 Catalyst 348
4.12 Tubes 349

4.13 Burners 349
4.14 Flow Distribution 350
4.15 Heat Recovery 350
5. Environmental Issues 351
   5.1 Flue Gas Emissions 351
   5.2 Process Condensate (Methanol and Ammonia) 352
   5.3 Wastewater 354
6. Monitoring Plant Performance 355
7. Plant Performance Improvements 357
8. Economics of Hydrogen Production 359
   8.1 Overall Hydrogen Production Cost 361
   8.2 Overall Production Cost Comparison 361
   8.3 Evaluation Basis 362
   8.4 Utilities 362
   8.5 Capital Cost 363
   8.6 "Life of the Plant" Economics 363
   8.7 Sensitivity to Economic Variables 364
   8.8 Feed and Fuel Prices 365
   8.9 Export Steam Credit 366
9. Conclusion 366
10. Additional Reading 367

26. Hydrogen: Under New Management
Nick Hallale, Ian Moore, and Dennis Vauk

1. Introduction 371
2. Assets and Liabilities 372
3. It's All About Balance 373
4. Put Needs Ahead of Wants 375
5. Beyond Pinch 382
   5.1 Multi-Component Methodology 383
   5.2 Hydrogen Network Optimization 384
6. You Don't Get Rich by Saving 388
7. Conclusions 391
8. References 392

27. Improving Refinery Planning Through Better Crude Quality Control
J. L. Peña-Díez

1. Introduction 393
2. Crude Oil Quality Control 394
3. New Technologies in Crude Oil Assay Evaluation 396
   3.1 Analytical Methods 397
   3.2 Chemometric Methods 397

3.3 Other Alternatives 398
4. Crude Assay Prediction Tool (CAPT) 398
   4.1 Model Description 398
   4.2 Potential Applications 402
   4.3 Model Results 403
5. Concluding Remarks 405
6. References 406

Index 409

Chapter 15

CONVENTIONAL LUBE BASESTOCK MANUFACTURING

B. E. Beasley, P. E.
ExxonMobil Research & Engineering Co.
Process Research Lab
Baton Rouge, LA 70821

This chapter reviews the basic unit processes in modern conventional lube manufacturing. As this is a large subject area, this chapter will focus on giving the reader an overview of the major processes most frequently found in the lube manufacturing plant. It will not cover all technologies or processes, nor will it discuss detailed plant design and operation as this would easily require another book. The reader should come away with a general understanding of the conventional lube manufacturing process and key factors affecting the unit processes.

1.   LUBE BASE STOCK MANUFACTURING

Lubes and specialties include a number of products that have a variety of end uses. Some end uses include:
- Automotive: engine oils, automatic transmission fluids (ATF's), gear oils.
- Industrial: machine oils, greases, electrical oils, gas turbine oils.
- Medicinal: food grade oils for ingestion, lining of food containers, baby oils.
- Specialty: food grade waxes, waxes for candles, fire logs, cardboard.

Lube manufacturing is complex and involves several processing steps. Crude is distilled and the bottoms, atmospheric resid, is sent to a vacuum distillation unit (VDU) sometimes called a vacuum pipestill (VPS) for further fractionation. Vacuum fractionation is used to separate the atmospheric resid into several feed streams or distillates. Conventional solvent processing uses selected solvents in physical processes to remove undesirable molecules (asphaltenes, aromatics, *n*-paraffins). Hydroprocessing is used to convert or remove the trace undesirables such as nitrogen, sulfur and multi-ring aromatics or to enhance base stock properties to make specialty, high quality products.

The manufacture of lubes and specialty products makes a significant contribution to refining profitability even though volumes are relatively small.

The business drivers of the lube business are for increased production to reduce per barrel costs, to reduce operating expense (OPEX) and for higher quality products to meet ever-increasing product quality standards.

Refiners produce **base stocks** or **base oils** and lube oil blenders produce **finished oils or formulated products**. See the American Petroleum Institute's API-1509.
- **base stocks** are products produced from the lube refinery without any additives in the oil
- **base oils** are blends of one or more base stock
- **finished oils** or **formulated** products are blends of baseoil with special additives

Lube Base stocks are given various names. Some of the common names include:
1. Neutrals - from virgin distillates ex. 100N, 150N, 600N, etc
2. Bright stock - from Deasphalted Oil (DAO), ex BS150
3. Grades - ex. SAE 5, 10, 30, etc.; ISO 22, 32, etc.

The most common name is **neutral** (N) which was derived in the days when the lube distillates were acid treated (sulfuric acid) followed by clay filtration. After clay treating the oil was acid free or **neutral.** The viscosity number in this example, 150 N, is the approximate viscosity of the base stock (Note: the ASTM viscosity classification refers to an industrial oil grade system, not the base stock viscosity system) expressed in Saybolt Seconds Universal (SSU) at 100°F.

Bright stock is a heavy lube grade that is made from deasphalted resid. The name refers to the "bright" appearance of the product as compared to the resid feed. Bright stocks are very viscous; a typical bright stock, BS150, has a viscosity of 150 SSU at 210°F.

Grades may refer to the actual viscosity. For example, ISO (International Standards Organizations) industrial oil grades = cSt at 40°C or the reference may be arbitrary such as SAE (Society of Automotive Engineers) engine oil grades.

There are many other grade names that are used to differentiate special products. These products may have special qualities that may make them very profitable even though they tend to be lower volume products.

Base stocks are assigned to five categories (see API-1509 Appendix E).
- Group I base stocks contain less than 90 percent saturates and/or greater than 0.03 percent sulfur and have a viscosity index greater than or equal to 80 and less than 120.

- Group II base stocks contain greater than or equal to 90 percent saturates and less than or equal to 0.03 percent sulfur and have a viscosity index greater than or equal to 80 and less than 120.
- Group III base stocks contain greater than or equal to 90 percent saturates and less than or equal to 0.03 percent sulfur and have a viscosity index greater than or equal to 120.
- Group IV base stocks are poly-alpha-olefins (PAO).
- Group V base stocks include all other base stocks not included in Groups I-IV.

2. KEY BASE STOCK PROPERTIES

Viscosity is a key lube oil property and is a measure of the fluidity of the oil. There are two measures of viscosity commonly used; **kinematic** and **dynamic**. The **kinematic** viscosity is flow due to gravity and ranges from approximately 3 to 20 cSt (centistokes) for solvent neutrals and about 30-34 cSt at 100°C for Bright stock. The **dynamic** viscosity is flow due to applied mechanical stress and is used to measure low temperature fluidity. Brookfield viscosity for automobile transmission fluids (ATF's) at -40°C and cold cranking simulator (CCS) viscosity for engine oils at -25°C are examples of dynamic viscosity measurements.

Lube oil **volatility** is a measure of oil loss due to evaporation. Noack volatility measures the actual evaporative loss which is grade dependent, and a function of molecular composition and the efficiency of the distillation step. The volatility is generally lower for higher viscosity and higher VI base stocks. The gas chromatographic distillation (GCD) can be used to measure the front end of the boiling point curve and may be used as an indication of volatility, e.g. 10% off at 375°C.

Viscosity index or **VI** is based on an arbitrary scale that is used to measure the change in viscosity as a function of temperature. The scale was first developed in 1928 and was based on the "best" and "worst" known lubes at the time. The best paraffinic lube was assigned a value of VI = 100 and the worst naphthenic was assigned a VI = 0. The quality of Base stock has been improved dramatically since 1928 with the VI of high quality Base stock in the 140+ range.

Pour point is the temperature at which the fluid ceases to pour and is nearly a solid. Typically the pour point ranges from -6 to -24°C for heavy to light neutrals.

The **cloud point** is the temperature at which wax crystals first appear.

Saturates, aromatics, naphthenes are measures of these molecular types present in the Base stock.

Color (appearance) and **Stability** are the measure of color and change in presence of light or heat.

Conradson carbon (CCR) or **Micro-Carbon Residue (MCR)** is a measure of the ash left after flame burning.

2.1   Lube Oil Feedstocks

Lube plant feedstocks are taken from the bottom of the crude barrel (see Figure 1).

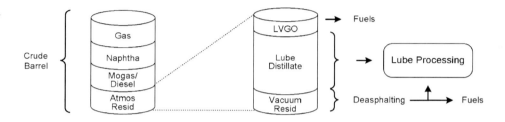

Figure 1. Lube Plant Feedstocks Are Taken from the Bottom of the Crude Barrel.

Lube crudes are generally paraffinic or naphthenic in composition. A paraffinic crude is characterized by a higher wax content. West Texas and Arab Light are good quality paraffinic crudes. Naphthenic crudes are characterized by their low wax content and they make base stocks with low viscosity index, e.g. Venezuelan and Californian.

In conventional solvent lubes the atmospheric resid (bottoms from the crude distillation tower) is upgraded to lube products through the following processes:
- vacuum distillation
- solvent extraction (N-methyl-2-pyrrolidone (NMP), furfural, phenol)
- solvent dewaxing (methyl ethyl ketone (MEK)/methyl isobutyl ketone (MIBK), MEK/toluene, propane)
- hydrofinishing (may be integrated with extraction)
- propane deasphalting
- hydroprocessing for higher quality

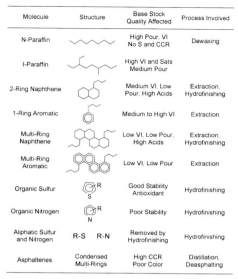

Molecule	Structure	Base Stock Quality Affected	Process Involved
N-Paraffin		High Pour, VI No S and CCR	Dewaxing
I-Paraffin		High VI and Sats Medium Pour	
2-Ring Naphthene		Medium VI, Low Pour, High Acids	Extraction, Hydrofinishing
1-Ring Aromatic		Medium to High VI	Extraction
Multi-Ring Naphthene		Low VI, Low Pour, High Acids	Extraction, Hydrofinishing
Multi-Ring Aromatic		Low VI, Low Pour	Extraction
Organic Sulfur		Good Stability Antioxidant	Hydrofinishing
Organic Nitrogen		Poor Stability	Hydrofinishing
Aliphatic Sulfur and Nitrogen	R-S R-N	Removed by Hydrofinishing	Hydrofinishing
Asphaltenes	Condensed Multi-Rings	High CCR Poor Color	Distillation, Deasphalting

Figure 2. Lube Oil Molecules Contribute to Lube Oil Properties

3.   BASE STOCK COMPOSITION

Lube oil is produced from a wide variety of crude oil molecules. The molecular types and effect on lube oil quality is summarized below along with the lube process that acts on them.

4.   TYPICAL CONVENTIONAL SOLVENT LUBE PROCESSES

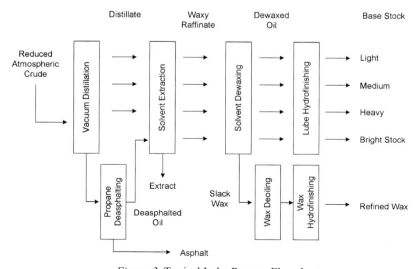

Figure 3. Typical Lube Process Flowchart

4.1 · Lube Vacuum Distillation Unit (VDU) or Vacuum Pipestill (VPS) - Viscosity and Volatility Control

The VPS is generally the first process unit. The VPS's goal is to fractionate the atmospheric resid or reduced crude so that the base stock will have the desired viscosity. The fractionation also controls the volatility and the flash point. The boiling point separation is accomplished by using high efficiency distillation/fractionation hardware. Secondary effects include asphalt segregation in the Vacuum Resid from the VPS (potential by-product), reduction in Conradson carbon and color improvement.

4.2   Solvent Extraction - Viscosity Index Control

Extraction is typically the second process although this is not always the case. The primary goal of extraction is to remove aromatics and polar molecules. This is accomplished through solvent extraction of the distillate using NMP, furfural, or phenol. By removing aromatics, the VI is raised. Secondary effects of extraction include reduction in the refractive index, reduction in density, reduction in Conradson carbon and improvement in color, color stability and oxidative stability.

4.3   Solvent Dewaxing - Pour Point Control

Conventional solvent dewaxing is an energy- and cost-intensive process, and you therefore want to operate on the fewest number of molecules consistent with a high product yield. Therefore you do extraction first to remove the non-lubes molecules and you do dewaxing last on the raffinate. But you optimize the total OPEX per volume through the entire process. If somehow it made better economics to do the DWX first, even though it is expensive, you would do it that way.

The primary goal of solvent dewaxing is to make the pour and cloud point requirements. This is accomplished by paraffin separation by solubility of non-paraffins in cold solvent, fractional crystallization, and filtering the solid paraffins from the slurry. This may be done in "ketone" units which use MEK, MEK/MIBK, MEK/Toluene solvents or in propane units which use liquefied propane as the solvent. Secondary effects include viscosity increase, density increase, sulfur increase, and reduction in VI.

4.4   Hydrofinishing - Stabilization

Hydrofinishing follows extraction or dewaxing. The primary goal is to improve appearance (color, color stability, and oxidative stability) and to remove impurities such as the solvent, nitrogen, acids and sulfur to meet the required specification. This is accomplished by hydrogen saturation and chain breakage that uses hydrogen at mild pressures and temperatures in the presence of a catalyst. Secondary effects include slight improvement in VI, slight improvement in saturates, reduction in viscosity, lower acidity, and reduction in Conradson Carbon.

4.5   Solvent Deasphalting

When used, it is always ahead of extraction. The primary goal is to remove asphaltenes, which could be a possible byproduct and to make the viscosity specification that is required. This is accomplished by asphaltenes separation by solubility of non-asphaltenes in a solvent and precipitation of asphaltenes using e.g. propane as a solvent. Secondary effects include Conradson Carbon reduced, metals reduced, saturates increased, viscosity index increased, and color improved.

4.6   Refined Wax Production

Wax deoiling and hydrofinishing follows the dewaxing unit. The primary goal is to reduce the oil content of the wax and to meet melting point and needle penetration requirements. This is accomplished by soft wax solubility and physical separation in the deoiling equipment. Hydrofinishing's primary goal is to saturate residual oxygenates and aromatics. Secondary effects include removal of impurities and color improvement.

A summary of the main and secondary lube qualities, by processing step, is shown in Figure 4.

	Distillation	Deasphalting	Extraction	Dewaxing	Hydrofinishing
Feed Stock	Atmospheric Resid	Vacuum Resid	Distillate or Deasphalted Oil	Raffinate	Raffinate or Dewaxed Oil
Lube Product	Distillate	Deasphalted Oil	Raffinate	Dewaxed Oil	H/F Oil
By-Products	Vacuum Resid	Asphalt	Extract	Wax	Naphtha
Main Lube Quality Improved	–	Con Carbon ↓	VI ↑	Pour Point ↓	Color
Other Quality Changes					
Density (SG)	Varies	↓	↓	↑	↓
Gravity (API)	Varies	↑	↑	↓	↑
Viscosity	Varies	↓	↓	↑	↓
Viscosity Index (VI)	–	↑	"↑"	↓	↑ (slightly)
Color (ASTM)	↓	↓	↓	–	"↓"
Pour Point	–	–	–	"↓"	–
Cloud Point	–	–	–	↓	–
Conradson Carbon	↓	"↓"	↓	↑	↓
Sulfur	–	↓	↓	↑	↓
Nitrogen	–	↓	↓	↑	↓
Saturates	–	↑	↑	↓	↑ (slightly)
Flash Point	↓	–	–	–	–
Refractive Index	–	↓	↓	↑	↓

Figure 4. Summary of Lube Process Impact on Product Quality

5.   KEY POINTS IN TYPICAL CONVENTIONAL SOLVENT LUBE PLANTS

- Majority of operations are "blocked operation" instead of "in-step". Blocked operation requires intermediate tankage between units and allows the optimum operation of each unit on each viscosity grade.
- The dewaxer is the most expensive unit to build, has the highest operating costs and is the most complex to operate. Therefore, you want to operate on the fewest number of molecules consistent with a high product yield.
- Bright stock is the most expensive conventional lube to manufacture and requires the addition of a deasphalting unit.
- Integration of extraction and hydrofinishing units saves energy, and the elimination of a hydrofiner furnace saves capital. However, this arrangement is less flexible than a standalone hydrotreater.

There are exceptions to the general flow. Some plants that process extremely high wax content crudes position dewaxing after vacuum distillation. Some plants position high-pressure hydrotreating upstream of dewaxing.

Hydroprocessed lubes will be covered in other chapters and includes:
- Lubes hydrocracking
- Wax isomerization
- White Oils hydrogenation
- Catalytic dewaxing

Other processes include:
- Clay Contacting or Acid Treating, both are older stabilization processes
- Duo-Sol, a process that combines propane deasphalting and solvent extraction

6.   BASE STOCK END USES

Conventional lube Base stocks are formulated into a multitude of finished products:
- Engine Oils
- Transmission Fluids
- Gear Oils
- Turbine Oils
- Hydraulic Oils

Metal Working (Cutting) Oils
- Greases
- Paper Machine Oils

Specialty products may include:

- White Oils: Foods, Pharmaceuticals, Cosmetics.
- Agricultural Oils: Orchard Spray Base Oils.
- Electrical Oils: Electrical Transformers (Heat Transfer).

7. LUBE BUSINESS OUTLOOK

Lubricant base stocks are produced in approximately 170 refineries worldwide that have a total capacity of over 900 kBD. The average capacity utilization is somewhere around 80%, to meet an industry demand of just over 700 kBD. About 75% of the total production is solvent-based refining, most making Group I quality base stocks. However, almost all new capacity is hydroprocessing-based, making Group II or Group III base stocks.

The lubricant market is roughly equally split between transportation lubricants (engine crankcase oils, transmission fluids, greases, etc.) and industrial process oils. Demand is growing at an average rate of only 1% / year, as robust growth in the developing economies (e.g. China, India) is being partially offset by declining demand in the mature markets (N. America, Europe) due to extended drain intervals for the higher quality engine oils. Engine builders tend to drive the transportation lubricant quality, as economic and environmental drivers push engine oils towards better oxidation stability, better low temperature properties, lower volatility, and lower viscosity. These desired characteristics drive formulators to favor hydroprocessed base stocks which have higher VI. However many other applications, such as most industrial and process oils, as well as older engine oils, still favor the characteristics of solvent-refined Group I base oils, which are expected to continue to play an important role in meeting the world's lubricant needs for year's to come.

8. FEEDSTOCK SELECTION

Crude selection is extremely important for the profitable production of lubes. Only a limited number of crudes contain a sufficient quantity of lubes quality molecules. Downstream unit operability is affected by crude selection, as are rates and yields. Typically, manufacturers would prefer operating at maximum throughput, thereby spreading costs over a larger volume. Poor crude selection can result in downstream bottlenecks reducing overall throughput.

8.1 Lube Crude Selection

Lube oil manufacturers may have a lube crude approval (LCA) process to assess the opportunity to manufacture Base stocks from crudes available in the marketplace. The LCA process defines the detailed steps to qualify a new crude for purchase by the refinery to make base stocks and / or wax products.

The first step entails identifying economically attractive crudes. These crudes are characterized, or assayed, to quantify their lube yield and qualities. The assay process includes subjecting a small sample of the crude to an atmospheric distillation, vacuum distillation, extraction and dewaxing to produce the desired base stock products. This information enables the manufacturer, through the use of modeling techniques, to predict the process response of the crude of interest to make the required Base stock products. These modeling techniques may also allow the manufacturer to investigate process variables and operating optimization for distillation, extraction, and dewaxing to assess manufacturing flexibility. Not all crudes are acceptable for Base stock manufacturing as yields may be too low or Base stock products may not meet requirements.

With an acceptable assessment of the new crude, the refiner may elect to validate the crude for Base stock manufacture. This may entail running a plant test to make Base stock products from the new crude. The products made from the plant test are typically blended into formulated oils and subjected to testing to demonstrate acceptable product performance.

Results of the plant test are reviewed with a focus on lube plant manufacturing performance and Base stock product quality to determine if the new crude can be approved for Base stock manufacture.

1) Lube plant manufacturing performance - actual rate, yield and operability. The actual operating conditions are compared to the predicted processing conditions to assess if the new crude processed as expected.

2) Base stock product quality - Plant testing protocol should be defined to ensure base stock products meet acceptable quality specifications. Care should be taken to avoid making base stocks that may not be representative of how the crude will typically be processed to make base stocks. The range of acceptable base stock qualities should be defined by the test protocol. Plant test product disposition may need to be defined as part of the plant test. Options may include blending the plant test products to dilute the new crude component or quarantining the product tank until product testing has been completed. Product testing failure will prevent the crude from being approved.

Results from the manufacturing test will determine if the crude will be accepted. The certification test must have been acceptable and the crude processed as expected. There must not be any evidence that Base stock quality is unacceptable. If the above is completed successfully, the crude may be approved and added to the manufacturer's list of approved crudes.

The approval protocol may require periodic re-evaluation of the crude in recognition that the crude may change.

9. LUBE CRUDE ASSAYS

A lube crude assay is a laboratory process to measure the lube processing response from crude to base oil. It is an important step in a manufacturer's lube crude selection. A crude assay will include process yields for desired base oils at their quality specifications. The manufacturer can use the assay data to predict the process response for their refinery and to assess the desirability of purchasing particular crude. The assay results may be used to calculate the impact on profitability.

Key steps to complete a typical lube assay include:

- Secure a representative sample of the crude. This may best be achieved by collecting a sample at a load port.
- Fractionate the crude into discreet components first to separate the light, non-lubes boiling material. The bottoms are then sent to a high vacuum distillation. The distillation produces several distillate blends for extraction. The distillates produced are sufficient to cover the Base stock viscosity range.
- The distillates are then extracted using a lab pilot unit and the preferred extraction solvent (ex. furfural, NMP or phenol). Waxy raffinates are produced from the extraction.
- The waxy raffinates are then dewaxed using solvents of interest (MEK, MEK/MIBK, MEK/toluene, etc.) to produce a dewaxed oil and a slack wax.
- The dewaxed oils will be characterized to quantify their properties and yields. This will enable an economic assessment to be made with respect to the crude's lube potential.

There are several lube assay objectives in distillation. One is to relate key lube properties such as viscosity, sulfur, density, refractive index, etc. to boiling point. A second is to determine the yield of material boiling in the lube range and a third is to determine the yield of material boiling in the asphalt range.

The objectives of the lube assay extraction are to generate data, which will determine the ability of the crude to produce base oil capable of meeting the base oil specifications. Obviously this is of great importance in the selection of lube crudes for the plant. Key lube oil qualities related to process response are determined over the full lube oil viscosity range. Yields are used in manufacturing economic calculations. All crudes were not created equal, although there may be similarities in a given region. For example, Middle Eastern crudes may contain high sulfur, high aromatics and high iso-paraffins while North Sea crude may be low in sulfur, contain high saturates, and have a medium wax content. There are always outliers in every region. Crude production from a given "field" may change over time. If so, this may require that the assay is repeated to update the crude's relevant information to remain current.

In summary, the lube assay will characterize the potential of a crude to produce a specific Base stock (viscosity, viscosity index, saturates, wax, sulfur, basic nitrogen, etc) and to determine the expected yields from distillation, extraction and dewaxing.

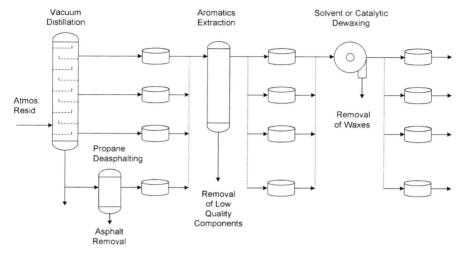

Figure 5. Lube Manufacturing Scheme

10.   VACUUM DISTILLATION

Vacuum distillation is used to fractionate the heavier molecules in the crude. In the majority of plants it is the beginning point for lube manufacturing. Vacuum distillation is applied to avoid the high temperature fractionation, which would lead to undesirable coking and loss of lube oil yield.

Crude oil was first distilled in batch distillation, like a lab technique, beginning in the 1850s. Advancements were made by increasing the size of the batch vessel. A continuous process was developed by using a series of batch stills - called a battery. The first continuous pipestill appeared in the 1920s and the "modern" pipestill came on the scene in the 1930s. A typical lube vacuum distillation unit is shown below (Fig 6).

Vacuum distillation equipment is often referred to as the vacuum distillation unit (VDU) or vacuum pipestill (VPS).

The objective of the VPS is to achieve on test product quality for viscosity, volatility and flash point. Maximizing the yield of the most valuable products requires using the right cutting schemes. Steady control will minimize distillate variability. Good fractionation makes for sharp separation which is beneficial to good performance in downstream equipment. VPS per barrel costs can be minimized by operating the VPS at high capacity with long run lengths and making the best use possible of utilities and chemicals per barrel.

Fractionation is the separation by boiling point of light and heavy components in the distillation tower achieved by intimate contact between hot rising vapor with the cooler falling liquid. The hot vapor strips the lighter components from the liquid and the cold liquid condenses heavier components from the vapor. Stripping requires heat in order to vaporize the lighter components and the condensation of heavy components releases heat. Good contact between the phases is essential to achieve maximum fractionation efficiency. In the presence of vapor the liquid may be carried upward in the form of a mist, foam or spray and may contaminate the desired distillates with heavy components. The contamination is known as entrainment and should be avoided.

The concept of a "theoretical stage" is a useful one and refers to the length of the VPS section required for the vapor and the liquid to reach equilibrium. The sharpness of the separation between adjacent streams may be measured in theoretical stages or minimum number of theoretical stages (Nm) which represents the number of theoretical stages at infinite reflux to effect the separations.

Sidestreams from the distillation tower are typically named from the top (lighter products) down to the bottom (heavier streams). Typical atmospheric and vacuum sidestream nomenclature for a typical atmospheric and vacuum tower is shown below (typical boiling point range of fractionated stream).

Table 1. Typical Distillation Tower Sidestream Names

Name	Description
AOH	atmospheric overhead (-30 to200 °C)
A1SS	atmospheric 1, or first, sidestream (150 to 210 °C)
A2SS	atmospheric 2, or second, sidestream (175 to 300 °C)
A3SS	atmospheric 3, or third, sidestream (190 to 400 °C)
LVGO	light vacuum gas oil, vacuum tower overhead (200 to 400 °C)
V1SS	vacuum 1, or first, sidestream (350 to 425 °C)
V2SS	vacuum 2, or second, sidestream (390 to 600 °C)
V3SS	vacuum 3, or third, sidestream (450 to 620 °C)
VRES	vacuum resid stream (500 to >900 °C)

Cut points are used to describe the pipestill product. Volume cut points are the cumulative yield on the crude and are expressed as a liquid volume percent of product. Temperature cut points are the boiling points that correspond to the volume cut point.

A key objective of the VPS is to set the viscosity of the final product. This basic product property is set in the distillation by setting the cut points of the product streams. Volatility, another key product specification is the amount of material removed at a certain temperature and is controlled in the distillation by cut point targets and front-end fractionation. It affects engine oil thickening and evaporative losses. The flash point is the ignition temperature above the liquid surface and affects engine oil thickening. It is a safety concern for storage of liquid product. Cut point targets and fractionation in the main tower and stripper are used to control the product flash point.

Distillate yields are affected by crude type, product viscosity and volatility specifications, the distillation tower cutting scheme, fractionation efficiency and the theoretical stages between the sidestreams. Poor fractionation efficiency can be caused by operating at feed rates above equipment design. If the feed is significantly lighter than the tower is designed to handle, fractionation efficiency may suffer. Mechanical damage such as dislodged or damaged internals, leaks or plugs in spray headers used to distribute liquids in the tower, leaking trays, etc. will degrade fractionation efficiency. Insufficient wash oil or reflux in the tower contributes to poor separation. Poor distribution of liquid or vapor reduces contact and leads to poor fractionation efficiency. Pumparounds are used to remove heat from the tower and to adjust the vapor-liquid flow in the tower. When pumparound duties get out of balance, fractionation efficiency is reduced. This can be because of reflux rates being above or below design specifications and also if flooding or entrainment is occurring in overloaded sections of the tower.

Poor tower fractionation efficiency may adversely affect downstream lube operations. Insufficient separation of light grade front ends may result in light oil carryover in extraction, increasing solvent ratio requirements, possibly reducing throughput and increasing energy usage. Dewaxing throughput and yields are adversely affected across all grades by the presence of "tail ends".

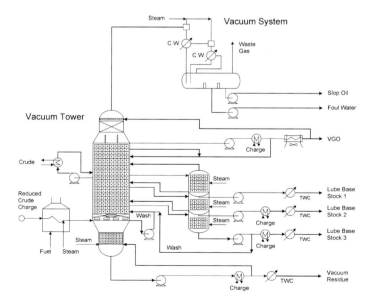

Figure 6. Typical Lube Vacuum Tower Design

10.1   Feed Preheat Exchangers

Feed preheat exchangers are used to recover heat from sidestreams and pumparounds and to make the overall distillation more energy efficient. Preheating minimizes the loss of heat to the atmosphere or cooling water. Heat integration reduces fuel consumption in the furnace and steam may be generated for stripping in the vacuum unit. Atmospheric and vacuum units may be heat integrated.

10.2   Pipestill Furnace

The furnace partially vaporizes the feed to the tower. A typical furnace has multiple parallel passes and the outlets are combined as feed to the distillation tower. Steam may be injected into the vacuum furnace coil to increase vaporization of feed at a lower temperature and to reduce the residence time. The vacuum cut points are set by the extent of the vaporization in the flash zone where temperatures may range from 390-420°C. Furnace firing is controlled to achieve the desired vacuum cut point.

10.3   Tower Flash Zone

The flash zone is a large area in the tower that allows for the disengagement of liquid and vapor. The height of the zone affects the separation. The flash zone is designed to facilitate disengagement. Internals in this section consist of annular rings or vapor horns and collector rings for the bottoms stripping inlet.

10.4   Tower Wash Section

The wash section cleans entrained liquids from the flash zone vapor phase. Vapor in excess of the amount needed to meet distillate requirements is referred to as overflash. The wash section condenses the overflash. It also provides some fractionation between the heavy lube sidestream and the vacuum resid stream.

The wash zone may include a Glitsch grid or random packing. An open structure gives a low-pressure drop while providing a high surface area to capture and retain resid. The resid is washed away by the wash oil that is applied through a spray header. The overflash, or spent wash, may be 40-50% resid and is removed and either sent to tankage as another sidestream or returned to the bottoms section for stripping. Maintaining wash oil flow is extremely important to efficient long-term operation. Loss of wash oil will result in rapid fouling.

10.5   Wash Oil

The wash oil that is used for de-entrainment is also important for improving the separation between the bottom side stream and the resid stream. Separation is enhanced by condensation of the overflash by vaporizing the bottom sidestream. The amount of overflash that is required is affected by packing type, depth and source of the wash oil. Overflash flow rates should be carefully monitored to make sure that there is no degradation in the bottom side stream fractionation, that there is no increase in pitch entrainment to the heavy solvent neutral stream and there are no major increases in coking in the wash bed zone.

10.6   Purpose of Pumparounds

Pumparounds are used to remove heat from the tower and to adjust the vapor-liquid flow in the tower. They condense vapors rising in the tower and create an internal reflux for the fractionation stages below the pumparound. They also reduce vapor loads in sections of the tower above the pumparound. A pumparound takes liquid from the tower, cools it, and returns it higher up in the tower. The liquid condenses the vapors in the pumparound section creating liquid reflux for fractionation lower in the tower. Vacuum pipestills do not use overhead reflux seen in other distillation towers, a top pumparound is used instead.

10.7   Tower Fractionation

As mentioned earlier, fractionation is used to generate the various product sidestreams off the tower by condensing rising hot vapor with falling colder liquid. At each stage in the fractionation section the highest boiling components are condensed, releasing heat that boils the lowest boiling point components, putting them into the vapor phase. Contacting between the phases is needed for the heat and mass transfer. Contacting equipment may include bubble cap trays, sieve trays, Glitsch grid, structured packing and many others. The number of theoretical stages between adjacent sidestreams typically varies from 1 to 3. The current trend is toward using packing.

10.8   Fractionation Packing

Packing used for fractionation can also reduce the pressure drop (Delta P) in a tower compared to trays. Tray designs are more limited in Delta P reduction. The packing surface allows intimate contact between vapor and liquid without having to have the vapor pass through the liquid. The liquid phase coats the packing surface as a film so the liquid phase movement is restricted only by the resistance of the packing surface. Packing has been used in high liquid loading service such as pumparounds and also in main fractionation sections. Packing is sensitive to liquid maldistribution so spray rates, pan level control and pumparound rate control are critical. A high quality liquid distributor is preferred.

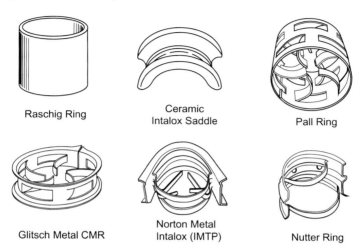

Raschig Ring   Ceramic Intalox Saddle   Pall Ring

Glitsch Metal CMR   Norton Metal Intalox (IMTP)   Nutter Ring

Figure 7. Various Types of Random Fractionation Packing (Drawing courtesy ExxonMobil Research and Engineering)

Structured Packing

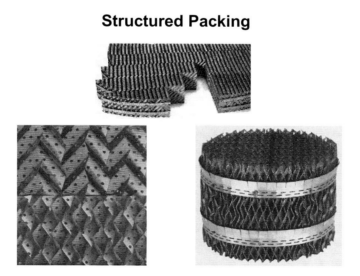

Figure 8. Various Types of Structured Fractionation Packing (photos by Ted Sideropoulos courtesy ExxonMobil Research and Engineering)

If the liquid flow to the packing is low and a spray distributor is used, the sprays can collapse so that the liquid contact with the vapor degrades causing low viscosity and poor volatility of the stream below. If the pumparound rate is too high, liquid may be atomized at the spray distributor and the small liquid droplets may be entrained upwards. If the liquid stream is from the pumparound, the product above the pumparound will be contaminated with heavy components. If there is leakage or overflow from the sidestream draw off pan, the falling liquid does not contact the vapor and it will reduce the viscosity of the stream below and also result in poor volatility.

10.9   Bottoms Stripping Section

The objective of the bottoms stripper is to strip distillate from the flash zone liquid, revaporize residual distillate that may be in the spent wash, and correction of bottoms flash. The bottom stripper typically has a design of 4 to 6 bubble cap trays or sieve trays. Some of the newer designs are using packing. Steam is used to reduce the hydrocarbon partial pressure to vaporize lighter molecules. A quench recycle is used to cool the stripped bottoms below 360°C (680°F) to reduce coking and cracking of the hydrocarbons.

10.10 Side Stream Strippers

Lube distillates are sent at their bubble points to side stream strippers. Steam is injected, reducing the partial pressure of hydrocarbons which effectively removes lighter hydrocarbons; improving volatility beyond that obtainable without side stripping. Ten to thirty percent of the stream may be removed in the stripper. If the heavier streams are not stripped this will reduce the yield of lighter lubes. Stripping is an important part of the overall operation to achieve the best separation and produce the desired products. A stripper typically consists of 4 to 6 sieve trays but packing may also be used.

10.11 Overhead Pressure

Pressure has a very large effect. Low pressure (15 to 100 mmHg overhead) is employed to reduce boiling points, allowing operation at temperatures low enough to minimize thermal degradation and cracking. The overhead vapors include steam, light hydrocarbons, and inerts. In the lower pressure design (15 to 50 mmHg) there is no precondenser before the first ejector. In higher pressure designs (40 to 100 mm Hg) a precondenser is employed and overhead pressure is dictated by condensing temperature (vapor pressure of water at the condensing temperature). Steam ejectors or vacuum pumps compress to atmospheric pressure and pump away the non-condensable hydrocarbons and inerts. Precondensers will reduce the overall load on the compression system. Ejectors use steam for compression in 2 or 3 stages. Each ejector typically has an intercondenser. Secondary and tertiary ejectors may sometimes be replaced by a liquid ring vacuum pump.

10.12 Tower Overhead Pressure With Precondensers

In a precondenser design, the lower the cooling water temperature the higher the achievable vacuum. The precondensers must operate below the water dew point to condense steam in the overheads. To achieve lowest tower pressure the tower should be operated at a low top temperature to minimize condensable hydrocarbons. Inerts should be minimized by reducing air egress and by keeping the bottoms temperature at or below 360°C to avoid excess thermal cracking and the formation of light gases. Cooling water flow should be sufficient to minimize the cooling water outlet temperature which sets the equilibrium conditions in the precondenser and therefore the achievable tower vacuum.

10.12a Tower Overhead Without Precondensers

In no precondenser designs, pressures can be lower than the vapor pressure of water at the condensing temperature. Lower pressure has both advantages and costs:

Advantages:
1. Higher distillate/resid cut-point
2. Less furnace coil and stripping steam required
3. Pressure is controlled at a constant value (vs. varying with cooling media temperature)

Costs:
1. Higher ejector steam rate
2. Larger diameter tower

10.13 Tower Pressure - Ejectors

The steam ejectors pump away the remaining vapor pressure of water, hydrocarbons and inerts. Ejector systems typically have two stages or three by 50% ejectors. Because of the criticality for tower operation most systems are overdesigned and it may be possible for the tower to operate with one 50% ejector in each stage. Intercondensers (1st stage) and after condensers (2nd stage) condense the steam from the ejectors, tower steam and condensable hydrocarbons. Motive steam flow must be maintained for best operation.

10.14 Factors Affecting Lube Distillate Production

- Crude Type
- Equipment Operation
    1. Cutting scheme selected
    2. Fractionation efficiency
    3. Pumparound heat removal capability
    4. Sidestream product stripper operation
    5. Equipment constraints
    6. Operational Stability
- Product Inspection Measurement precision

Table 2. Nominal Lube Product Boiling Range

	Two Product Sidestreams , °C	Three Product Sidestreams, °C	Two Product Sidestreams, °F	Three Product Sidestreams, °F
Vacuum Gas Oil	345 - 385	345 - 370	650 - 725	650 – 700
Light Neutral	385 - 455	370 - 425	725 - 850	700 – 800
Medium Neutral	-----	425 - 490	-----	800 – 915
Heavy Neutral	455 - 540	490 - 550	850 - 1005	915 – 1025
Overflash	540 - 580	550 - 580	1005 - 1075	1025 – 1075
Vacuum Resid	580+	580+	1075+	1075+

Table 3. 2-Sidestream vs. 3-Sidestream Product Comparison

Two Sidestream Products		Three Sidestream Products	
Viscosity (SSU at 100°F/38°C)	Yield on Crude (Vol%)	Viscosity (SSU at 100°F/38°C)	Yield on Crude (Vol%)
150	9.8	100	9.2
450	8.5	300	7.7
		700	5.3
Total	18.3	Total	22.2
Resid (1000 + °F)	18.3	Resid (1075 + °F)	16.8
(538 + °C)		(579 + °C)	

11.   PIPESTILL TROUBLESHOOTING

11.1 Material Balance and Viscosity Measurements

1. Tabulate rates of crude feed, reduced crude, overhead condensate rate, and all VPS sidestream and bottoms rates for material balance calculations.
2. Take a sample of VGO and each sidestream and measure viscosities (100°C) of all the VPS products.
3. Calculate yields, cumulative yield ranges and mid yield points for all the VPS products and combine with measured viscosities.
4. Compare viscosity to yield for each product. Compare to assay or lab generated distillation cuts and viscosities.

11.2   Tower Pressure Survey

- Use the same vacuum gauge to measure tower absolute pressures. (Pressures are low and different gauges can have calibration offset. Best results are obtained by moving the same gauge to the desired locations.)
  1. Make pressure measurement at flash zone, tower top and points in between.
  2. Determine Delta P across the strippers, both with steam on and with steam off.
  3. Measure ejector inter-stage pressures and condenser Delta P, noting temperature differences between condenser liquid and vapor outlet as well as Delta T across the condensers.
  4. Determine pressure drop across spray nozzles.
  5. Measure the transfer line pressure drop.
  6. Measure the ejector motive steam pressure.
  7. Measure the steam source pressure.
- Inferences based on findings
  1. If overall **tower pressure** drop is too low from the flash zone to the top then there may be damage to the tower internals or hydraulic problems. If the pressure drop is too high then flooding, plugging or internal damage may be indicated.
  2. **Pumparounds** typically have higher pressure drop than the fractionating sections.
  3. No or low Delta P in a **tower section** may indicate missing trays or absence of liquid. High Delta P in a tower section may indicate that the drawoff is partially restricted or blocked, that may be due to high liquid rates in that section of the tower, flooding, or too much stripping steam.
  4. Check **spray nozzle** Delta P, actual vs. expected. If higher than expected, the spray nozzle may be plugged. If lower than expected, the header may be leaking or missing a nozzle(s).
  5. If **Stripper** Delta P is too high then this may be an indication that too much steam is being used. If the Delta P is too low there may be too little steam being used of the trays or packing are damaged.
  6. If the **precondenser** Delta P is too high this may be an indication of poor design, flooding or fouling. If too low, equipment damage may be indicated.
  7. Review **ejector interstage** pressures vs. design. Low interstage pressure may be an indication of 2nd stage overload.
  8. **Tower pressure cycling** may be due to steam ejector underload and high ejector discharge pressure.
  9. **Condenser** liquid and vapor temperatures should be about 3°C apart. If the temperature difference is greater than this it may be an indication of bypassing.

10. Increase in cooling water temperature (in vs. out) should be about 5-8°C. If too low this may indicate fouling or bypassing. If too high cooling water rate may be too low.
- Comparison to Design
  1. Cooling Water - Higher / lower rate than design
  2. Vacuum system - Higher / lower rate than design
  3. Steam Injection - Higher / lower rate than design

12.   SOLVENT EXTRACTION

The properties of the lube oil that are set by the extraction process are the viscosity index (VI), oxidation stability and thermal stability. These properties are related to aromatics, aliphatic sulfur, total sulfur and nitrogen levels present in the base stock.

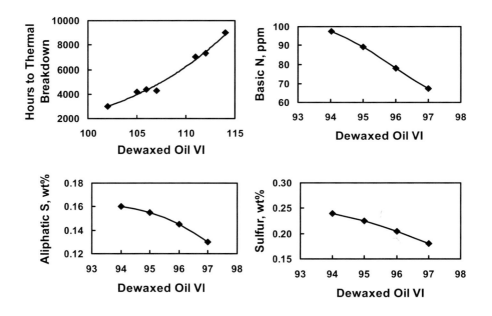

Figure 9. Typical impact Of Extraction On VI And Lube Oil Properties

Base stock VI has historically been used as a performance indicator for the base stock. The VI specification sets the extraction severity required to achieve the target. VI is also an indicator of relative stability from the same feed. VI is crude sensitive under constant extraction conditions.

Table 4. Impact of Molecular Type on Lube Oil VI and Stability

	VI	Stability
Paraffins	Excellent	Good
Mono-Naphthenes	Good	Fair
Poly-Naphthenes	Fair	Fair
Mono-Aromatics	Good	Fair
Poly-Aromatics	Poor	Poor

Molecular structure affects Lube quality. Solvent extraction and dewaxing processes preferentially separate the molecules as shown in Figure 10. Extraction separates *n*-paraffins, *i*-paraffins, naphthenes and some aromatics from the distillate into the raffinate phase. Dewaxing rejects the n-paraffins and some *i*-paraffins from the raffinate to produce a dewaxed oil or base stock. The dewaxed oil will contain the "slice" of molecular types as shown in Figure 10.

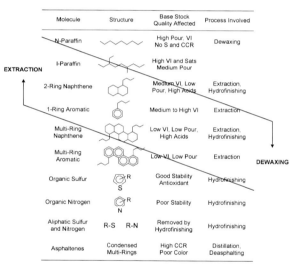

Figure 10. Principal Molecular Types And Their Effect On Lube Quality

The extraction process is a physical separation that is used in all conventional lube plants. The solvent is added to the distillate and then separated to produce a raffinate (the desired product) and an extract that contains a higher percentage of aromatics and impurities. Typical solvents used are N-methyl-2-pyrrolidone, furfural, and phenol. Properties of the solvents are shown in Figure 11.

Solvent	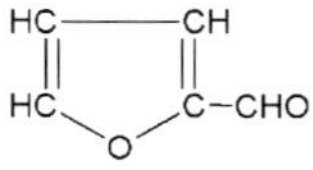 Furfural	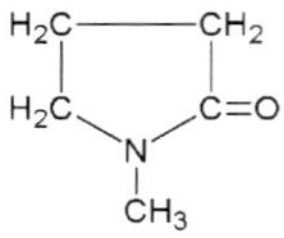 N-Methyl-2-pyrrolidone	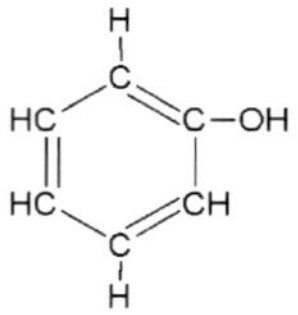 Phenol
Molecular Weight	96	99.1	94.1
Specific Heat @ 130°F	0.42	0.47	0.56
Boiling Point @ 1 atm., °F	323	399	359
Flash Point, °F	137	187	175
Viscosity @ 140°F, Cp	0.95	1.02	2.58
Melting Point, °F	-37	-11.6	105.6
Latent Heat @ b.p., Btu/lb	194	187	206
Toxicity	moderate	low	high
Specific Gravity @ 20°C	1.16	1.03	1.07

Figure 11. Physical Properties Of Typical Extraction Solvents

12.1   The Characteristics of a Good Extraction Solvent

A good extraction solvent will have a high selectivity for the undesirable components of the distillate stream. The solvent must also have good solvent power so that a low solvent to feed ratio may be used in the extraction plant. The solvent promotes rapid mass transfer. The solvent partitions between the raffinate and extract phases and must be recovered. Easy recovery via distillation is desired. A high density is also a characteristic of a good extraction solvent as this allows rapid separation of the oil and solvent phases. High demulsibility is needed for a rapid separation of the oil and solvent. The solvent must be chemically and thermally stable or inert in the lube extraction and recovery equipment. The ideal solvent would work for a wide range of feed stocks that the refiner might process. Solvent must be available at a reasonable cost and be non-corrosive to conventional materials of construction and it must be environmentally safe.

Table 5. NMP Relative To Furfural

NMP Solvent Property	
Thermally Stable	Heat integration with no measurable solvent decomposition
More Selective	Higher yields at lower solvent treats
Lower Latent Heat	Requires less energy for solvent recovery
Chemically Stable	Eliminates the need for feed deaerator for removal of oxygen
Higher Boiling Point	More efficient heat integration

Table 6. NMP Relative to Phenol

NMP Solvent Property	
Lower toxicity	Much safer
More selective	Higher yields and/or lower solvent treats
Lower Latent heat	Less energy required for solvent recovery
Higher Boiling point	More efficient heat integration
Lower Melting point	Less steam tracing required, less chance of solidifying in piping
No hydrogen bonding effect with the oil	More efficient stripping, easier to achieve low solvent concentration in product
No azeotrope	Simplifies water recovery

12.2 Extraction Process

Distillate is brought in contact with the solvent, and aromatics and polars are preferentially dissolved in the solvent phase. Saturates do not dissolve and remain in the hydrocarbon or dispersed phase. The hydrocarbon phase is lower in density than the solvent phase and rises as bubbles through the continuous phase. After separation the raffinate and extract solution are sent to their respective solvent recovery sections. Integration of a hydrofiner on the raffinate product is in some lube plants for heat integration because this eliminates the need for an additional hydrofiner furnace.

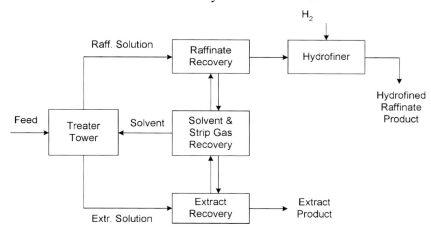

Figure 12. Simplified Extraction Flow Diagram

There are several types of continuous treater tower designs used in conventional lube plants. These include trayed towers, packed towers and rotating disc contactors (see Figure 13). The treater tower internals are designed to promote contact and separation of the oil and the solvent phases.

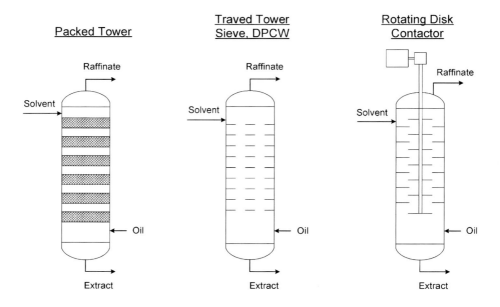

Figure 13. Types of Continuous Extractors

An example of a tray design is shown below.

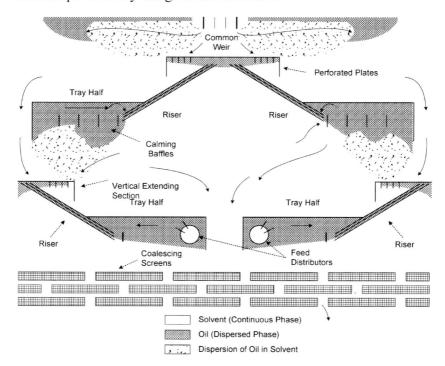

Figure 14. ExxonMobil® Patented Dual Pass Cascade Weir Tray

An example of a typical rotating disc contactor is shown below.

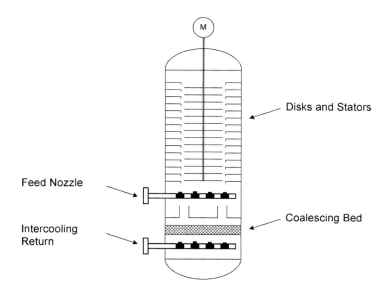

Figure 15. Typical Rotating Disc Contactor

There are several factors that affect extraction efficiency and in general the efficiency depends on the mixing/settling and coalescence characteristics of the system. Important factors include:

1. Hardware and staging
2. Throughput
3. Viscosity/Gravity of Oil/Solvent
4. Solvent dosage and composition
5. Temperature and temperature gradient
6. Solvent quality
7. Dispersion energy

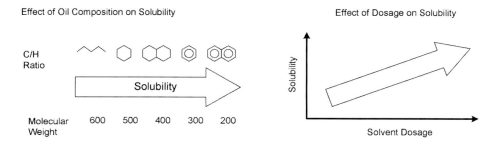

Figure 16. Effect On Oil Composition And Dosage On Solubility

12.3   Extraction Process Variables

INDEPENDENT variables are controlled by the operator and include:
* Treat Ratio - the volume ratio of solvent to feed
* Solvent Composition (water for NMP, Phenol)
* Bottom and Top treater temperature

DEPENDENT variables rely on the independent variables.
* Raffinate Quality
  - Viscosity Index
  - Saturates content as an indicator of the degree of aromatic removal
  - Sulfur
* Raffinate Yield - primarily dependent on the treat ratio at constant VI

12.4   Solvent Contaminants

Water from steam stripping in the solvent recovery section must be removed. In the furfural solvent system water is removed for process effectiveness and product quality. Water contamination in furfural reduces DWO VI and leads to furfural degradation. In NMP and phenol systems excess water is removed for process control.

Oil in the solvent results from incomplete solvent-oil separation and may be due to entrainment from flash vessels, volatilization or stripper flooding. Characterization of the solvent contamination by GCD can be used to determine if contamination is occurring by light or heavy oil fractions. A light oil contamination suggests that the accumulation of distillate in the front end. Presence of heavy oil suggests entrainment oil in the solvent, which can reduce raffinate yield and increase the treat rate required.

12.5   Solvent Recovery

The objectives of solvent recovery sections are to:
* Recover furfural/NMP/phenol from product streams
* Purify furfural/NMP/phenol for recycle
* Maximize energy efficiency while recovering solvent
  Simplified recovery sections are shown below

12.5.1 Raffinate Recovery

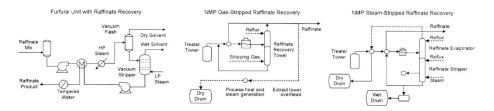

Figure 17. Typical Raffinate Recovery Diagrams

12.5.2 Extract Recovery

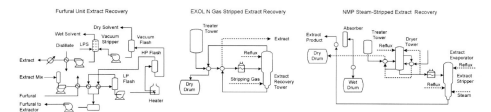

Figure 18. Typical Extract Recovery Diagrams

12.6 Minimizing Solvent Losses

12.6.1 Recovery Sections

NMP concentration in the product can be readily reduced to <20 ppm and furfural to <50 ppm, but keeping solvent loss low requires energy and the final concentration is an optimization problem. The furnace coil outlet temperature is a function of the pressure. Typical furnace COT's are higher for NMP than furfural due to solvent boiling points and the concern for furfural degradation. If fuel gas failure occurs at the furnace the plant immediately goes on oil recycle. The solvent concentration in the product can rise to several percent if the COT is too low. Too high of a furnace temperature leads to an increase in solvent decomposition and to premature furnace coking.

12.6.2 Other Contributors to Solvent Losses

- Some solvent is lost through line flushing during product sampling. To reduce losses, send sample line flushings back to the unit.
- Address miscellaneous leaks at flanges.

- The start-up and shutdown operation typically will increase solvent losses in the year the shutdown is taken. Consider ways to minimize solvent left in vessels.
- Check valves on bypass lines that may not hold and may allow bypassing of solvent.
- Water sent to the wastewater treatment unit might contain solvent.
- Phenol will oxidize and form deposits in the unit.

Effective solvent loss performance is considered to be:

NMP   = <0.03 Lbs/Bbl feed,
Phenol = <0.1   Lbs/Bbl feed,
Furfural = <0.2   Lbs/Bbl feed

Furfural will decompose when exposed to oxygen or excessive temperatures.

Sources of oxygen ingress include:

- Pump suctions
- Vacuum systems
- Oil feed poorly deaerated
- Poor/non-existent inert gas blanketing

13.   CORROSION IN NMP PLANTS

NMP, by itself, is not corrosive to carbon steel. However, because of NMP's high dielectric constant, other corrosive compounds will readily ionize in NMP and become very aggressive. The NMP condensing circuit may be at risk to accelerated corrosion from accumulated corrosive elements or corrosion/erosion from high velocities.

The refiner can take the following corrective actions. Avoid corrosive species ingress into the circuit. Remove H_2S from recycle stripping gas with a ZnO bed. Neutralize acids with neutralizing additives or caustic injection. Consider alloy upgrade in areas of known corrosion or areas of known impingement or very high velocity, such as 316 SS impingement baffles in exchangers. Insulate the overhead line to the first condenser to avoid premature condensation leading to impingement issues.

14.   EXTRACTION ANALYTICAL TESTS

There are numerous analytical tests that are used to assess the extraction operation and to help optimize the treater tower. A few of the most important tests are listed below in Table 7.

Table 7. Typical Analytical Tests for Extraction

TEST	ASTM Test No.	APPLICATION
Refractive Index @ T°C	D 1218	Correlation with dewaxed oil quality: VI, saturates etc
		Better for yield calculations than relying on process flow meters
Density at 15°C	D 4052	Better for yield calculations than relying on process flow meters
Water in Solvent by Karl Fischer	D 6304-3, E 203-1	Process optimization (NMP, Phenol)
Treater Carryunder		Measures the amount of distillate bypassing the treater and being downgraded from lubes to fuel
Oil in solvent 1. Percent 2. Characterization		Oil in solvent adversely affects raffinate yield
DEWAXED OIL		
Viscosity	D 445	Primary base stock specification
VI	D 2270	
Molecular Analysis	D 2007, D2887	Troubleshooting
	D 2140, D2501	Base stock quality

15. DEWAXING

Waxy raffinates from extraction are not useful as lubes because they contain too much wax. Referring to our molecular drawing (Figure 10) we can see that the objective of dewaxing is to remove the paraffins from the raffinate to produce a final Dewaxed Oil base stock that when additized becomes the finished lubricant.

Dewaxing is a physical process that adds solvent to a raffinate (or distillate) and once the mixture is cooled, the n-paraffins drop out of solution as solid wax crystals. The slurry is filtered to remove the wax crystals and produce a dewaxed oil or base stock and a valuable wax by-product. Waxes may be further refined to make hard waxes by melting and separating the soft wax. Hard waxes may meet FDA standards for use in direct or indirect contact with food.

The majority of dewaxing processes today use Methyl Ethyl Ketone (MEK), Methyl IsoButyl Ketone (MIBK), mixtures of MEK and MIBK, or mixtures of MEK and Toluene or propane. There are advantages and disadvantages to each solvent system.

Dewaxing sets the primary properties shown in Table 8.

Table 8. Properties Set by Dewaxing

Dewaxed Oil Properties	Wax Properties
Pour Point	Oil Content
Cloud Point	Melting Point
Low Temperature Fluidity	Needle Penetration

A general flow plan for a dewaxing plant is shown below.

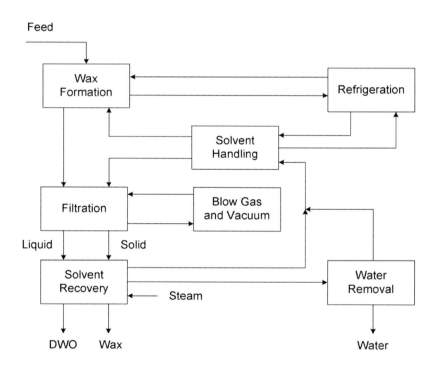

Figure 19. Simplified Dewaxing Flow Diagram

16.   THE ROLE OF SOLVENT IN DEWAXING

As solvent is added to the waxy raffinate the oil is diluted and the viscosity of the oil solvent mixture decreases allowing filtration to take place more easily. The polarity of the oil-solvent mixture increases, decreasing the solubility of the wax and promoting the formation of more compact wax crystals. But as solvent is added the resulting filtrate becomes more dilute, loading up filtrate pumps and solvent recovery facilities.

Properties to Consider in Selecting a Dewaxing Solvent:

1. Solubility
2. Selectivity
3. Solvent boiling point lower than the boiling point of the oil
4. Low heat capacity
5. Heat of vaporization
6. Low viscosity
7. Non-Toxic
8. Non-corrosive
9. Low freezing point
10. Inexpensive
11. Readily Available

Ketone units typically use a dual solvent system consisting of MEK and either MIBK or Toluene. The MEK acts as an antisolvent to reject wax molecules from solution. This reduces refrigeration requirements but excessive MEK may cause oil phase separation. The second solvent keeps the oil in solution but also dissolves some wax. MIBK and toluene act as prosolvents.

Solvent properties are compared in the table below.

Table 9. Typical Dewaxing Solvent Properties

Solvent	Wax Solubility g/100 ml	Viscosity @ 0°C, cSt	BP, °C	Latent Heat of Vaporization, cal/g	Specific Heat, cal/g-°C
MEK	0.25	0.40	80	106	0.55
MIBK	0.90	0.61	116	87	0.46
Toluene	13.0	0.61	111	99	0.41

MEK/MIBK refrigeration requirements are lower than MEK/Toluene because the Pour-Filter spread is smaller due to the lower wax solubility. The Pour-Filter spread is the difference between the Dewaxed Oil pour point and the filtration temperature required to meet the Dewaxed Oil pour point specification. Wax has a higher solubility in Toluene than MIBK and MEK/Toluene systems will require a lower filtration temperature to achieve the same pour point. MEK/MIBK solvent mixture viscosity is lower than MEK/Toluene. Filtration rates are higher for MEK/MIBK. Toluene costs less than MIBK.

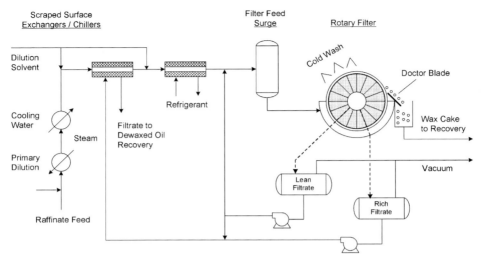

Figure 20. Simplified Incremental Dilution Dewaxing

17. KETONE DEWAXING PROCESSES

17.1 Incremental Ketone Dewaxing Plant

Incremental ketone dewaxing plants (see Figure 20) use a combination of Scraped Surface Exchangers that use cold filtrate for cooling the slurry followed by Scraped Surface Chillers, which use a refrigerant (propane, propylene, ammonia) to cool the slurry to the filtration temperature. Solvent may be added at the beginning or along the train in "increments".

Solvent may be mixed with raffinate before the scraped service exchanger as primary dilution. The slurry passes through a feed heat exchanger to melt any crystals that may have formed in tankage. The slurry temperature is then reduced in a feed precooler. The slurry flows to the Scraped Surface Exchangers and through the tube side of the exchanger. Cold filtrate from the filters is used to cool the feed below the cloud point and initiate crystallization. Solvent may be added in increments to reduce the slurry viscosity and enhance heat transfer. Slurry flows to Scraped Surface Chillers and through the tubeside. Propane, propylene or ammonia are typical refrigerants used on the shell side. The slurry exits the last scraped surface temperature at filtration temperature and enters the filter feed drum. The slurry flows by gravity to the filters where the wax crystals are filtered across a rotary vacuum filter. Wax is removed from the filter and sent to the wax recovery section. Oil and solvent — filtrate — is collected and sent through the shell side of the Scraped Surface Exchangers on its way to the Dewaxed Oil recovery section. Filtrate may also be recycled back to the slurry to adjust the final dilution before filtering.

17.2 DILCHILL™ Dewaxing

In DILution CHILLing the Scraper Surface Exchangers (SSE) are replaced with a multistage crystallizer. Cold Solvent is added at each stage and the slurry is mixed with an impeller. The crystallizer replaces all the SSEs with a single mixer.

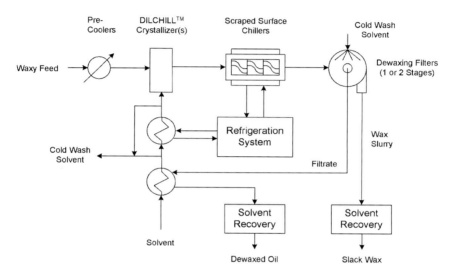

Figure 21. Simplified DILCHILL™ Dewaxing

The key features of DICHILL™ are summarized below:
* Less oil is occluded in the wax crystal due to the vigorous "micromixing" in the crystallizer
* Higher filtration rates and lower oil-in-wax are achieved as a results of compact, spherical crystals
* Lower dilution solvent
* Lower operating costs
* Easier Operation
* Higher Service Factor.

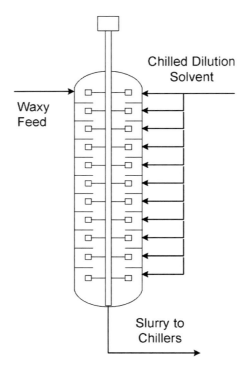

Waxy Feed

Chilled Dilution Solvent

Slurry to Chillers

Figure 22. Simplified DILCHILL™ Tower

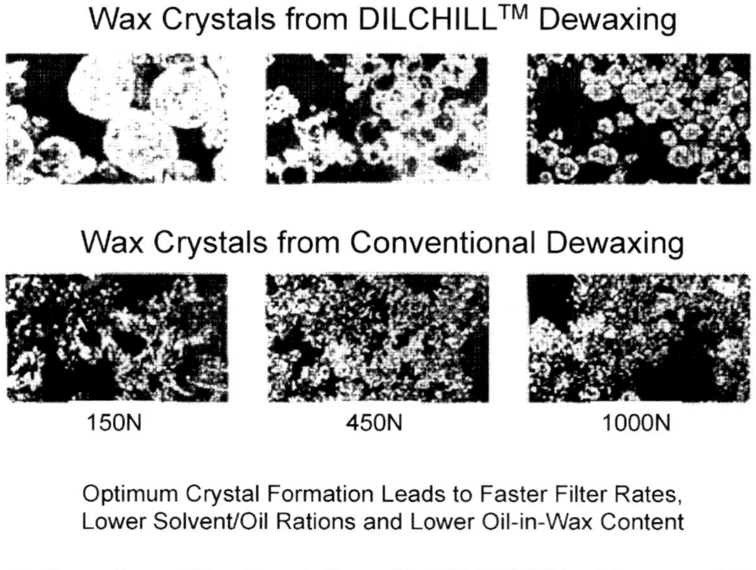

Wax Crystals from DILCHILL™ Dewaxing

Wax Crystals from Conventional Dewaxing

150N   450N   1000N

Optimum Crystal Formation Leads to Faster Filter Rates,
Lower Solvent/Oil Rations and Lower Oil-in-Wax Content

Figure 23. Comparison of Wax Crystals Formed in DILCHILL™ and Incremental Dewaxing

17.3 Dewaxing Process Variables

INDEPENDENT variables are controlled by the operator and include:
- Charge Rate
- Solvent Composition
- Precooler Outlet Temperature
- Dilution Ratio (Increments)
- Chilling Rate
- Filter Wash Solvent Ratio
- Wash Temperature
- Mixing Energy (DILCHILL™ Only)
- Filtration Temperature
- Filter Speed
- Dewaxing Aids

DEPENDENT variables rely on the independent variables:
- Filter Feed Rate
- Dewaxed Oil Quality
  - Cloud Point
  - Pour Point
  - Low Temperature Properties
- Dewaxed oil yield
- Wax Yield
  - Oil in Wax
  - Melting Point
  - Needle Penetration

18. PROCESS VARIABLE EFFECTS

18.1 Crude Source Affects Dewaxed Oil Yield

As was mentioned earlier, crude selection can have an influence on downstream units. The table below shows the impact of crude source on dewaxed oil yield.

Table 10. Typical Basestock Yields for Two Different Crudes

Basestock		Arab Light	Statford
	Minimum Fluid Point, °C	Yield, Vol%	Yield, Vol%
100 SUS	-18	81	79
300 SUS	-9	81	79
700 SUS	-7	85	80
Bright stock	-7	85	78

19.   SOLVENT COMPOSITION

19.1   Miscible and Immiscible Operation

MEK is an antisolvent for the wax and helps to reduce its solubility. If the MEK content is too high the Basestock may become insoluble and a phase separation will occur. MIBK or Toluene is added to help solubilize the oil. Both of these prosolvents have a higher affinity for wax molecules than MEK. The higher the concentration of prosolvent the more wax stays in solution, and ends up in the filtrate. This raises the pour point of the dewaxed oil and since the manufacturer must meet dewaxed oil pour point specification the manufacturer is forced to reduce the filtration temperature to remove more wax. The reduction in filtration temperature increases the viscosity of the slurry and filtration rates are slower and oil removal from the wax cake becomes more difficult. Thus the objective is to use the maximum amount of MEK without having a phase separation. A plot of the phase separation temperature or miscibility temperatures vs. solvent composition may be used to help set the optimum solvent composition.

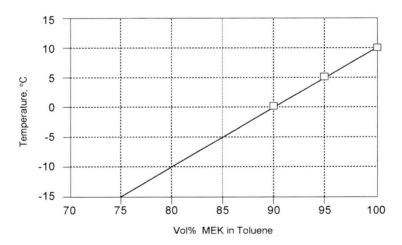

Figure 24. Typical Heavy Neutral Miscibility Curve

The ideal would be to operate as close to the miscibility curve as possible. However, this curve shifts depending on the Basestock. Wax molecules in heavier grades come out of solution earlier and this has the effect of shifting the miscibility curve to the left. Plants equipped with solvent splitters to separate the MEK from the MIBK or Toluene after it has been recovered in the DWO and Wax recovery sections may blend to the optimum solvent composition for each Basestock. Manufacturers without the capability to change solvent composition will set the plant solvent composition based on the heaviest grade to avoid immiscible operation. This increases the pour filter spread, because the prosolvent composition is too high for the lighter grades. This will also affect plant processing capacity.

The shift from miscible to immiscible operation occurs over a narrow range of solvent composition. When the plant moves deep into immiscible operation and phase separation occurs, the oily phase, containing the desirable high VI molecules, will hang up in the filter cake and be very difficult to remove, increasing the oil in wax content of the wax and reducing dewaxed oil yield. Dewaxed oil properties may be adversely affected, and VI may decrease. Interestingly, in severely immiscible operation the dewaxed oil pour point may actually decrease as the waxy molecules remain in the wax cake.

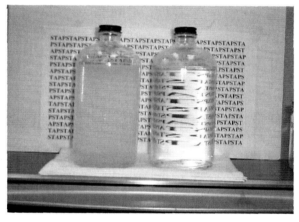

Figure 25. Comparison of Immiscible and Miscible Filtrates (*photo by B.E. Beasley courtesy ExxonMobil Process Research*)

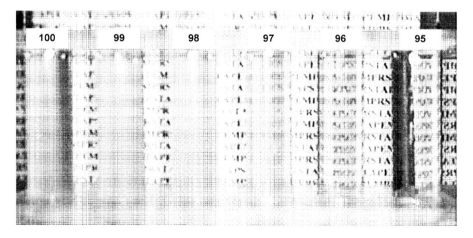

Figure 26. Impact Of Change In Solvent Composition On Miscibility (*photo by B.E. Beasley courtesy ExxonMobil Process Research*)

19.2   Effect of Viscosity on Filtration Rate

Higher viscosity Base stocks will filter more slowly. Increasing the MEK concentration in the solvent reduces the slurry viscosity but the maximum amount of MEK is limited by miscibility considerations.

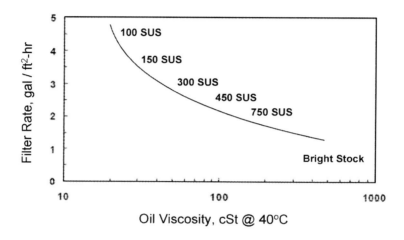

Figure 27. Effect Of Oil Viscosity On Filtration Rate

19.3   Effect of Chilling Rate On Filtration Rate and Dewaxed Oil Yield

Increasing the chilling rate forces wax molecules to come out of solution quicker. When the wax molecules come out of solution they may either form a separate nucleus (nucleation) or attach onto an existing nucleus (growth). When the chilling rate is high, nucleation is favored over growth with the result that the average crystal size is smaller, the wax is more difficult to filter, and the wax retains a higher oil content.

Table 11. Example of Effect of Chilling Rate on Yield and Filtration Rate

Chilling Rate, °C/min	DWO Yield, Vol%	Filter Rate, Gal/ft‾-hr
2.2	85.3	2.00
4.4	82.0	1.94
6.7	81.3	1.92
8.9	79.8	1.90

19.4   Effect of Temperature Profile

The preferred temperature profile for incremental and DILCHILL™ dewaxing is a linear profile. Both methods of dewaxing add solvent, either in increments along the SSE/SSC train or in stages in the crystallizer. Thus the flow rate increase from the feed to the filters. Other temperature profiles are possible. Filtration rates in plants with a convex temperature profile are typically not as good as those with a linear profile. This may be due to the very high chilling rates in the SSC or last few DILCHILL™ stages, required to reach the filtration temperature, and which favors nucleation at the expense of growth and leads to smaller average crystals. Typically concave profiles offer no advantage over linear. In the rare case where a concave profile has produced improved filtration rates, it has been surmised that a high chilling rate at the beginning of the train, which will favor nucleation, was needed to establish "seed" crystals for growth.

19.5   Effect of Solvent Dilution Ratio

19.5.1 Filtration Rate

Solvent may be added in increments along the SSE/SSC train or in the DILCHILL™ Crystallizer. The solvent reduces the slurry viscosity and facilitates filtration. When the dilution is too low, the slurry viscosity will be too high, and the slurry filtration will be reduced. This is often referred to as VISCOSITY limited. When the dilution is too high, the volume of liquid to be filtered exceeds the capacity of the filter (filter cake resistance is limiting) and the filtration is said to be hydraulically limited. This balance results in an optimum dilution ratio dependent on the Basestock being processed, the solvent composition being used and the crystallization, which sets the filter cake resistance (see Figure 28). Light Basestock operations are more likely to be Chilling or Refrigeration limited rather than Filtration limited because the Light Basestock pour points are typically lower than the heavy grades while the average wax crystal size is larger, reducing wax cake resistance.

19.5.2 DWO Yield

Increasing the dilution ratio will reduce the Oil in Wax and increase the dewaxed oil yield.

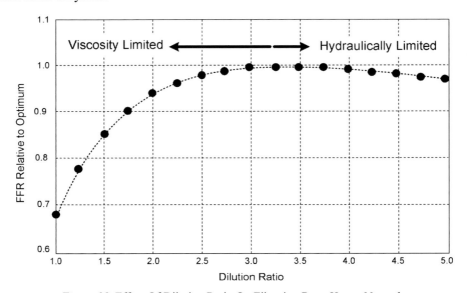

Figure 28. Effect Of Dilution Ratio On Filtration Rate, Heavy Neutral

19.6   Effect of Water

Water may enter with the feed from the extraction process or from tankage. It may also enter the system from leaks in water coolers or condensers that are used in the dewaxing plant, or it may reenter the solvent system through inadequate performance of the dehydrator section in the plant. Water is a strong anti-solvent, and reduces the solubility of wax and oil molecules so that they come out of solution at a higher temperature. This has the effect of shifting the miscibility curve to the left. Typically the increase in water concentration from the normal levels is an indication of a process/equipment problem and the root cause must be found and dealt with quickly.

Excessive water may drop out of solution and form ice in the slurry or on the inside of the scraped surface equipment. It may also flow downstream to the filters where it may be captured in the wax cake or on the filter cloth.

19.7   Effect of Increased Raffinate VI

Raffinate VI may be increased in extraction through more severe extraction. This increases wax concentration in the raffinate stream. These waxy molecules will come out of solution in the dewaxer at a higher temperature, resulting in the miscibility curve shifting to the left. This will require the manufacturer to add more prosolvent to avoid immiscible operation.

19.8   Effect of Pour Point Giveaway on Product Quality and Dewaxed Oil Yield

When a Basestock produces a dewaxed oil with a lower pour point than the specification, the plant is "giving away" this product quality. The economic penalty for pour point give away is lower filtration rate and lower throughput and decreased dewaxed oil yield. Lower pour point also means that more wax has been removed and this will lower the dewaxed oil VI.

Table 12. Examples Of Quality Giveaway

Base Stock	Pour Point Reduction, °C	Typical VI Loss	Typical Yield Loss, vol%
Solvent Neutrals	3	1	1.0
Bright Stock	3	0.5	1.0

20.   SCRAPED SURFACE EQUIPMENT

SSEs and SSCs are used in incremental dewaxing and SSCs in DILCHILL™ dewaxing. SSEs and SSCs are double pipe exchangers with slurry inside the central tube and the cooling media (filtrate or refrigerant) in the annular area around the inside tube (pipe). Tubes are typically stacked in groups of 8, 10 or 12 and internal flow may be in parallel or series. Internal pipe diameters are typically 6, 8, 10, 12 inches.

A shaft extends the length of the inner pipe. Spider bearings are used for shaft support. Each shaft is fitted with a scraping blade; there are two leading designs. The shaft extends through a seal or packing at the drive end. A sprocket is attached to the shaft outside the tube and the shaft is turned by a lubricated chain that is connected to all the sprockets in a tube back and to a fixed speed drive motor mounted above the tube bank. Shaft speeds range from 2-30 rpm.

Figure 29. Scraped Surface Equipment (photos reprinted courtesy Borsig®)

Borsig GmbH and Armstrong International, Inc. are the leading designers of scraper internals. Both use blades that are softer than the pipe, so that the blades will wear down first. Each uses a spring to apply pressure on the scraper blade to push it against the tube wall so that it scrapes the wax off the wall as the internal shaft is turned.

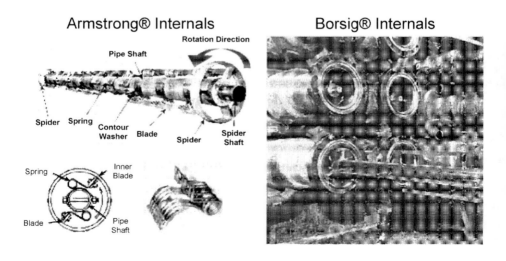

Figure 30. Scraped Surface Equipment (Photos reprinted courtesy Armstrong® and Borsig®)

Both designs attempt to accomplish wax removal from the wall by scraping while maximizing open cross sectional area in the tube. Wax will tend to accumulate on cold internals like the Spider bearing and other surfaces affixed to the shaft. The build up of wax will increase the pressure drop. When pressure drop becomes limiting, the equipment will either be taken off-line and cleaned with hot solvent or cleaned on-line. Various methods have evolved to optimize the cleaning with minimum impact on the process.

SSEs and SSCs require very high maintenance support in comparison to other equipment in the plant. Shear pins protecting the internal shaft from damage will fail and the shaft will stop turning. This reduces the effective heat transfer, which can limit the plant throughput if the plant is near a chilling or refrigeration limit. Seals and packing will leak, contributing to solvent loss in the plant. Poorly lubricated chains will break or jump off the sprockets. SSEs and SSCs overall heat transfer coefficients are low. Performance should be monitored, which is typically not an easy task. Internals must be overhauled when performance has permanently degraded.

21.   FILTERS

21.1   Filter Operation/Description

Filters are used to separate the wax crystals from the slurry. The slack wax and filtrate are collected and sent to their respective recovery sections in the plant to recover and recycle solvent.

Rotary vacuum filters are used in ketone dewaxers and rotary pressure filters are used in propane plants. The principles of operation are the same. A typical filter is shown in Figure 30.

The filter drum rotates at speeds from 0.2 - 1.6 rpm. The surface of the drum is divided into segments that run the length of the filter. Newer designs have 30 segments. Each segment has a lead and trail pipe connected to the segment in several locations running the length of the drum. The segment lead and trail pipes combine so that there are 30 lead and trail pipes, one for each segment, that carry filtrate to the master valve, which is also called the trunnion valve. The master valve is stationary while the filter drum rotates. The master valve has internal bridge blocks that effectively segregate the filtrate collected from the filter into as many as four distinct compartments or "pick-ups".

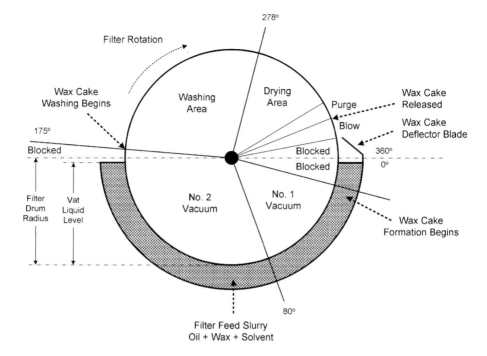

Figure 31. Typical Filter Wash - Dry Segments

Figure 32. Filter Master Valve (*photos by B.E. Beasley courtesy ExxonMobil Process Research*)

Starting from the deflector blade (see Figure 31), the drum rotates into the slurry in the vat and filtration begins at the first "wet pick up" port or "No. 1 vacuum". The filter segment continues its rotation into the "2nd wet pick up" or "No. 2 vacuum". The rotation continues and the filter moves into the wash area or "first dry pick-up" and finally into the "drying area" or "second wet pickup". The drying area may also be washed with cold solvent. The filter rotates into the Purge zone where N_2 blow gas is applied to the trail pipe while the lead pipe stays under vacuum. The purge zone clears out the filtrate in the pipes and increases the dewaxed oil yield by preventing the filtrate in the pipes from being blown back through the pipes and into the wax scroll with the wax. Finally the segment moves into the blow gas zone and blow gas is applied to the lead and trail pipes and the cake "pops" off of the filter and slides across the deflector blade and into the wax scroll.

The scroll is an Archimedes screw that moves slack wax from the outer ends of the filter to a center pipe. The slack wax falls down the pipe and may either proceed down a slide where it may be combined with slack wax from other filters into a slack wax drum or into a wax "boot" or vessel. This is a small drum dedicated to the filter. From the large wax drum or the wax boot the slack wax is pumped to either wax recovery, the second stage of dewaxing or deoiling depending on the plant configuration.

Filter Drum   Solvent wash sprays and drip pipe   Wax scroll

Figure 33. Typical Rotary Vacuum Filter (*photos by B.E. Beasley courtesy ExxonMobil Process Research*)

21.2   Filter Media

The filter drums used in ketone dewaxing (rotary vacuum) and in propane dewaxing (Rotary pressure) have filter cloths secured to the drum surface. The filter cloth retains the wax crystals while allowing the filtrate to pass through the filter cake, through the filter cloth and into the internals of the filter drum and eventually to the filtrate receiver drum.

Many different types of material have been used over the years. These include woven and non-woven materials, single and dual layer cloths. A few examples are shown below.

SATIN WEAVE

TWILL WEAVE

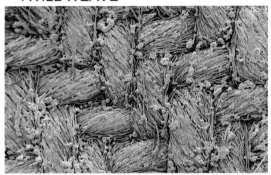

PLAIN WEAVE

NON-WOVEN

Figure 34. SEM of Typical Filter Medium (photos reprinted courtesy Madison U.S. Filter)

The requirements of a good filter cloth are:
1. Low Interfacial Resistance for high filtration rates

2. High filtration rates lead to high wash acceptance or lower filter speeds for increased dewaxed oil yields
3. Elimination of wax bleed through and haze formation
4. Thermally stable at hot wash temperature
5. Chemically stable in ketone and propane service

A comparison of typical filter cloth materials is included below.

Additional considerations include the mechanical stresses generated in vacuum and pressure filters that lead to fabric stretch and wear associated with filter blow gas. The filter cloth may suffer mechanical damage during installation, or from exposed deflector blades or other sharp or raised edges. The cost of the filter cloth is minor compared to the lost production time incurred when a filter has to be re-clothed or when the dewaxed oil quality does not meet product specification.

Fiber selection, fabrication technique (woven, non-woven) and cloth finishing all play an important role in determining filter cloth performance.

The ExxonMobil® Patented proprietary filter cloth has been used in ketone and propane dewaxing service for many years. It has higher resistance to fouling due to several unique design features and can achieve >15% improvement in filtration relative to woven cloths.

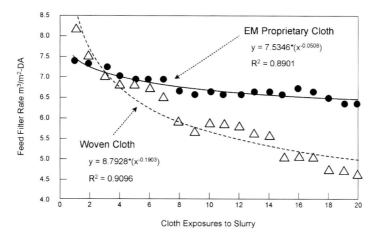

Figure 35. Typical Filter Cloth Fouling

Filter cloth is cut the length of the filter and depending on the width, several sections are required. The cloth sections are held in place on the surface of the drum by a caulking bar which is hammered into the caulking bar groove during the cloth installation. A steel wire is wrapped, under tension, around the drum to further secure the cloth. The wire will tend to migrate away from the drive end and many filters have installed a take-up tensioner device to extend the length of time before wire breakage occurs.

Upper left: Caulking bars hammered into caulking bar groove
Upper right: Removal of excess cloth. A small cloth strip is used in all grooves that do not have
cloth overlap to provide tight fit.
Lower left: Technicians tamp down the caulking bar as the wire is wound onto the drum.
Lower right: Wire tension take up device.

Figure 36. Filter Cloth Installation (*photos by B.E. Beasley courtesy ExxonMobil Process Research*)

22.   COLD WASH DISTRIBUTION

Cold ketone wash is applied to the filters through either sprays or drip pipes or both. There are advantages and disadvantages to both designs and both are in practice today. Cold wash is essential to reduce the Oil in wax, increase dewaxed oil yield and increase the wax product value.

Good distribution is necessary to spread the wash fully over the wax cake and avoid dry sections. The wash must not impinge with excessive velocity onto the wax cake in order to avoid knocking it off or digging channels or grooves. The wash system should be resistant to fouling. Spray nozzles and drip pipes are the major wash distribution systems in use today. Both are acceptable provided the system is properly designed and operated within the design parameters.

ExxonMobil patented Drip Pipes provide excellent coverage and are capable of good turndown. Wash falls by gravity onto the wax cake minimizing potential dislodging of the cake by wash impingement. Stainless steel construction and upstream filtration prevent fouling.

- Poor wash distribution can result from collapsed sprays creating a wide area without coverage.
- If cold wash flow from the spray nozzles is too high wax may actually be blown off the filter cake. Evidence of this may be seen as an accumulation of wax on the filter windows.
- Conventional (not ExxonMobil) drip pipes may not be balanced, resulting in some areas with no flow and some areas with too much flow.
- Excessive wash flow rates through the spray nozzles may cut grooves in the cake and wash wax and solvent back into the vat, resulting in recycling.
- Combinations of too high a flow through conventional drip pipes will collapse spray cones. Excess accumulation in a local area may cut grooves in the wax cake.

Poor Distribution from Conventional Drip Pipe

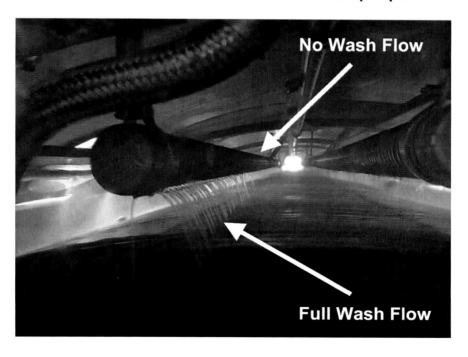

Figure 37. Example Of Poor Filter Cold Wash Distribution (*photo by D. S. Sinclair courtesy ExxonMobil Process Research*)

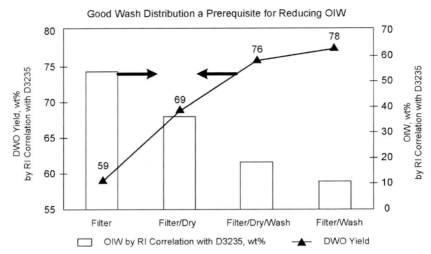

Figure 38. Effect Of Filter-Wash-Dry Sequence On Oil Yield (Lab) (*photo by B.E. Beasley courtesy ExxonMobil Process Research*)

Impact of Wash Application on Filter Feed Rate (FFR)

Drying out / compressing wax cake due to lack of wash over part of top pickup zone affects both DWO yield and FFR

* Recent plant tests redistributed wash solvent to prevent large "dry" area on descending side of filter, gave 15% increase in FFR
* Graph shows effect of continuing to apply vacuum without wash and then reapplying wash. FFR improved but did not come back up to previous level

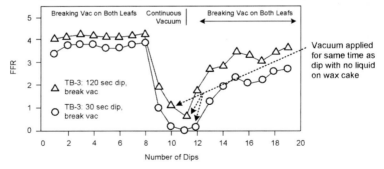

Figure 39. Impact of Wash Application on Feed Filtration Rate (FFR)

23.   WASH ACCEPTANCE

Wash acceptance is the amount of wash that can be accepted through the wax cake before it begins to spill or run off the wax cake. The volume of filtrate collected as a function of time decreases in a square root relationship, for non-compressible wax cakes, until the end of the filtration step. At that point filtration is complete, the wax cake formation has stopped and the wash is applied to the cake. The flow of wash through the cake is at the same rate as the last increment of filtrate just at the final moment of filtration (ignoring for the moment slight differences in liquid viscosity and wax cake compressibility). If too little wash is applied the Oil in Wax will be higher than optimum and yield will be lower. If excess wash is applied the excess wash will roll off into the filter vat or into the scroll.

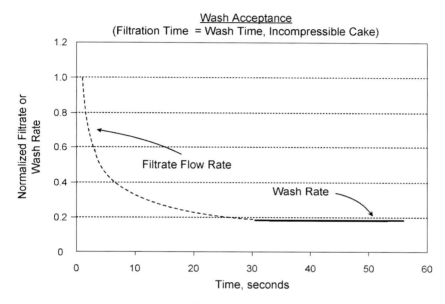

Figure 40. Filtrate Volume Versus Time

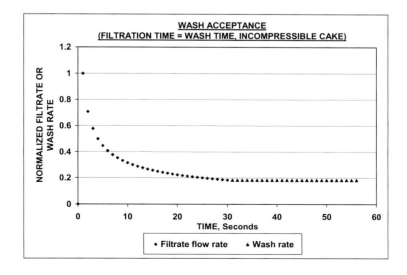

Figure 41. Normalized Filtrate Or Wash Rate Versus Time

24.   WASH EFFICIENCY

Cold wash, distributed over the wax cake by drip pipes or sprays, will displace the cake liquids, reducing the Oil in Wax and increasing the yield. This occurs in two steps. The first step is a piston displacement where the wash liquid pushes out the cake liquids. In the second step, oil from within the wax crystal diffuses into the low oil concentration wash liquid. The theoretical reduction in oil content may be predicted by the Butler equation [1].

The wash efficiency is defined as the actual oil in wax obtained for a given amount of wash applied relative to that predicted by the Butler curve. Efficiencies may be less than 100%. This may result from:
1.   Poor spray distribution
2.   Fouled, leaking or missing spray nozzles
3.   Poor distribution along the drip pipe, "out of level" on older designs
4.   Fouled drip pipes
5.   Area of excessively fouled filter cloth may occur when "drying" is practiced
6.   Cracked filter cake, solvent flows through crack instead of cake
7.   Wash rate exceeding wash acceptance
8.   Wash temperatures significantly lower than filtration temperature, seals off wax crystals
     Wash efficiencies may be found to exceed 100%. This may be due to:
1.   Hydraulic compression of the wax cake, "squeezing" oil out of the cake without using any wash solvent such as may occur in the drying section of the filter
2.   Higher crystal or wax cake compressibility in the presence of wash.

Some plants will dedicate a significant portion of the filter area to "drying" and will block off sprays or drips in this area. The intent is to reduce the Oil in Wax to increase yield and possibly debottleneck wax recovery. An unintended consequence is that the vacuum that remains on in this section will pull the wax cake into the filter cloth, accelerating the filter fouling which will reduce wash acceptance and increase oil in wax. It will also require more frequent filter washing which will reduce plant throughput. These competing factors must be balanced by the manufacturer to achieve the optimum filter performance.

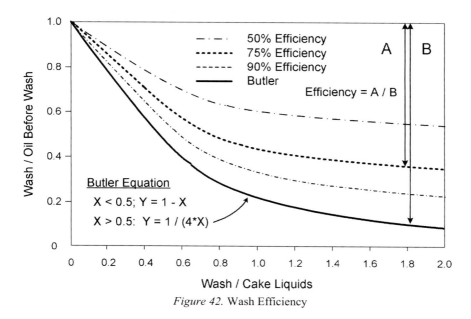

Figure 42. Wash Efficiency

25.   FILTER HOT WASHING

The filtration rate of a filter decays over time due to plugging of the filter cloth by small wax crystals. A typical decay curve is shown below. The feed rate measured by flow meter is plotted against the number of "DIPS" or exposures of the filter cloth to the wax slurry. The shape of the decay curve depends on the filter media, in this case an ExxonMobil proprietary cloth.

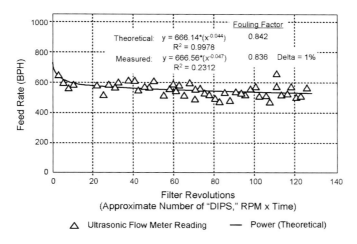

Figure 43. Feed Rate Decay Curve (Light Neutral)

As the filtration rate decays, throughput is lost and at some point the filter must be taken offline and washed to restore the filtration rate. The time that the filter is off-line also represents lost production. The optimum time between washes is the economic balance between production lost due to the decay curve compared to the time to wash the filter and bring it back on line. An example of a series of decays and washes are shown below. If the shape of the decay curve is known and the wash time is known the optimum wash time may be found analytically.

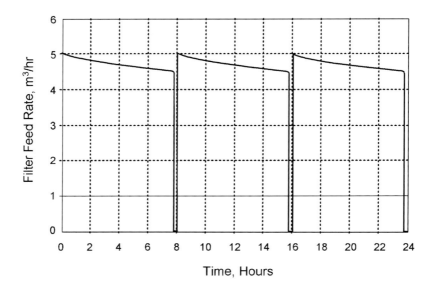

Figure 44. Filter Feed Rate Decay Curves And Filter Washes

Hot ketone solvent is used to wash the filters. Cold wash solvent is blocked off and hot solvent flows through the same sprays or drip pipes, over the filter cloth, melting the wax crystals, cleaning the cloth. Filter washings are collected in a filter washing drum. The washings may be:
1. Injected with the feed as "predilution"
2. Injected between 1st and 2nd filter stages
3. Injected into the appropriate stage of the DILCHILL™ crystallizer
4. Sent to wax recovery.

Hot wash temperature is controlled. If the temperature is too low it may not adequately melt the wax crystals. If the wash temperature is too high wire migration may occur resulting in reduced wire life and/or an increase in the frequency of wire retensioning.

26. DEWAXED OIL/WAX-SOLVENT RECOVERY

The objective of the Dewaxed oil and wax recovery section is to remove solvent from the product stream for recycle back to the chilling train and the filters. This is done using low and high pressure flashes and a high temperature flash followed by steam stripping and vacuum drying (dewaxed oil). Solvent concentrations are reduced to meet the product specification. High concentrations in the product represent avoidable solvent losses and increase operating expenses. In some cases excessive high temperature operation in the recovery section can lead to light oil vaporization and accumulation in the solvent. This represents avoidable excess energy costs. The operation of the solvent recovery system becomes an economic optimization problem.

Dry "Waxy Solvent" from wax recovery should be segregated from the Dry "Clean" solvent recovered from the Dewaxed Oil recovery section. "Waxy solvent" tends to have wax in it and cannot be chilled to the same temperature as that of the "Clean" solvent from the dewaxed oil recovery. It may be cooled and temperature blended with the "Clean" and used as filter wash. Flow restrictions may be set on flow to wax recovery to prevent wax carryover and downstream fouling.

Wet solvent from the stripper and solvent from the low pressure flash is sent to the dehydrator for solvent recovery.

A typical recovery section is shown below.

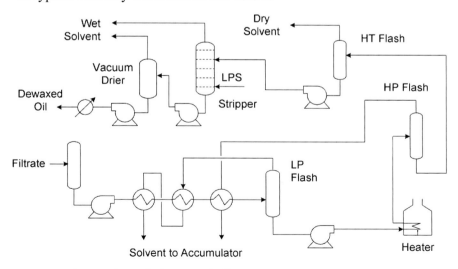

Figure 45. Typical Dewaxed Oil Recovery Section Flow Diagram

27.   SOLVENT DEHYDRATION

Solvent from the low pressure flashes are sent to the dehydrator tower. Overheads from the dehydrator tower are combined with stripper overheads and the overhead from the water tower and sent to the decanter. The solvent rich phase from the decanter is sent back to the dehydrator. The water rich phase to the water tower. Dry solvent from the dehydrator bottoms is returned to the unit. The water from the water tower is sent to waste water treatment.

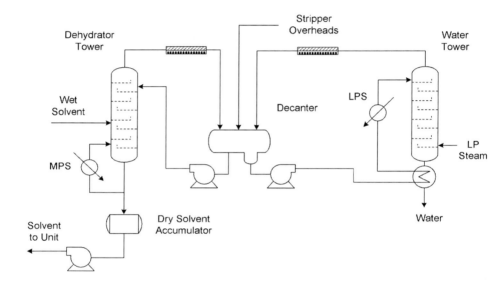

Figure 46. Typical Solvent Dehydration Flow Diagram

28.   SOLVENT SPLITTER

The solvent splitter allows the manufacturer to separate the solvent mixture back into MEK and MIBK or Toluene which can then be added back to the solvent mixture in the plant to optimize the solvent composition and minimize the "pour - filter" temperature spread to achieve maximum throughput.

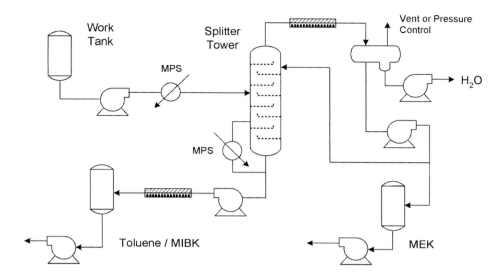

Figure 47. Typical Solvent Splitter Flow Diagram

29.   2-STAGE DEWAXING

A second stage of filtration can be used to reduce the oil in wax and to increase the dewaxed oil yield. Solvent (called repulp or repuddle) is added to the wax from the first stage, typically in the wax scroll or wax boot depending upon the plant design, and pumped to a filter feed drum that feeds the second stage of filters.

Filtrate from the second stage is "lean" in oil and can be used in first stage operation. It may be blended with "fresh" solvent and used as first stage wash, or it may be added as the final increment of dilution in an incremental plant. Recycling the 2nd stage filtrate reduces the overall solvent usage.

30.   DEOILING

The deoiling process is used to produce a hard wax containing a very low oil content. Waxes produced in deoiling have melting point and needle penetration specifications. Waxes intended for food grade use must also meet UV absorption specifications and require wax hydrotreating.

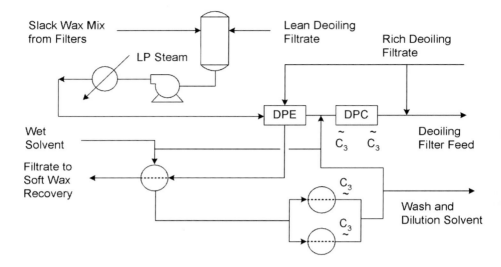

Figure 48. Typical Chilling Section In Recrystallization Deoiling

In recrystallization deoiling, lean solvent is added to the wax from the dewaxing plant and the resulting slurry is pumped through a heat exchanger where the wax crystals are melted. The wax is then recrystallized in SSEs and SSCs. Typically two stages of deoiling are used to meet the low Oil in Wax specification.

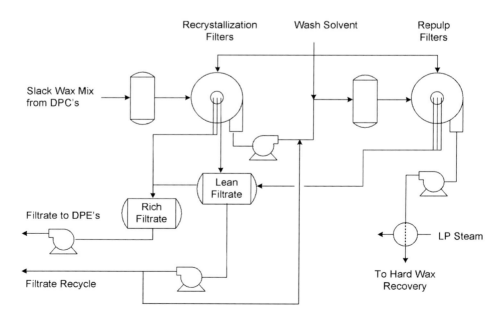

Figure 49. Typical Filtration Section In Recrystallization Deoiling

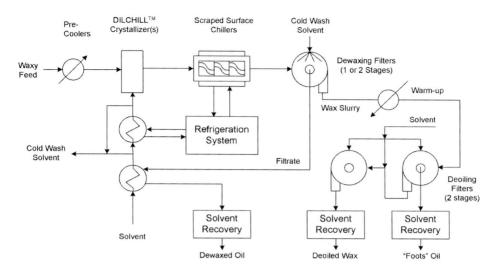

Figure 50. Simplified DILCHILL™ Dewaxing/Warm-up Deoiling Flow Diagram

DILCHILL™ dewaxing is ideally suited for Warm-up deoiling. The spherical nature of the crystals facilitates the melting of the softwax and re-crystallization is not required.

The Deoiling (Filtration) temperature in the deoiling plant is key to meeting final wax properties. As the temperature is increased the soft wax melts and is removed, leaving behind the higher melting, harder wax. Needle penetration decreases while melting point increases. The melting point and needle penetrations define the specification "box" for the manufacturer. Usually the "box" is wide enough that a range of Deoiling temperatures may be used that will produce a hard wax that meets all product specifications. In this case, the manufacturer will select the lowest deoiling temperature that will achieve the desired product properties, because this will give the manufacturer the maximum yield.

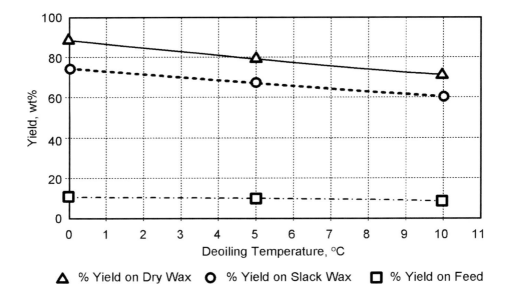

Figure 51. Typical Light Neutral Refined Wax Yields Versus Deoiling Temperature

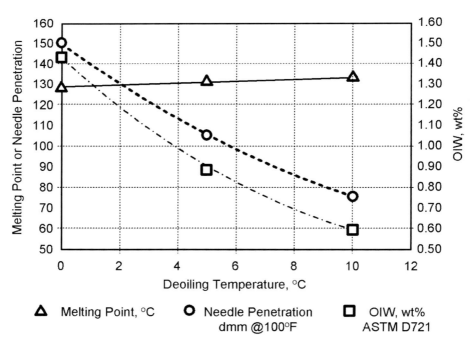

Figure 52. Typical Light Neutral Melting Point and Needle Penetration Versus Deoiling Temperature

Table 13. Comparison of Dewaxing, 2-Stage Dewaxing and Deoiling Heavy Neutral

Yields, vol%	1-Stage Dewaxing	2-Stage Dewaxing	1-Stage Dewaxing 2-Stage Deoiling
Dewaxed oil	68	73	68
Slack Wax	32	27	0
Soft Wax (Foots Oil)	0	0	19
Finished Wax	0	0	13

31. PROPANE DEWAXING

Propane dewaxing uses liquid propane as the solvent. Propane is normally a gas at ambient temperature and the vessels in the unit must be pressure vessels. This also includes the filter. Filtration in propane dewaxing is pressure filtration vs. ketone dewaxing which uses vacuum filtration. The propane temperature depends on the pressure, so that it is of paramount importance that the pressure be controlled. It is often said that in propane dewaxing propane is always either flashing or condensing. This adds an additional level of complexity that is not present in ketone dewaxing.

There are advantages and disadvantages to ketone vs. propane dewaxing. A brief comparison is shown below.

Table 14. Propane Vs. Ketone Dewaxing

	Propane	Ketone
Higher filter rates		
150N	30-50	7-9
600N	18-30	4-5
Bright stock	10-15	2-3
Lower dilution ratios		
150N	1.2-1.6	2
600N	1.4-2.0	3
Bright stock	2-2.2	4
Economics		
Investment	-40%	base
Utilities	Less	base
Operating costs	More	base
Pour point	Limited to -15°C	
2-stage	yes	yes
Deoiling	yes	yes

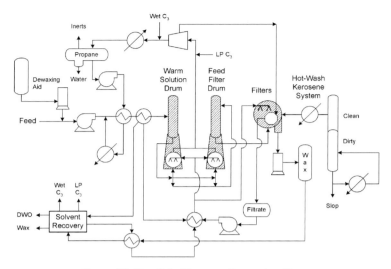

Figure 53. Simplified Propane Dewaxing Plan

Warm liquid propane is added to waxy raffinate and the slurry is prechilled in a shell and tube prechiller, with the slurry in the tube side. The Warm Dilution Ratio is controlled by the operator. Like the feed precooler in ketone dewaxing the outlet temperature of the prechillers is maintained just above the cloud point of the slurry to avoid fouling in the prechiller. Some newer designs use a prechiller tower.

The feed solution passes through the prechillers and into the warm solution drum. Up to this point the propane dewaxing process is a continuous process. The warm solution drum alternates feeding one of two chillers and the process now becomes a batch process.

Large batch chillers cylindrical (older) or spherical in design accept prechilled slurry from the warm solution drum and batch chilling begins. Pressure on the chiller is slowly released and the liquid propane evaporates, cooling the batch of slurry in the chiller.

In older plants control of the "vent gas" sets the chilling rate. This may be done with two pressure control valves (chiller vent valves). Advanced control and valve design has been successfully applied to allow adequate pressure control using a single valve.

Make-up propane is added to replace the vented propane so that at the end of the cycle the dilution in the slurry, the cold dilution ratio (CDR) is at the desired target. Similar to ketone dewaxing where the filtration rate depends on the solvent dilution to the filters, the filtration rate in propane dewaxing is dependent on the CDR that is also the dilution that will be seen at the filters. The CDR also has a big influence on the Pour point. Wax molecules are highly soluble in propane, more than Toluene. Increasing the CDR will carry more wax molecules into the filtrate and increase the pour point of the dewaxed oil. This requires the manufacturer to reduce the filtration temperature (by reducing the filtration pressure) to compensate. Because of the large pour-filter spread that exists in propane dewaxing the base stock pour point is typically limited to -15. This is a drawback for the propane dewaxing process.

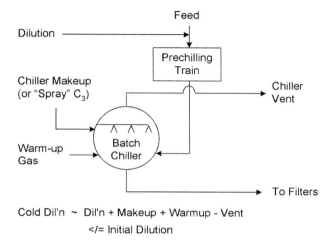

Cold Dil'n ~ Dil'n + Makeup + Warmup - Vent

</= Initial Dilution

Figure 54. Simplified Propane Dewaxer Chilling Section

Slurry in the chiller is cooled to the target conditions and then the bath is transferred to the filter feed drum that provides holding capacity for the filters.

Under ideal conditions the transfer of the cold slurry would occur the moment after the chiller target temperature has been reached. In practice two conditions may exist. If the chiller final chill temperature has been reached and the filters are not ready to accept the new batch, then the slurry is held in the chiller and the slurry "waits" for the filters. This is referred to as "wait time". If the filters are ready to accept the next batch of feed but the chiller has not reached it's final chill temperature so that it is ready for transfer the filter levels will drop and the filters will be starved for feed. This is referred to as "starve time". Chilled slurry is transferred to the filters. After transfer has been completed the chiller is warmed by pressuring with propane, which condenses on the walls of the chiller. The wall temperature is "warmed-up" to a high enough temperature so that when the prechilled slurry from the warm solution drum is fed to the chiller shock chilling at the wall will not occur. Warm-up takes time which limits production. The steps in the chiller cycle are:

1. Fill with prechilled slurry
2. Chill slurry to target temperature by venting propane, add make-up to make target CDR
3. Wait time for filters
4. Transfer slurry to filter feed drum
5. Warm-up

Total cycle times may range from 20 to 30 minutes.

Rotary Pressure filters are used to filter the slurry. Filtration rate is a function of the pressure drop across the filter. The temperature of the slurry also depends on the pressure. This requires carefully controlled pressure balance from the filter feed drum to the filters and to the filtrate drum. If the pressure in the filtrate drum is too high the temperature will be high and the resulting dewaxed oil pour point will be too high. If the filter pressure is not increased to maintain the Delta P then filtration rate will also be reduced.

Wax and propane and oil and propane are sent from the filters to their respective recovery sections. High pressure flash, low pressure flash, stripper and drier vessels are used to recover propane from the product.

31.1   Effect of Water

Solvent "Drip Pipes" which are used to distribute the liquid propane wash over the wax cake in the filters, will foul if the water content in the propane is too high. The water freezes and will reduce the wash flow and/or may change the wash distribution. The Oil in Wax will increase and the overall dewaxed oil yield will drop. Methanol or acetone may be added to the propane to "de-ice" the drip pipes.

32.   2 -STAGE PROPANE DEWAXING

Two stage dewaxing may be used to reduce the Oil in Wax and increase Dewaxed oil yield and to reduce solvent requirements. Repulp/repuddle propane is added to the slack wax from the first dewaxing stage. The slurry is pumped to the second stage filters. Control of the pressure balance is critical to avoid shock chilling and second stage bog-downs that will limit plant throughput. Second stage filtrate may be used as first stage wash or as dilution.

32.1   Propane Deoiling

Propane deoiling is accomplished using high pressure filters that are required to handle the higher temperatures that must be used to melt the softwax. The finished wax produced from deoiling is a valuable by-product and margins may at times exceed dewaxed oil margins.

32.2   Propane Filter Washing with Hot Kerosene

Temperatures required to successfully melt the wax from the filter cloth do not allow the use of liquefied propane due to the very high pressure required.

Instead kerosene is used to wash the filters. Kero wash temperature is typically controlled. If the wash temperature is too low the wax crystals will not be melted, and if the wash temperature is too high wire migration will increase leading to reduced wire life and/or more frequent wire retensioning.

33.   DEWAXING AIDS

DeWaxing Aids (DWA) may be used in ketone dewaxing but are always used in propane dewaxing. Typically the DWA doses used in propane plants are 2-3 times higher than in ketone plants. While DWAs are economically justified for use on all grades in the propane plant, performance on light neutrals in ketone plants has typically not been economically justified.

DWAs are expensive and may represent the single largest controllable operating expense in the propane plant. Filtration rates in the propane plant may be improved from a level that is almost inoperable to several times the highest rate experienced in the ketone dewaxing plant. Oil in wax may be greatly reduced, increasing overall yield and enhancing the wax product value.

DeWaxing Aids (DWA) are required to achieve high filtration rates in propane dewaxing. Typically a dewaxing aid consists of a polymer backbone with alkyl side chains. Factors affecting the DWA performance include:
1.   Raffinate Feed
2.   Polymer "backbone" chemistry
3.   Molecular weight, number distribution
4.   Side chain distribution
5.   Active Ingredient
6.   DWA ratio (when combined with other additives) in blend
7.   DWA dose, the concentration used
8.   Asphaltene contamination
9.   Regulatory Requirements (FDA approval)
     Major DWA Polymer Backbone Chemistries include:
1.   Poly Alkyl Methyl Acrylate (PAMA)
2.   Poly Alkyl Acrylate (PAA)
3.   Co-Polymerization of PAMA, PAA
4.   Di Alkyl Fumerate Vinyl Acetate (DAFVA)
5.   Ethylene Vinyl Acetate (EVA)
6.   Wax Naphthalene Condensate

Molecular weights of the DWA may vary from 10,000 to 1,000,000. Molecular number, the "branchiness" may range from 7,000 to 300,000. Side chain lengths may vary from C_{14}-C_{26} and with various distributions.

Active ingredient may range from 10-100% with 15-30 most typical. Dewaxed oil (Light or Heavy Neutral) or toluene may be used as diluent. DWA viscosity will depend on the active ingredient, and the diluent and the viscosities can be quite high and may affect pumpability.

Typically DWA is stored at elevated temperatures. Positive displacement or centrifugal pumps are used for injection and various meters may be used to monitor the flow rate. Mass flow meters have been used with great success in the application.

34.   DWA MECHANISM

The exact mechanism of how the Dewaxing aid works is still being studied. Leading theories include:
1. Co-crystallization of the DWA into the wax crystal matrix that changes growth direction of crystal planes
2. Agglomeration Mechanism
3. Wax Crystal Modifier that associates with the surface to change crystal growth plane

Studies have shown that DWAs may be combined to give synergistic performance. Currently, because of the large number of variables (feed, DWA chemistry, etc.) it is not possible to predict DWA performance a-priori and DWAs must be tested in a pilot plant that can simulate the propane dewaxing process. Lab and analytical tests typically over predict DWA performance due to favorable conditions of the lab and inability to simulate the plant.

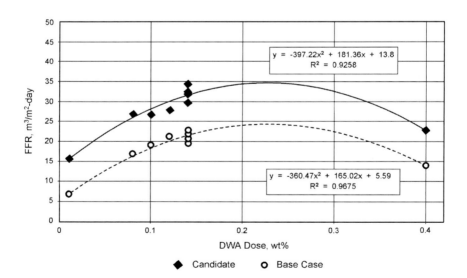

Figure 55. Light Neutral DWA Dose Response Curve, DWA Ratio 1:1

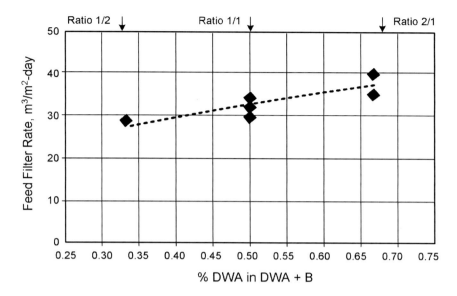

Figure 56. Effect of DWA Ratio on Light Neutral

35. ASPHALTENE CONTAMINATION

The presence of asphaltenes in the feed to the propane dewaxer or in the propane will significantly reduce the filtration rate and must be avoided. Asphaltenes may be present in the heavy neutral due to entrainment from the VPS section below the heavy sidestream drawoff. Contamination may occur if the lube deasphalting unit (LDU) and propane dewaxer share propane.

36. REGULATORY REQUIREMENTS

The manufacturer using DWAs must determine if a FDA approved DWA is required in their product market. The number of DWAs meeting FDA technical requirements is small and these DWAs command a premium in the marketplace. The qualification of new DWAs is expensive and takes a long time. DWA chemistry type, Molecular weight (Mw), Molecular number (Mn), PolyDIspersity (PDI = Mw/Mn), unreacted monomer concentration and concentration in the final product are some of the major factors the FDA considers and the candidate must pass technical specifications in these areas in order to obtain approval.

Table 15. Analytical Tests in Dewaxing

TEST	ASTM Test No	APPLICATION
Pour Point	D 97, D 5949, D 5950, D 5985, D 6749	Key specification for dewaxed oil
Oil-in-wax %	D 721, D 3235	Wax specification, measure of yield performance
RI @ T °C	D 1218	Yield calculation, solvent composition
Miscibility Curve		Used to set solvent composition for each stock to avoid immiscible operation
Gas Chromatographic Distillation	D 2887	GCDs may be for troubleshooting and predicting dewaxing performance
Feed Cloud Point	D 2500, D5551 D 5771, D 5772, D 7773	Helps set feed precooler outlet temperature
Water-in-Solvent	D 6304-03, E 203-1	Excess water may lead to immiscible operation and icing of solvent chillers

37. GLOSSARY

Aromatic — An unsaturated ring compound having a basic 6-carbon-atom ring with either a hydrogen atom or a chain joined to each carbon atom.

API — American Petroleum Institute

ATF — Automatic transmission fluid

Base oil — "A base oil is the base stock or blend of base stocks used in API licensed oil " (See API 1509)

Base stock — "A base stock is a lubricant component that is produced by a single manufacturer to the same specifications (independent of feed source or manufacturer's location); that meets the same manufacturer's specification; and that is identified by a unique formula, product identification number, or both." (See API 1509)

Base stock slate — "A base stock slate is a product line of base stocks that have different viscosities but are in the same base stock grouping and from the same manufacturer." (See API 1509)

Batch — Any quantity of material handled or considered as a unit in processing.

Batch treat — A treatment of a limited quantity of material with chemicals or solvent to improve quality.

Blowback — The term applied to blowgas usage. Blowback Gas used to "blow" the wax cake from the filter cloth by being applied under pressure beneath the cloth

Bright Stock	A Vacuum Distillation Tower Bottoms that has been deasphalted and extracted or hydrotreated and probably dewaxed as a heavy Base stock
Caulking Grooves	"U-shape" piece of metal welded to the drum deck. The resultant "grooves" are used with "caulking bars" that hold the filter cloth in place on the surface of the filter drum. The grooves are slightly slanted inward at the top to provide tension on the bar and help hold the caulking bar in the groove. The upper edges of the U shape are lipped to provide an edge to retain the grid on which the filter cloth rests.
CCR	Conradson Carbon Residue
Controlled Chillers	SSE using cold filtrate to cool the slurry
Carryover	Term in solvent extratic which refers to entrainment of heavy phase into the light phase and out with the raffinate solution (essentially solvent at the tower top).
Carryunder	Term in solvent extratic which refers to entrainment of light phase which has not settled from the heavy phase and exits with the extract solution (essentially feed at the tower bottom).
Centistoke	The worldwide unit of kinematic viscosity.
Coalescence Rate	Relative rate at which fine, small droplets form larger droplets, which settle faster, forming a clarified oil layer.
Continuous Phase	Heavy, NMP-rich phase which forms an unbroken stream through the treater in solvent extraction.
Corrosion	The gradual eating away of metallic surfaces as the result of oxidation or other chemical action. It is caused by acids or other corrosive agents.
Crystallization	May be defined as a phase change in which a crystalline product is obtained from a solution
Cut	The portion or fraction of a crude oil boiling within certain temperature limits.
Cut point	The temperature limit of a cut or fraction, usually but not limited to a true boiling point basis.
DAO	Deasphalted Oil: The extract or residual oil from which asphalt and resins have been removed by an extractive precipitation process called deasphalting.
DAU	DeAsphalting Unit: A process for removing asphalt from reduced crude or vacuum residua (residual oil) which utilizes the different solubilities of asphaltic and nonasphaltic constituents in light hydrocarbon liquids, e.g., liquid propane.
Density	The mass of a unit volume of a substance.

Deck Pipes	Each segment of a dewaxing filter is drained by a series of pipes located along side each caulking groove
DILCHILL™	DILution CHILLing. ExxonMobil process where chilled solvent is added stagewise to an agitated tower. Mixing, cooling and crystallization occur in the tower
Dilution Ratio	Term used in solvent dewaxing to refer to the ratio of solvent to waxy raffinate feed (vol/vol)
Dispersed Phase	Light, oil-rich phase which "bubbles" through continuous phase, then settles, and repeats the process on each tray in solvent extraction.
Distillate	Product of distillation collected by passing vapors through a condenser.
DPC	Double Pipe Chiller
DPE	Double Pipe Exchanger
Drum	The horizontal drum on which the filter cloth is applied to its circumference surface in solvent dewaxing.
Dry Wax	The solid component present in a waxy feed at a certain oil pour point if all oil were removed from it
DWO	DeWaxed Oil
Entrainment	Entrainment is the carryover of liquid by the vapor phase. Liquid may be in the form of a spray, foam or mist.
EXOLFINING™	An ExxonMobil® extraction process used to extract aromatics from lube feedstocks to improve the viscosity index and quality of lubricating oil base stocks, integrated with HYDROFINING
Extract	In solvent refining processes, that portion of the oil which is dissolved in and removed by the selective solvent; the solvent rich phase. Contains a low percentage of oil which is typically high in aromatic content.
Extraction	The process of separating a material, by means of a solvent, into a fraction soluble in the solvent and an insoluble residue.
Feed Retention Drum	Feed solution hold up drum
Feed Solution	Waxy feed plus warm propane in propane dewaxing.
FFR	Feed Filter Rate (M^3/day/M^2 of filter cloth). Refers to feed only in solvent dewaxing.
Filter Cloth Blinding	Term applied to the reduction in filter rate due to wax or ice particles plugging up the flow paths in the filter cloth
Filter Hood	The top part of the filter outer enclosure

Filtrate (or Wet) Port   The section or nozzle on the master valve that collects filtrate from the vat slurry

Filtrate   Liquid product (oil plus solvent) in solvent dewaxing.

Final Dilution   The last increment of solvent added after the shock chillers in incremental dilution solvent dewaxing.

Foots Oil   The soft wax melted or dissolved/washed away during wax deoiling. In the old days, it was the oil drawn from the bottom (foot) of a pan separator for the wax formed in a wax "sweater".

Fractionation   Fractionation is the separation of light and heavy fractions in the distillation tower.

Furfural   An aldehyde obtained from corn shucks, wheat, or oat hulls, used in an extraction process for removing aromatic, naphthenic, olefinic and unstable hydrocarbons from a lubricating oil charge.

Haze   Visible, uncombined or flocculated, wax or water in Dewaxed Oil

Hot Wash   The technique or the hot solvent used to periodically melt and dissolve wax out of the filter cloth

Immiscible   Two separate phases, one oil rich and one solvent rich, which will not mutually dissolve. Normally not desirable in solvent dewaxing

Incremental Dilution   Refers to the addition of solvent, in increments to the waxy raffinate feed as it flows through the heat exchanger equipment leading to the solvent dewaxing filters.

Internal Reflux   Portion of the light phase which springs from the heavy phase due to lower temperature at treater bottoms, and recycles back up the tower in the dispersed phase of solvent extraction.

Kinematic viscosity   The ratio of the absolute viscosity of a liquid to its specific gravity at the temperature at which the viscosity is measured.

L/S   Liquids to solids ratio in the wax cake. The liquids consist of oil and solvent OR liquids to solids in the filter vat

LHU   Lube Hydrofining unit to improve color stability of lube oils using hydrogen

Lead Pipes   The pipes on the leading side of each segment as the drum rotates

Lubricant   Any substance interposed between two surfaces in relation to motion for the purpose of reducing the friction and/or the wear between them.

LPS   Low Pressure Steam

LVPS   Lubes Vacuum PipeStill

Master Valve   Trunnion Valve, The stationary sectioned casting in contact with the rotating end of the drum (i.e., the wear plate) at which the lead and trail pipe manifolds are dead-ended in solvent dewaxing filtration.

MEK   Methyl Ethyl Ketone or loose term for ketone dewaxing unit

MEK OIW Test   Oil content of wax using ASTM D921 (MEK solvent at -25°F)

MIBK   Methyl IsoButyl Ketone

Miscible   Two or more liquids which mix to form one homogeneous phase. Usually preferred in solvent dewaxing. Refers to oil and solvent mixture

Miscibility temperature   Temperature at which solvent and feed are completely miscible - all feed dissolved in solvent - and there is no phase separation in treater tower or in dewaxer slurry.

MPS   Medium Pressure Steam

NMP   N-methyl-2-pyrrolidone, a solvent used as an alternate to furfural and phenol for the extraction of lubricating oil fractions.

Naphthene   A group of cyclic hydrocarbons also termed cycloparaffins. Polycyclic members are also found in the higher boiling fractions.

Naphthenic crudes   Class designation of crude oils containing predominantly naphthenes or asphaltic compounds.

Neutral   A VPS distillate, extracted and dewaxed, made into a Base stock

Normal paraffin   A straight chain hydrocarbon in which no carbon atom is united with more than two other carbon atoms.

Pale oil   A petroleum lubricating or process oil refined until its color is straw to pale yellow.

Paraffin-base crudes   Crude containing paraffin wax and practically no asphalt or naphthenes.

Paraffinic   Describing the paraffin nature or composition of crude petroleum or products therefrom.

Paraffins   A homologous series of open-chain saturated hydrocarbons of the general formula C_nH_{2n+2} of which methane (CH_4) is the first member.

Paraffin wax   A colorless wax extracted from paraffin-base lubricating oils. Typically solid at room temperature.

PDU   Propane Dewaxing Unit to remove wax from oil using propane as a solvent

PNA	PolyNuclear Aromatic. A compound composed of two or more aromatic rings (see aromatic). These compounds may impart unwanted color, cause stability problems and are classified as carcinogens.
Polarity	A measure of the asymmetric distribution of a molecule's electrical charge. MEK>MIBK>H_2O
Pour Point	The temperature at which oil will not pour
Predilution	Solvent addition occurring before the precoolers. Sometimes called primary dilution.
Prechillers	In propane dewaxing, Shell and Tube exchangers that chill the feed solution before batch chillers
Raffinate	In solvent-refining practice, that portion of the oil which remains undissolved and is not removed by the selective solvent; the solvent lean phase. Contains a low percentage of solvent and typically a low aromatic content oil.
Rectification	Rectification is the removal of heavy material from the vapor phase in fractionation.
Saybolt Universal Seconds (SUS)	A measure of kinematic viscosity, expressed as the time in seconds for 60 ml of fluid to flow through a standard Saybolt Universal viscometer at a specified temperature. ASTM Method D-88 describes the method and apparatus.
Scale Wax	Wax with an oil content of 1-5%, e.g. Wax from a 2-stage dewaxer
Segment	Each solvent dewaxing filter drum segment is made up of a pair of the caulking grooves running longitudinally on the drum deck with the grid in between
Selectivity	A measure of the ability of a solvent to separate compounds of different structure, e.g. aromatics from paraffins from naphthenes.
Service factor	A quantity which relates the actual on-stream time of a process unit to the total time available for use of the unit. Frequently a ratio of the number of actual operating days divided by 365.
Settling Rate	Rate at which droplets of the dispersed phase rise through and separate from the continuous phase.
Shock Chillers	Double pipe exchangers (see SSE) that use propane or propylene as a refrigerant in solvent dewaxing. Rapid cooling rates (>2%/min) "shock" the slurry resulting in smaller crystals and lower FFR.

Shock Chilling	Rapid uncontrolled crystallization, characterized by high nucleation rates producing very small average crystal size.
Slack Wax	Wax with an oil content >5%, e.g. Wax from a 1-Stage Dewaxer
Solubility	Degree to which the oil (especially the aromatic fraction) dissolves in the solvent. This is a function of aromatic type and concentration, changes from one grade to another as well as for the same grade from one crude to another, for a given solvent.
Solvent neutral oil (SNO)	A paraffinic base oil which has been solvent refined, dewaxed, and finished and is ready to be used in blending or compounding.
SSC	Scraped Surface Chiller. Usually a double pipe design of 6, 8, 12 inch diameter with internal scrapers to remove wax from the cold walls. Typically a refrigerant such as propane, propylene or ammonia is used on shell side. Used in conventional dewaxing
SSE	Scraped Surface Exchanger. Usually a double pipe design of 6, 8, 12 inch diameter with internal scrapers to remove wax from the cold walls. Typically cold filtrate is used on shell side. Used in conventional dewaxing
SSU	See Saybolt Universal Seconds.
Stability	Resistance to chemical change.
Stripping	Stripping is the removal of light material from the liquid phase.
TAN	Total Acid Number
Third Phasing	Immiscible condition in a dewaxer where you have two liquid phases and a solid (wax phase)
Trail (or Lag) Pipes	The pipes on the trailing side of each segment as the drum rotates
Trunnion Valve	Master Valve, the stationary sectioned casting in contact with the rotating end of the drum (i.e., the wear plate) at which the lead and trail pipe manifolds are dead-ended
Vat	The bottom part of the filter outer enclosure into which the filter feed slurry flows
Viscosity Index	A measure of the change in viscosity with temperature; ASTM D-2270.
VPS (VDU)	Fractionation equipment, a vacuum pipestill (VPS) or vacuum distillation unit (VDU) is used to distill atmospheric bottoms into gas oil or lube distillate cuts.

VTB	Vacuum tower (VPS/VDU) bottoms, or vacuum residue or vacuum resid
Warm-up	A step in the batch chiller sequence during which the chiller internal metal wall is warmed in preparation of accepting feed, and avoiding shock chilling at the wall
Warm Solution Drum	Drum containing prechilled feed solution
Wash	Generally the term used for the cold solvent applied to the wax cake in solvent dewaxing.
Wax Cake	A cake formed on the surface of the filter in solvent dewaxing. Typically consists of a Liquids (solvent plus oil) to Solids (oil free wax) ratio of 2-10. Wax Doctor Knife dislodges the cake.
Wax, Refined	Wax of, usually, 0.5-2% oil-in-wax produced from deoiling
Wax, Scale	Wax of, usually, 2-10% oil-in-wax produced from two stage dewaxing.
Wax Scroll	The screw conveyor rotating in the wax trough. All scrolls in dewaxing service are the center- discharge type
Wax, Slack	Wax of, usually, <10% oil-in-wax produced from one stage dewaxing.
Wax Trough	The semi-circular channel that the wax cake falls into after blow gas dislodges it and into which the wax doctor knife deflects it

38. ACKNOWLEDGEMENTS

A Special Editorial thanks to Mike Davis and XB Cox at ExxonMobil Research and Engineering Co., for their comprehensive detailed review and recommendations.

Special thanks to ExxonMobil employees and annuitants: Bob Aupperlee, Doug Boate, Joe Boyle, Barry Deane, Sasha Glivicky, Dave Mentzer, Dominick Mazzone, Chuck Quinlan, Ken Del Rossi, Evelino Ruibal, Dave Sinclair, Bernie Slade, and Howard Spencer.

39. REFERENCES AND ADDITIONAL READINGS

1. Butler, R. M.; Tiedje, J. L. "The Washing of Wax Filter Cakes," *Can. J. Technol.* **1957**, *3(1)*, 455-467.
2. Citarella, V. A.; Ruibal, E. A.; Zaczepinski, S.; Beasley, B. E. "Crystallization Technique to Simplify Dewaxing", *Pet. Technol. Quartely*, Winter **1999/2000**; pp. 37-43.
3. Klamann, D. *Lubricants and Related Products*, Verlag Chemie Gmbh, D-6940: Weinheim, 1984.

4.   Sequeira, A. "Lubricant Baseoil and Wax Production", Marcel Dekker: New York, 1994.
5.   Fiocco R.J. "Development of the Cascade Weir Tray for Extraction", *AIChemE Symposium Series, New Developments in Liquid-Liquid Extraction*, ISEC 83, **1984**, *80*, 89-93.
6.   Sankey, B. M. et al. "EXOL N:New Lubricants Extraction Process", *Proc. Tenth World Petroleum Congress,* **1979**, *4*, 407-14.
7.   Bushnell, J.D.; Fiocco, R.J. "Engineering Aspects of the Exol N Lube Extraction Process", *Proc. - Refining Department American Petroleum Institute*, **1980**, *59*, 159-67.
8.   Sankey, B.M. "A New Lubricants Extraction Process", *Can. J. Chem. Eng.*, **1985**, *63*, 3-7.
9.   Davis, M. B. et al. "The EXOL N Extraction Process - Flexibility and High Efficiency to Meet Modern Lubes Product Requirements", *Advances in Production and Application of Lube Base Stocks,* Indian Institute of Petroleum, Nov. 23-5, 1994, Tata McGraw Hill, New Delhi, India, pp. 24-32.
10. Gudelis et al. "Improvements in Dewaxing Technology", *American Petroleum Institute Proceedings - Division of Refining,* **1973**, *53*, 725-37.
11. Bushnell, J. D.; Eagan, J. F., "Commercial Experience with DILCHILL™ Dewaxing", paper F&L-75-50, 1975, NPRA Fuels and Lubricants Meeting, Sept., 11-12, 1975, Houston, Texas,
12. Gudelis, D. A. et al, "New Route to Better Wax", *Hydrocarbon Process.*, **1973**, *52(9)*, 141-6.
13. Wax HYDROFINING™, *Pet./Chem. Eng*, **1970**, *42(9)*, p. 36.
14. *The Exxon Wax HYDROFINING™ Process*, Exxon Research and Engineering Co, Technology Licensing Division, February, 1986.
15. Eagen, J.F., et al., "Successful Development of Two New Lubricating Oil Dewaxing Processes", *Proc. Ninth World Petroleum Congress*, Vol. 5, Applied Science: London, 1975; pp. 345-357.
16. *Engine Oil Licensing and Certification System*, American Petroleum Institute, API 1509, Fifteenth Edition, April 2002.
17. R.R. Savory, "Chaper 11: Base Oil Processes", *Modern Petroleum Technology, Volume 2 Downstream,* The Institute of Petroleum, John Wiley and Sons, 2000.
18. B.C. Deane, "Chapter 25: Base Oil Quality", *Modern Petroleum Technology, Volume 2 Downstream,* The Institute of Petroleum, John Wiley and Sons, 2000.

Chapter 16

SELECTIVE HYDROPROCESSING FOR NEW LUBRICANT STANDARDS

I. A. Cody
ExxonMobil Research and Engineering Co
Process Research Laboratories
Baton Rouge, LA 70805

1. INTRODUCTION

World basestock demand is expected to remain steady or rise only slightly in the next decade, but the nature of basestocks is anticipated to change dramatically, primarily driven by tougher specifications for automotive lubricants.[1,2] Automobile manufacturers continue to seek lubricants that provide better fuel economy, lower emissions and longer life; for the refiner this means producing basestocks with lower viscosity, lower volatility and higher saturates content. Lower viscosity of the formulated oil can improve both cold start performance and fuel economy by reducing friction in the engine. Lower volatility directly lowers emissions and minimizes oil thickening, extending oil change intervals.[3,4] Also, with less volatiles there is less stress on catalytic converters. Higher saturates can help make oils more stable by lessening oxidation, thereby extending service. These trends in formulated oil performance present difficulties to the refiner in making suitable basestocks for these formulations using conventional processing strategies. For example, lower viscosity cannot be achieved merely by cutting a lighter stock because that would increase volatility. Conversely, cutting deeper to resolve volatility adversely affects fluidity and fuel economy. Separations-based processing alone can be inadequate because the hydrocarbon molecules with the required properties are not abundant in many of the mineral crudes refined today. Accordingly, many lube refiners have elected to incorporate new conversion technologies[5] to boost populations of the desired molecules by selectively transforming the indigenous hydrocarbons. At the same time, some of the largest oil companies are poised to invest in "Gas-to Liquids" (GTL) technologies that can produce basestocks with excellent properties. This chapter describes some of the current and future hydroprocessing technology that can generate basestocks capable of meeting the new standards in lubricant quality.

Until recently, improvements in standards for passenger vehicle lubricants (PVL's) and commercial vehicle lubricants (CVL's) were achieved largely with the use of better additives, such as anti-oxidants, antiwear agents, detergents, and viscosity improvers. In the 1990's that began to change as the equipment manufacturers accelerated efforts to improve automotive performance (see Table 1). Additives alone have not been able to address the new requirements, resulting in dramatic changes to basestock properties and consequent refining strategies.

Table 1. Evolution of Multi-Grade PVL's

1930	API/SA	Mineral Oil + castor oil
		No/few additives
		No performance requirements
1950s	API/SB	Antioxidants, antiwear introduced
		Take head off to "de-coke" engine after 20,000 miles
1960s	API/SC	Detergency to control high temperature deposits
		Multi-grade introduced
1964-1980s	API/SC to SG	Tripartite of ASTM.SAE/API control quality
		Requirements (new additive technology)
1990-present	GF, CF Categories	Rapid change, Equipment manufacturers control quality targets (better basestocks for e.g. fuel efficiency, reduced emissions)

A simple measure of the impact on basestocks resulting from changing engine oil specifications is the empirical property of viscosity index (VI). For example, in North America in the past decade, SAE 5W-30 oils have required the light basestock VI target to increase from about 100 to 115 due to the progressively tougher ILSAC, GF-1, GF-2 and GF-3 standards. This target is achievable only in low yield from most crudes by the conventional, separations based, processing steps of vacuum distillation, solvent extraction, and solvent dewaxing.   Similar trends have occurred with ACEA requirements in Europe.

VI is a convenient guide to low temperature viscosity and volatility, properties that really underpin automotive oil performance. This is illustrated in Figure 1. Basestocks meeting the 5W specification have a kinematic viscosity of about 4 to 4.5 cSt at 100c, and, until the 1990's, the acceptable limit of formulated oil volatility for this grade was as high as 30 wt%, as measured by the Noack volatility test.[6] Basestocks with about 95 VI could be blended to this target and at the same time stay below the maximum allowable low temperature dynamic viscosity limit for a 5W- oil. However, in subsequent years, the industry impetus to reduce volatility while maintaining low kinematic and dynamic viscosity has resulted in ever higher VI basestocks to achieve GF-1, then GF-2, and now GF-3. Today, with a 5W formulated oil volatility of 15 wt % Noack, basestocks suitable for GF-3 require about 115 VI. Further changes in basestock quality are expected in 2003 with the introduction of GF-4.

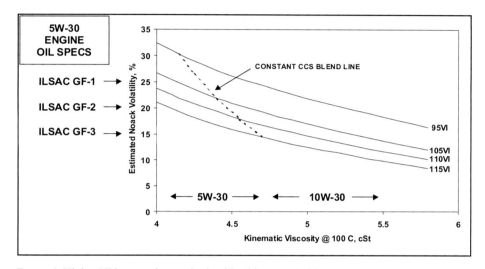

Figure 1. Higher VI basestocks required to blend lower volatility engine oils (from reference 3)

Basestocks with higher VI provide value to the formulator in two ways: the "volatility@viscosity" relationship is better and at the same time the low temperature viscometric properties are improved. The sloping blendline in Figure 1 shows that the same low temperature Cold Cranking Simulator (CCS) viscosity value is achieved at higher kinematic viscosity when the basestock VI is higher. This allows the refiner the additional flexibility to cut a basestock with higher kinematic viscosity, still meeting the CCS target and lowering volatility even further.

Refiners have only two ways to tailor basestock volatility while holding viscosity constant. One is tighter fractionation: by narrowing the boiling range distribution of a basestock, populations of low boiling hydrocarbons are minimized and volatility is accordingly improved. Since the mid-boiling point remains constant, the volatility@viscosity relationship is also improved. However, this strategy has limited value because volatility improvements are ultimately dictated by the nature of the molecules within the boiling range envelope. Step improvements in volatility@viscosity can only be achieved by raising VI.

This is evident in Figure 2, which shows the relationship between boiling point (and the associated Noack volatility) for a basestock with a typical boiling range distribution versus kinematic viscosity. Each curve represents a class of lube hydrocarbons that have been assayed into fractions with viscosities spanning the typical lube range of 10 to 1000 cSt at 40°C (corresponding to light basestocks to Brightstocks derived from vacuum tower bottoms). Boiling point rises (and volatility falls) as viscosity increases for a given hydrocarbon class. But to raise boiling point while maintaining viscosity requires a shift to a new class, linked to a higher VI. For example, at 30 cSt @ 40°C a basestock with 100 VI has a mid boiling point around 425°C, i.e. about 15 wt% Noack volatility, whereas a basestock of the same 40°C viscosity with 140 VI has a mid-boiling point of about 465°C, fully 40°C higher boiling point. This corresponds to a Noack volatility of less than 12 wt%.

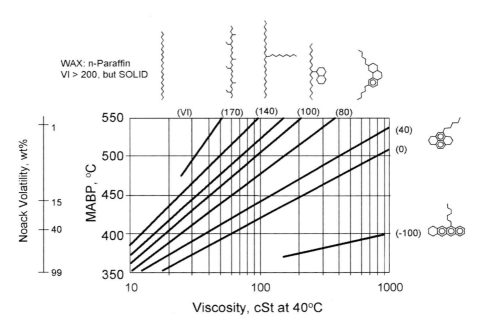

Figure 2. Isoparaffin molecules needed for better volatility (from reference 10)

Viewed in the context of the specification changes discussed above, while 100 VI was sufficient to achieve 15 wt% Noack volatility in a 10W type PVL, i.e. with basestock viscosity around 30 cSt, 115 VI or greater is needed to achieve that same volatility standard in a 5W formulation at 25 cSt. The molecular implications of VI improvement associated with increasing boiling point while fixing viscosity are shown in the perimeter of the chart: the basestocks must comprise hydrocarbons that have regions of more paraffinic character with fewer ring clusters, either naphthenic or aromatic. Ultimately, straight chain normal paraffins have the best volatility@viscosity characteristic, but, since they also have high melting points (i.e. are waxes), they are unsuitable for use in automotive lubricants across the full temperature range of application. From a purely viscometric and boiling point perspective, the most attractive hydrocarbons for lubrication are the lightly branched isoparaffins.

In large part, hydroprocessing technology has been introduced to make basestocks with higher boiling point for a given viscosity by selectively boosting the population of isoparaffins with catalytic methods that eliminate or convert ring species and/or that isomerize normal paraffins. An associated benefit of hydroprocessing is that aromatic ring species become saturated, contributing to an overall improvement in oxidation performance of the finished oil.

Hydroprocessing is the means by which refiners are now producing basestocks with higher saturates and VI, typified by the API basestock categories, Group II and III, see Table 2.

Group IV basestock, typified by polyalphaolefins, derived from the oligomerization of selected olefins, represent the pinnacle of basestock performance, but, because of their high cost to manufacture, have so far occupied only a small niche in top tier.

Lower cost Group III basestocks, particularly those derived from GTL, are expected to partially displace PAO's.

Table 2. American Petroleum Institute (API) Basestock Categories

API Group	Sats		Sulfur	VI	Typical Manufacturing Process
I	<90%	and/or	>0.03%	80-119	Solvent Processing
II	90+%	and	<0.03%	80-119	Hydroprocessing
III	90+%	and	<0.03%	120+	GTL; Wax Isom.; Severe hydroprocessing
IV					Polyalphaolefins (PAO)
V					All other basestocks

2. HYDROPROCESSING APPROACHES

Today, hydroprocessing plays many roles in the manufacture of basestocks and specialties. In some cases all-catalytic technologies have replaced separations based processing, while in others, refiners have inserted hydroprocessing into the pre-existing solvent-based technology train. The first strategy involves higher capital cost but can provide lower overall operating expense; the second offers a lower capital cost route that can leverage the existing facilities.

A composite of several possible schemes for basestock and specialties manufacture (that no single lube plant employs in total) is illustrated in Figure 3.

Group I basestocks are generally made by the traditional separations-based sequence of distillation, solvent extraction, and solvent dewaxing. These steps tailor the volatility, viscosity and low temperature properties by shaping the boiling range, and then removing the most aromatic species (by extraction with e.g. furfural, phenol or N-methyl pyrollidone) followed by fractional crystallization (to separate the highest melting species, generally paraffins) using ketones. An alternative to solvent dewaxing of raffinates is catalytic dewaxing in which the wax is converted to lower melting products by either boiling range conversion or isomerization. This step is lower cost than solvent dewaxing and is used in areas where wax is valued as fuel only. An example of this technology is ExxonMobil's MLDW™ process.[7]

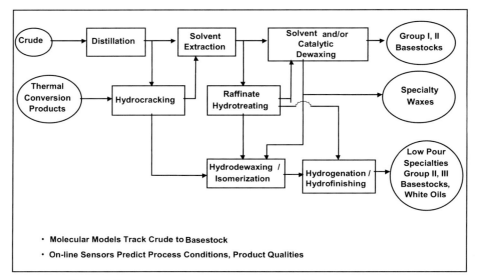

Figure 3. Hydroprocessing technology developed for entire process train

Finally, a hydrofinishing step is often used to improve color and stability by saturation of trace multiring species and by removing some residual polar hydrocarbons, e.g. those containing nitrogen. Basestocks from typical crudes have VI ranging from 95 to 105, with saturates contents 65 to 85% and sulfur levels usually above 0.1 wt%.

An "all catalytic" approach to basestock manufacture is to feed virgin gas oils (from vacuum distillation of crude) and/or thermally converted feeds, such as coker gas oils, through a sequence of hydrocracking, hydrodewaxing, and hydrofinishing (hydrogenation).[8] In the first step, a high pressure, high temperature, hydrocracker operation generates species that have increasing paraffinic character. Next, the lube boiling fractions are block processed through a moderate pressure hydrodewaxing step, in which the most highly paraffinic hydrocarbons are hydroisomerized into isoparaffins. Finally, the hydrodewaxed streams cascade over a saturation catalyst operating at moderate to high pressures to hydrogenate most of the remaining aromatics into naphthenes. Depending on the gas oil properties (such as wax content) and the processing severity, basestocks meeting Group II and III requirements, i.e. with saturates in the high 90's and VI's ranging all the way to 150, can be produced. Such processes are more flexible than separations based refining approaches in that they can utilize a wider range of crudes and achieve unique properties. For example, basestocks may have excellent low temperature viscosity, features attractive for use as refrigerator oils and transformer oils; others may be deeply saturated to meet the most stringent specifications for use as "white" oils for application in the food and cosmetics industries.

Some refiners use combinations of separation and catalytic processes to retain specialties like waxes, a by-product of the solvent dewaxing step, and/or to leverage the hydroprocessing step. A raffinate hydroprocessing step inserted between extraction and solvent dewaxing can be designed to achieve the same kind of basestock upgrading achieved by a VGO hydrocracker but at milder conditions using less hydrogen. At the same time the extraction step preceding the hydroprocessor can operate at higher yields and rates because it is not needed to achieve final basestock property targets like VI. ExxonMobil's Raffinate Hydroconversion (RHC™) process is a recent example of this approach.[9,10]

An alternative process may involve a sequence of raffinate hydrotreating, hydrodewaxing and a hydrofinishing step. In either case, a range of Group II and III basestocks and specialties with excellent properties are feasible.

To optimize overall processing, comprehensive models are increasingly used to relate crude and vacuum distillate properties to the required process conditions needed for the catalysts to mediate the desired reactions. In particular, a compositional modeling technology, "Structure-Oriented Lumping" (SOL) developed by ExxonMobil, provides a way to describe crudes, intermediates, and finished oils mathmatically by combining high-detail hydrocarbon analysis and reaction rules.[11] SOL, in combination with refinery process models, makes it possible to evaluate the economics of any lube crude and predict refining yields and product performance. Supplementing the models are on-line sensors that provide immediate feedback on the approach to target.

3.   CHEMICAL TRANSFORMATIONS

Three classes of catalyzed chemical transformation characterize how modern hydroprocessing is used to improve basestock properties-two concerning ring structures and the other paraffins.

3.1   Ring Conversion

The most facile reactions in a hydrotreating or hydrocracking environment involve the removal of sulfur and nitrogen. These species are present in both aliphatic and ring structures, and at moderate severity conditions essentially complete ejection can be achieved of even the most refractory of these heteroatoms, such as those embedded in structurally hindered rings. For example on a medium viscosity raffinate feed (5 cSt at 100C), conditions around 350°C and about 40 bar H_2 partial pressure and at 1.0 space velocity on a contemporary hydrotreating catalyst comprising Co, Ni, Mo or W on an alumina support may be sufficient to convert most liquid phase sulfur and nitrogen into H_2S and NH_3. Their removal opens out the heterocycle into a more aliphatic arrangement, resulting in a modest increase in basestock VI. Generally, these transformations alone are insufficient to improve basestock properties to the level needed to meet current and future automotive standards. Instead, higher severity processing is required.

At the next level, the reaction that contributes most to VI improvement with contemporary catalysts at hydotreating conditions is ring de-alkylation of multiring hydrocarbons. As side chains are cleaved by a beta scission step, both the side chain and ring fragments may fall below the front of the lube boiling range, leaving behind more paraffinic hydrocarbons in the lube range. Conditions to achieve this transformation must also be conducive for ring saturation, requiring moderate to high pressures. Accordingly, the remaining rings in the lube range become progressively more naphthenic. The "saturation-de-alkylation" zone typically occurs at reactor temperatures from 350°C to 380°C (assuming 1.0 space velocity) which requires pressures to be higher than about 60 bar H_2 in order for the aromatic-naphthene equilibrium to favor saturation. (See below for further discussion of saturation strategies).

At the end of the hydroconversion scale, both ring and paraffin type hydrocarbons undergo hydrogenolysis as hydrotreating gives way to hydrocracking, resulting in a significant shift of lube range hydrocarbons to lower boiling species. Hydrocracking is generally mediated by catalysts with an acid function to promote the formation of carbenium ions, leading to carbon-carbon cleavage.[12] These reactions are prominent above about 380°C (at 1.0 space velocity), and require pressures of 100 bar H_2 or higher to sustain saturation rather than aromatics formation mitigating both coke on the catalyst (associated with catalyst deactivation) and poorer basestock oxidation stability.

The overall change in composition of the remaining lube range (343°C+) hydrocarbons resulting from the progressive hydroconversion of a heavy lube distillate is illustrated in Figure 4. Looking at the composition on a dewaxed oil basis, it is evident that paraffins and mononaphthenes grow in population at the expense of aromatics and polynaphthenes. Viscosity declines, as does the average boiling point of the 343°C+ envelope. However, the net gain in paraffinic character of the basestock means that viscosity must decline faster than the mid-boiling point, corresponding to a VI rise. That is, for the highly hydroconverted product, the mid-boiling point rises substantially over the mid-boiling point of the feed at a given viscosity.

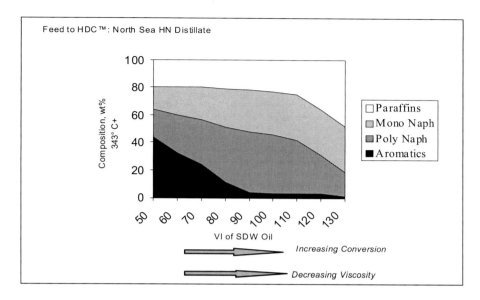

Feed to HDC ™: North Sea HN Distillate

Composition, wt% 343° C+

Paraffins
Mono Naph
Poly Naph
Aromatics

VI of SDW Oil

Increasing Conversion

Decreasing Viscosity

Figure 4. Hydrocracking impact on base stock composition

Selective ring conversion of a raffinate can also be a very effective route to improved basestock properties. This is dramatically illustrated by comparing the profiles of a Group I basestock made from conventional separations-based processing with a Group II basestock derived from a similar raffinate, but converted by a moderate severity hydrotreating type process, RHC™.

Both the conventional basestock (100VI) and the RHC™ derived basestock (116 VI) were separated in a thermal diffusion column, see Figure 5. Thermal diffusion is an analytical technique that applies a thermal gradient across a narrow annular wall, driving the basestock sample in this space to move in a convection cycle.[13] After about one week, the sample separates by density with the least dense, paraffinic species, at the top and most dense, clustered ring species, at the bottom, as shown in the right-hand illustration. Samples from each of ten ports can be characterized both for basestock properties and composition.

The plot of VI against port number reveals that the Group I basestock has a very wide range of properties, with VI's spanning from +170 to -160; port 10 comprises a high concentration of multiring species. However, these species are mostly absent in the Group II basestock; the combination of extraction and RHC™ has selectively eliminated the "port 10" components found in the feed, with little else altered. As a consequence, the basestock "volatility@viscosity" is greatly improved.

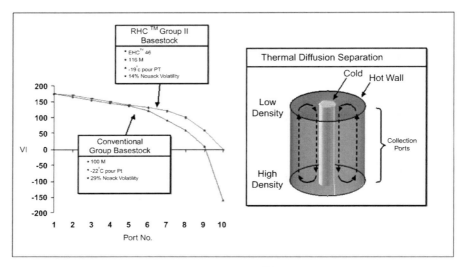

Figure 5. Basestock property improvement in RHC^{TM} is by ring conversion (from ref. 10)

3.2   Paraffin Conversion

Contemporary catalysts used for lubes hydrotreating or hydrocracking do not typically exploit the high VI potential associated with the waxy paraffinic component of distillates or raffinates. Paraffins either go through these processes little changed, or, at more severe hydrocracking conditions, are converted along with other non-parafffins into lower valued fuels.

In general, catalysts that function well for boiling range conversion and ring conversion processing are not suitable for selective paraffin conversion. Paraffin conversion can be mediated by amorphous surface catalysts such as platinum on fluorided-alumina, but the most effective catalysts are those that possess shape selectivity.   These catalysts are able to selectively process paraffins in an environment that excludes most other non paraffins. On such catalysts, access of hydrocarbons to the catalytic conversion sites is governed by the molecular cross section of the molecule rather than adsorption energy. Ring containing species that may preferentially adsorb onto the amorphous surface of an alumina supported hydrotreating catalyst on account of their greater bulk and polarity, are mostly denied entry to the regular microporous surfaces that characterizes modern "dewaxing" catalysts.

Two distinct kinds of   "Catalytic Dewaxing" are in use today. One involves the boiling point conversion of the paraffin components of waxy feeds   by   selective   hydrocracking   and   the   other   by   selective hydroisomerization.

An example of the first kind of process that has been in use since the late 1970's is MLDW™, a very durable technology that handles feedstocks across the entire lube boiling range, utilizing the zeolite ZSM-5.[14] This unique material has also been used in chemicals and fuels applications: non-equilibrium product distributions can be forced by limiting access and diffusion of reactants into the micropores and by preventing certain products from forming because of internal pore geometry constraints. The microporous network of ZSM-5 consists of two intersecting channels with pores of size and shape to admit normal and monobranched paraffins, but little else. One channel runs sinusoidally along the crystals with a nearly circular pore (5.5 Å diameter), whereas the crossing channel is straight and has an elliptical section (5.1, 5.5 Å). Several generations of this catalyst have been employed since its inception, resulting in catalysts today that have a higher resistance to coking with consequent better activity and longevity.

MLDW™ technology offers interesting options to the refiner. It has a lower operating cost process than solvent dewaxing and can lead to basestocks with excellent low temperature properties. Also, for higher viscosity grades, particularly Brightstock, yield and VI at the target pour can match or exceed that of the comparable solvent dewaxed basestock. Countering this, the potentially valuable wax component of the waxy feed is downgraded to light fuel, and on medium and low viscosity lube grades, yield and VI are generally lower than that of solvent dewaxed basestocks with the same pour point.

Alternate approaches to MLDW™ emerged in the 1990's in the form of Chevron's IsoDewaxing™[15, 16] and ExxonMobil's MSDW™ process[17] as the industry trend toward greater hydroprocessing continued to grow. The new technologies are based on the hydroisomerization of paraffins rather than selective hydrocracking, and are applied downstream of a hydroprocessing step such as a lube hydrocracker or hydrotreater. In this environment, the feeds to the hydrodewaxer are "cleaner", with fewer coke precursors and polars, allowing highly shape selective microporous materials to be used. Yields and properties of the basestocks from these processes are boosted because the wax component of the feed is transformed selectively into high VI isomerate.

These new processes differ from earlier commercial forms of hydroisomerization technologies that utilized amorphous bifunctional acidic, noble metal catalysts supported on fluorided aluminas and silica-aluminas. The amorphous catalyst processes were the first to illustrate that hydroprocessed waxy feeds could be manipulated into valuable isoparaffins with exceptional properties.[18]

For example, thermal diffusion of a wax isomerate derived from the hydroisomerization of a heavy slack wax using a fluorided alumina catalyst reveals the profile shown in Table 4. Notably, there are samples from most ports with both high VI and low pour point, testifying to molecules present that are not common in mineral basestocks derived from separations processes or even in basestocks made from hydroprocessing by only ring conversion methods.

Table 3. Thermal Diffusion Illustrates Unique Properties of Wax Isomerates (from Ref. 18)

Sample Port	VI	Pour Point, °C
(whole sample)	143	-21
1	173	-6
3	161	-21
5	146	-34
7	134	-46
10	95	-31

Feed: heavy grade slack wax; Catalyst: Pt-fluorided alumina

Ultimately the shortcoming of amorphous catalysts is that, unlike microporous catalysts, they allow access to all the molecules in the feed, making it difficult to achieve an adequate reduction in paraffin melting point without also over-converting (by hydrocracking) the isomerates in the lube range to lower boiling fuels. Figure 6 illustrates in a simplified way how the relative kinetics for hydroisomerization and hydrocracking dictate the selectivity of the process.

Paramount to an effective hydroisomerization process is that the relative rate (k1) of paraffin (wax) conversion to isoparaffin well exceeds the rate of isoparaffin conversion into fuels (k2) or of paraffin conversion directly to fuels (k3). As conversion of wax progresses, e.g. with increasing reactor temperature or by extending residence time, the yield of isomerate at first rises sharply, reaches a maximum, and then declines as more of the isomerate produced is hydrocracked to fuels. For example, with k2 at about one-fifth of k1 (k3 is generally very low and can be ignored in this discussion), yields of isomerate may reach 60%, meaning that 60% of the wax component of the feed has converted to isomerate. However, at this maximum there is still a component of feed with melting point above the target, i.e. counted as wax, that remains and must be removed by an additional step, such as solvent dewaxing, for the isomerate basestock to meet all the specified properties. In this process a balance has to be struck to achieve sufficient hydroisomerization to lube isomerate, minimizing fuel production, yet ensuring that there is not too much residual wax in the isomerate that might impose a limit in the downstream solvent dewaxer. Overall, this is a relatively inefficient process because the additional solvent dewaxing step can add significantly to the manufacturing cost.

With the advent of microporous hydroisomerization catalysts, it became possible for the first time to achieve the desired pour point in one step. As shown in Figure 6, MSDW technology achieves complete wax disappearance (to the target pour point). A high percentage of the high melting paraffins are converted into desirable isoparaffins with minimal conversion to fuels products, corresponding to a very high k1/k2. This is believed to be a consequence of the isoparaffins being shape constrained against further reaction, diffusing intact from the pores of the catalyst.[17]

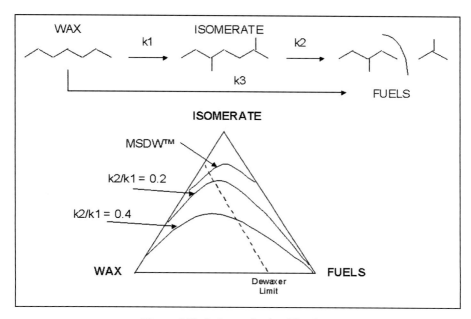

Figure 6. Hydroisomerization Kinetics

A specific example of the upgrading power of selective hydroisomerization technology is shown in Figure 7; a hydrotreated heavy grade slack wax with 30% entrained oil in wax is converted to an isomerate with 150 VI at 0°F pour point at a yield of greater than 70%.

These technologies are effective on a wide range of hydroprocessed stocks of varying wax content from light grades to Brightstock. Basestocks from these processes typically have superb low temperature viscometric properties.

With this level of hydrodewaxing capability, some refiners have moved to "all catalytic" processes with a ring conversion hydroprocessing step complemented by a single step paraffin conversion process that achieves the desired low temperature specifications by completely converting residual wax components, primarily into lube range isomerate.[19] (See further discussion under "Process Combinations", below.)

3.3   Saturation

While ring and paraffin conversion technologies are effective in creating more paraffinic streams with excellent volatility@viscosity, as well as low temperature property characteristics, they may not necessarily provide basestocks or finished oils that are also oxidatively stable.

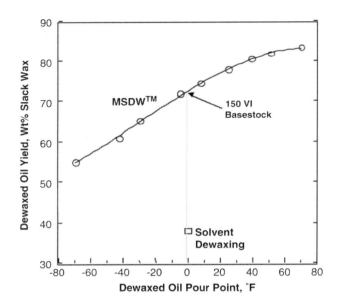

Figure 7. MSDW™ selectively converts wax to high VI isomerate

A factor that directly relates to the oxidative stability of basestocks, particularly Group II and III basestocks, is the extent of ring hydrogenation (saturation). At first sight, saturation appears straightforward. If there is enough driving force to reach equilibrium (i.e. sufficient reactor temperature, catalyst activity, and residence time), then the bulk measure of saturates associated with ring structures will simply be a function of pressure. This is shown in Figure 8, representing the saturates equilibrium of a medium grade raffinate; at higher H_2 pressure and lower reactor temperature, conditions are favorable for producing highly saturated basestocks.

In reality, particularly with conventional hydrotreating type processes, the driving force may be insufficient to approach equilibrium. For example, the catalyst mediating the saturation reaction may have low activity for hydrogenation and/or the residence time for the reaction may be too brief, resulting in a basestock that falls well short of the equilibrium lines defined by the process H_2 pressure. Furthermore, polar compounds present in the feed may hamper the inherent saturation activity by poisoning sites that mediate hydrogenation.

The properties of basestocks cannot be fully described by focusing on bulk saturates alone. Within the envelope of saturates are one, two, three, and higher ring classes, each of which behaves differently, both in terms of the kinetics and equilibrium for saturation. A crucial point is that the conditions favoring an increased rate of saturation of a particular ring class may promote

unfavorable equilibrium (i.e. the formation of aromatics) in another class.[20] This is illustrated in Figure 9.

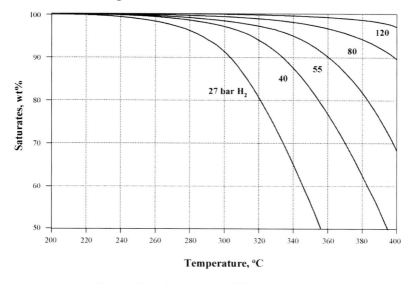

Figure 8. Estimate of Equilibrium Saturates

Single aromatic rings, whether linked to an aliphatic chain or a naphthene ring, have the characteristic of being kinetically slow to saturate, yet favored thermodynamically. By contrast, clustered aromatic rings may saturate rapidly, yet be unfavored thermodynamically. These conflicting trends can make it difficult to achieve complete saturation at one set of conditions.

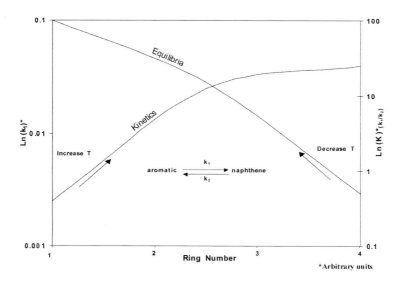

Figure 9. Kinetics and equilibria differ with ring class

For example, by increasing the reactor temperature, the rate of saturation of an isolated one or two-ring aromatic will increase because the forward reaction is well favoured over the reverse reaction (ring dehydrogenation), resulting in an overall increase in the population of naphthenes. Since most ring species in a typical lube raffinate or distillate feed are one or two ring aromatics, the net result of increasing temperatures in a hydrotreater or hydrocracker is to increase the overall saturates content. However, this action will have the reverse effect on aromatics centered in a cluster. At higher temperatures, complete saturation of a four ring aromatic cluster is not favored thermodynamically, and any increase in temperature only makes saturation of this ring class more difficult by tipping the equilibrium further in favor of dehydrogenation. So, while the majority of aromatic species may become saturated at moderate temperatures and pressures in a hydrotreater, the aromaticity of, e.g., three-plus ring species may increase, resulting in basestocks that are colored and/or exhibit higher UV absorbances and oxidative instability.

Two primary strategies are used to achieve nearly complete saturation across all ring classes. An effective but relatively costly approach is to use higher pressure. This works well because it forces the aromatic/naphthenic equilibrium toward naphthenic molecules, allowing the higher ring class species to remain mostly saturated at temperatures needed to kinetically drive the lower ring classes into naphthenes.

An alternate strategy, broadly applied, is temperature-staged processing. In this approach, a lower temperature stage is used following first stage processing such as hydrotreating, hydrocracking, or hydrodewaxing. Provided that process H_2 pressures are not too low, e.g. 40 bar or greater, the higher temperature step can do an effective job of converting mono-aromatics into naphthenes, even though there may be considerable reversal of higher ring classes back toward aromaticity. Then, as product from this reactor cascades to the lower temperature step, nearly complete saturation of all ring classes may be achieved; the higher ring class aromatics convert rapidly to naphthenes because the reaction is still kinetically viable and the conditions more thermodynamically favorable.

Temperature staging is used in the RHC™ process[10] to achieve a highly saturated, oxidatively stable, colorless basestock. This process incorporates three reactors, the first two (referred to as stage 1) operating at higher temperatures to achieve ring conversion for desired volatility@viscosity improvements and to drive saturation of mono-aromatics. Product from these reactors cascades to a lower temperature third reactor (stage 2), comprising a sulfur tolerant catalyst. At such conditions, equilibrium favors near complete saturation of clustered rings, resulting in excellent color and UV stability. In Figure 10, the equilibrium curve for higher ring classes (i.e. 3 ring and greater) aromatics is shown. With the catalyst and conditions chosen for the first stage, liquid effluent from the first stage can have residual aromaticity above the targeted level needed to ensure subsequent basestock stability. But, this is overcome in the second stage by operating at mild conditions. Even at short residence times, a temperature can be chosen that balances kinetics and equilibrium such that the residual clustered ring aromatics are readily saturated to a low level, ensuring excellent stability.

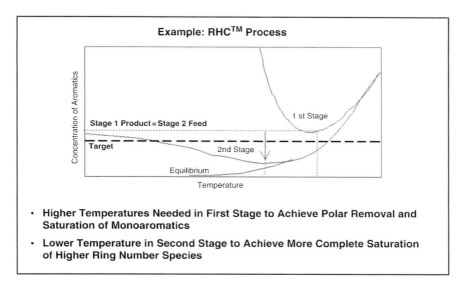

Figure 10. Temperature Staging for More Complete Saturation

The conflicting tug of the kinetics of saturation versus equilibrium has a direct bearing on process conditions and choice of catalyst to mediate the reaction. In some scenarios, a base metal, sulfided catalyst (such as Ni and Mo sulfides on alumina) is preferred, even though this type of catalyst has only moderately active hydrogenation sites compared to noble metal catalysts. This is because the gas stream cascading from the first stage is sour, containing H_2S and NH_3 that may significantly poison noble metal sites and erode catalyst performance. Base metal sulfided catalysts are much less susceptible to poisoning in this environment, making them a cost effective candidate in a process of this type *providing* there is sufficient pressure to offset the higher temperatures needed to drive saturation with a less active hydrogenation catalyst (per the guidelines shown in Figure 8). Conversely, in an environment where gas and liquid feeding to a saturation unit have only low levels of polars, noble metal catalysts may be preferred because they offer the greatest inherent activity for hydrogenation, allowing lower reactor temperatures and pressures to be employed while achieving near-complete saturation of all ring classes.

4.   PROCESS COMBINATIONS

4.1   Ring Conversion-Hydroisomerization-Hydrofinishing

Since 1965, a prevailing reason for using hydroconversion technologies has been feed flexibility; the envelope of crudes viable for lubes applications is considerably broadened when there is a hydroprocessing step in the manufacturing sequence. For example, crudes with high aromaticity such as Maya and Alaska North Slope can be upgraded by ring conversion to make moderate quality basestocks that would not be possible by solvent processing. However, as basestock specifications have continued to tighten, and VI requirements increase, processes such as lube hydrocracking that utilize only a ring conversion step incur high yield losses even on good crudes.

Figure 11 illustrates that, relative to VGO from a good lube crude such as Arab Light, the additional conversion required to upgrade VGO's from poorer crudes grows considerably as the basestock target VI increases. On the same 370°C+ basis, yields from poorer quality VGO's may be lower by a factor of three when the basestock VI target is above 110. The lower slope for VI improvement of the poorer feed is indicative of a lack of molecules amenable to upgrading by de-alkylation, a sign that much of the ring population consists of clusters with only short aliphatic side chains.

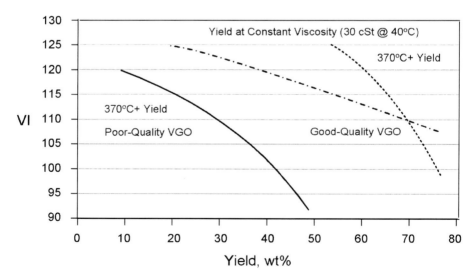

Figure 11. Yield loss is high with ring conversion alone (from reference 21)

Note though that even "good" VGO feeds can be reduced to low yields of basestock when a high VI target is fixed on a particular viscosity grade, e.g. 30 cSt @ 40C. Clearly the kind of ring conversion technology that typifies many hydroprocessing operations today is not very efficient at achieving basestock properties needed for modern automotive use. Above 110, additional improvements in VI are very costly, causing at least a 3% loss for each unit of VI improvement at a fixed viscosity.

The situation improves considerably when the hydroconversion step is combined with a selective hydrodewaxing step, employing a catalyst such as MSDW-2™. Since the paraffinic components of the VGO can be selectively converted to very high VI isomerate, this permits hydrocracking severity and conversion to be reduced while achieving the same nominal basestock properties. For example, a VGO with 20% wax may have an additional uplift of 5 to 6 VI points associated with the selective paraffin conversion step. This can result in an overall yield of basestock at an improved 115 VI that is 15 to 20 wt% greater than that achieved by the hydrocracking step alone. In this way, selective hydrodewaxing can be a very effective lever to increase quality and/or basestock volumes, as well as to further extend feed flexibility.[21]

The layout of an integrated Lube Hydrocracking-Hydrodewaxing Complex, used by ExxonMobil in Singapore, is shown in Figure 12.

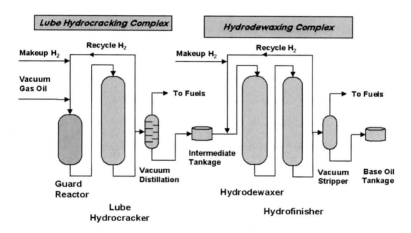

Figure 12. Hydrocracking-Hydrodewaxing Complex

This combination of technologies opens up new opportunities to the refiner. For example, waxier crudes become viable since there is no longer a (potentially rate limiting) solvent dewaxing step involved. With increasing wax content in the VGO feed there has to be a rebalancing of the upgrading load from the hydrocracker, placing a greater burden on the hydrodewaxer. For the same target basestock properties, feed to the hydrodewaxer will now have lower VI, potentially more aromatics and residual polars, as well as more wax to convert to isomerate. This can be a problem for some hydrodewaxing catalysts, because, despite their inherently excellent selectivity for converting polar free paraffinic feeds into isomerate, they can be intolerant of even quite low levels of residual nitrogen leading to much less selective performance. Some degree of polar tolerance in the hydrodewaxing step has significant ramifications, because if the refiner can operate the hydrocracker to a VI target rather than to a residual polars target, the overall yield from the process can be greatly improved.

This is because the hydrocracker lubes fraction VI target and an essentially polar free lubes fraction target may be separated by considerable additional conversion, see Figure 13. For example, the extra conversion required to progress from 10 ppm residual nitrogen in a hydrocracked VGO down to 1 ppm nitrogen may exceed 30 wt%! Fortunately, the latest generation of hydrodewaxing catalysts are more tolerant, permitting much greater process flexibility.

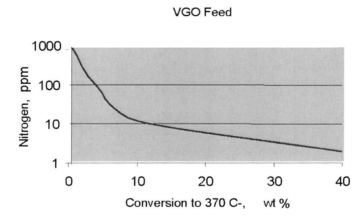

Figure 13. Hydrocracker yields strongly influenced by nitrogen target (from reference 21)

A caveat here is that for the overall process scheme of Figure 12 to be robust, the hydrofinisher must also be more tolerant of polars since the effluent gas and liquid flow directly to it from the hydrodewaxer. These hydrodewaxing and hydrofinishing catalysts need to operate in tandem to achieve a wax free, highly saturated, stable basestock. Here too, the most recent versions of hydrofinishing catalyst, exemplified by Exxonmobil's MAXSAT™, have a corresponding resilience to polars while retaining a high activity for saturation of all ring classes, Figure 14.[21]

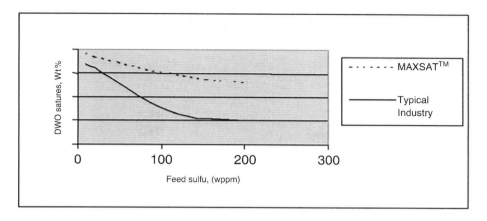

Figure 14. MAXSAT^{TM} polar tolerance is outstanding (from reference 21).

4.2 Extraction-Hydroconversion

Another process combination that is very effective in meeting modern basestock standards is linking extraction with a hydroconversion step. Both processes help reduce populations of low VI ring structures to improve volatility at a given viscosity, but there are optimal ways to operate the combined process by playing on the strength of each. A good example is the RHC™ process, illustrated in Figure 15.

An important aspect of this process is the use of raffinate feeds that are "under-extracted" relative to the raffinates required to meet typical Group I basestock properties using extraction only. In the combined process this lower severity mode takes best advantage of the extraction step because a higher percentage of the species removed into the extract phase are clustered aromatic ring structures with few aliphatic characteristics. In other words, the species that have the *least* potential to be converted into more highly paraffinic structures in the subsequent hydroconversion step are removed more selectively when the extraction treat is mild.

Furthermore, the increased raffinate yield associated with under-extraction more than offsets yield loss in the subsequent hydroprocessing step. There are some trade-offs, like higher hydrogen consumption in the hydroconversion step and possibly shortened catalyst life, but the overall advantage weighs heavily toward using under-extracted raffinate feeds. The RHC™ process has been found to be most effective with raffinate feeds having a VI (on a dewaxed oil basis) in the range of 80 to 95.

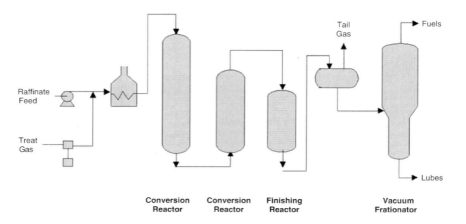

Tail
Gas

Fuels

Raffinate
Feed

Treat
Gas

Lubes

| Conversion Reactor | Conversion Reactor | Finishing Reactor | Vacuum Frationator |

Figure 15. Raffinate Hydroconversion Process™

The combination of extraction and hydroprocessing is a very efficient route to basestocks needed for GF-3 quality. Extraction alone is inappropriate because of an inability to selectively remove multiring naphthenes (these tend to be split evenly between the raffinate and extract phases). Yields by extraction to the same basestock property levels may be less than half of that achieved by RHC™. Also, VGO hydrocracking, i.e. with no pre-extraction step, requires more severe conditions and has potentially lower yields than RHC™ because of the higher conversion needed to offset the highly negative VI characteristics of refractory multi-ring species present in the distillate feed.

The RHC™ process exploits the two factors that influence volatility@viscosity, basestock VI and boiling range distribution. This occurs because the hydroprocessing step not only raises VI by ring conversion, but is also effective in shifting the highest boiling species to the middle of the boiling range. After vacuum stripping, a narrower boiling fraction is formed, effectively lowering volatility at the viscosity target. In fact, this works better when more of the upgrading workload is placed on the hydroprocessing step and less on extraction, that is, using under-extracted feeds. Figure 16 illustrates how this benefit is achieved on a raffinate feed derived from a 250N lube distillate.

In this example, the nominal target properties for the basestock from the RHC™ process were set at 112 VI, 7% Noack volatility, and, following vacuum stripping and solvent dewaxing, a pour point of -18°C. The base operation is shown with a raffinate extracted to achieve typical Group I basestock properties (103VI, on a dewaxed basis) then hydroprocessed to 112 VI. From this platform, as feed VI declines, the basestock yield at target properties rises, a testament to the greater ability of hydroconversion to raise VI, versus extraction capability in this range. The extra advantage of using lower VI raffinates in the hydroprocessing step is that the resultant basestock has *lower* viscosity for the same VI and volatility (by as much as 0.1 cSt, associated with using a raffinate feed with 85 VI rather than 95 VI). This credit can be taken as improved low temperature viscometric properties and/or as a yield increase resulting from being able to operate to a lower VI target, achieved by lowering conversion in the hydroprocessing step.

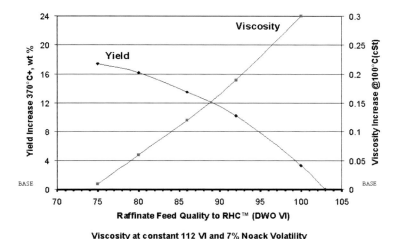

Figure 16. Yield and property improvements with under-extracted feeds (from Ref. 10)

A further embellishment to this process results from combining extraction and hydroconversion with a selective hydrodewaxing step, as discussed above.

5. NEXT GENERATION TECHNOLOGY

The next steps in technologies for better basestocks are likely to include conversion of natural gas into high quality fuels, and, potentially, basestocks and specialties. Although Fischer-Tropsch (FT) chemistry has been around since the catalysis discovery in 1923, it has had only minor impact on the traditional petroleum market, and none at all on the world of lubricants. But now, advances in catalysts and process design have significantly reduced costs and improved flexibility to the point where major oil companies such as Shell Gas and Power, ExxonMobil Corp., and Sasol Ltd. are technically ready for larger scale plants. While a major part of product from GTL will be diesel, it is expected that some basestocks can be also produced with properties ranging from those of the best Group III's made today from petroleum waxes, to the current benchmark for excellence, synthetic Group IV basestocks. The difference is that GTL basestocks volumes could well exceed the relatively small market for today's Group III and PAO applications and be produced at much lower operating cost than any currently made basestock. They have the potential to revolutionize the basestock business.[22]

The nominal sequence for making GTL basestocks as shown in Figure 17 can provide a reference for this if needed. First, natural gas (predominantly methane) is mixed with oxygen and converted into synthesis gas, a 2:1 mixture of H_2 and CO. Synthesis gas is then knitted into extended -CH_2- links using highly selective FT catalysts, forming paraffins with carbon numbers ranging from one to hundreds.

In the last step, the fraction above about C_{20} is selectively hydroisomerized into iso-paraffins, yielding basestocks with viscosities ranging from about 3 to 10 cSt @ 100°C (or higher, depending on the technology). Catalysts used in preliminary demonstrations are in the same family as those used today for hydrodewaxing of hydroprocessed petroleum streams. They create very stable isoparaffinic basestocks with exceptional high and low temperature viscometric properties and low volatility.

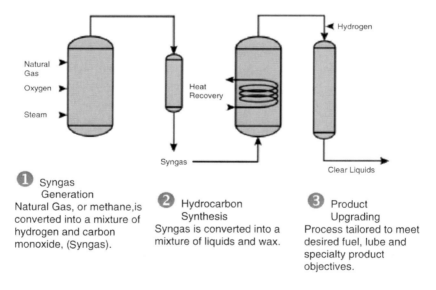

Figure 17. AGC-21 Process

Looking into the next decade, there should be even greater opportunities in lube processing to manipulate basestock properties to meet ever tightening specifications. More streams will be available for upgrading that are relatively free of polar and aromatic species, permitting highly selective lubes catalysis. It may be possible, for example, to open rings more efficiently and to be more precise with placement and type of branch created in hydroisomerization processes.

As more is learned about how properties and performance relate to hydrocarbon structure, hydroprocessing will continue to evolve, providing ever more subtle routes to superior lubricants.

6.   REFERENCES

1. Carnes, K., *Lubr. World*, **2001**, *11(10)*, 16-20.
2. Carnes, K., *Lubr. World*, **2002**, *12(2)*, 16-20.
3. Deane, B.C.; Choi, E.; Crosthwait, K., The Relationship of Base Oil Volatility to Oil Loss in Automotive Applications, 1997 NPRA Annual Meeting, March 16-18, 1997, San Antonio, (AM-97-40)
4. Choi, E.; Deane, B.C., Base Oil Volatility and Oil Consumption, 1998 NPRA Annual Meeting, March 16-17, 1998, San Antonio, (AM-98-47)
5. Rousmaniere, J., Group I Base Oils Decline in North America, *Lubr. World*, **2000**, *10(11)*, 14-16.
6. Noack, K., *Angewardte. Chemie*, **1936**, *49*, 385.
7. Smith, K.W.; Chen, N.Y., New Processes for Dewaxing Lube Basestocks, *Oil Gas J.*, **1970***, 78(21)*, 75, May 1970.
8. Wuest, R.G.; Anthes, R.J.; Hanlon, R.T.; Jacob, S.M.; Loke, L.; Tan, C.T., An Early Change, *Lubr. World*, **2000**, *10(2)*, 12-15..
9. Gallagher, J.E.; Cody, I.A.; Claxton, A.A., Raffinate Hydroconversion, *Lubr. World*, **2000**, *10(2)*, 16-19.
10. Cody, I.A.; Deane, B.C.; Claxton, A.A.; Gallagher, J.E.; May, C.J., Efficient Raffinate Hydroconversion Process for High Quality Lube Basestocks, 16th World Petroleum Congress, Calgary, Canada, June 11-15, 2000.
11. Jacob, S.M.; Banta, F.X.; Hoo, T.M.; McGuiness, M.P.; Quann, R.J.; Sanchez, E.; Staffeld, P.O.; Wells, M.E.; Wuest, R.G., Compositional Modelling of Lube Processes and Lube Base Oil Processing for the 21st Century, AIChE Spring National Meeting, March 10, 1998 (Paper 25c).
12. Scherzer, J.; Gruia, A.J., *Hydrocracking Science and Technology*, Marcel Dekker Inc., 1996
13. White, J.R.; Fellows, A.T., Thermal Diffusion Efficiency and Separation of Liquid Petroleum Fractions, *Ind. Eng. Chem.*, **1957**, *49(9)*, 1409-1418.
14. Kokotailo, G.T., *Nature*, **1978**, *272*, 437.
15. Lok, B.K.; King, R.R.; Lee, S.K.; Wilson, M.M.; Lopez, J., Cost Effective Mineral Oils with Synthetic Performance, AIChE Spring Meeting, February 25-29, 1998, New Orleans; Paper No. 67(d).
16. Miller, S., Wax Isomerization for Improved Lube Quality, AIChE, Spring National Meeting, March 8-12, 1998, New Orleans; Paper No. 25(b)
17. Helton, T.E.; Degnan, T.F.; Mazzone, D.N.; McGuiness, M.P.; Jacob, S.M.; Dougherty, R.C., Mobil's Lube Hydroprocessing Technologies --A Legacy of Catalytic Innovation and Commercial Success, AIChE Spring National Meeting, March 10, 1998 New Orleans; Paper No. 25(d).
18. Cody, I.A.; Ball, K.J.; Murphy, W.J.; Ryan, D.G.; Silbernagel, B.G., Exxsyn 6 --Exxon's New Synthetic Basestock, ACS National Meeting, Base Oil Symposium, San Diego, March 16,17, 1994.
19. Wuest, R.G.; Anthes, R.J.; Hanlon, R.T.; Jacob, S.M.; Loke, L.; Tan, C.T., Commercialization of Mobil's All Catalytic Lube Technology and Next Generation Improvements, AIChE Spring National Meeting, March 14-18, 1999, Houston, TX, Session 31(e).
20. Frye, C.G., Equilibrium Hydrogenation of Multiring Aromatics, *J. Chem. Eng. Data*, **1969**, *14*, 372.

21. Hilbert, T.; Cody, I.A.; Hantzer, S., Process Options for Producing Higher Quality Basestocks, NPRA, Lubricants and Waxes Meeting, November 8-9, 2001, Houston (LW-01-128).
22. Cox, X.B.; Burbach, E.R.; Lahn, G.C., The Outlook for GTL and Other High Quality Lube Basestocks, 2001 NPRA National Meeting, May 25, 2001, New Orleans, LA. (AM-01-64).

Chapter 17

SYNTHETIC LUBRICANT BASE STOCK PROCESSES AND PRODUCTS

Margaret M. Wu[a], Suzzy C. Ho[b], and T. Rig Forbus[c]
(a) ExxonMobil Research & Engineering Co. Annandale, NJ 08801
(b) ExxonMobil Chemical Co. Synthetic Division, Edison , NJ 08818
(c) The Valvoline Co. of Ashland, Inc., Lexington, KY 40512

1.   INTRODUCTION

This chapter reviews the product and process for synthetic base stocks produced from chemicals of well-defined chemical structures and in processes tailored to optimize important properties and performance features. These synthetic base stocks are critical components used in the formulation of many synthetic lubricants. (In this chapter, we use "synthetic base stock" to represent the base fluid and "synthetic lubricant" to represent formulated, finished lubricant product.)

At the start of this chapter, we briefly discuss the background and the driving force for using synthetic lubricants. The major part of the chapter discusses the key synthetic base stocks - chemistry, synthesis processes, properties, their applications in synthetic lubricant formulation and advantages compared to petroleum-derived base stocks.

Many U.S. base oil manufacturers and formulators include some Group II+ and Group III base stocks as synthetic, as their manufacturing process includes varying degrees of chemical transformation. These base stocks are usually produced by hydroprocessing or hydroisomerization, which is typically part of a refining process[1]. Discussion of these hydroprocessed base stocks can be found in the previous chapter. In this chapter, we limit discussion to those synthetic base stocks produced from chemicals of well-defined composition and structure.

1.1   Why Use Synthetic Lubricants?

Synthetic lubricants are used for two major reasons:
* When equipment demands specific performance features that can not be met with conventional mineral oil-based lubricants. Examples are extreme high or low operating temperature, stability under extreme conditions and long service life.
* When synthetic lubricants can offer economic benefits for overall operation, such as reduced energy consumption, reduced maintenance and increased power output, etc.

Conventional lubricants are formulated based on mineral oils derived from petroleum. Mineral oil contains many classes of chemical components, including paraffins, naphthenes, aromatics, hetero-atom species, etc. Its compositions are pre-determined by the crude source. Modern oil refining processes remove and/or modify the molecular structures to improve the lubricant properties, but are limited in their ability to substantially alter the initial oil composition to fully optimize the hydrocarbon structures and composition. Mineral oils of such complex compositions are good for general-purpose lubrication, but are not optimized for any specific performance feature. The major advantages for mineral oils are their low cost, long history and user's familiarity. But this paradigm is now changing.

The trend with modern machines and equipment is to operate under increasingly more severe conditions, to last longer, to require less maintenance and to improve energy efficiency. In order to maximize machine performance, there is a need for optimized and higher performance lubricants. Synthetic lubricants are designed to maximize lubricant performance to match the high demands of modern machines and equipment, and to offer tangible performance and economic benefits.

1.2   What Is a Synthetic Base Stock?

Synthetic lubricants differ from conventional lubricants in the type of components used in the formulation. The major component in a synthetic lubricant is the synthetic base stock. Synthetic base stocks are produced from carefully-chosen and well-defined chemical compounds and by specific chemical reactions. The final base stocks are designed to have optimized properties and significantly improved performance features meeting specific equipment demands. The most commonly optimized properties are:
-    **Viscosity Index (VI).** VI is a number used to gauge an oil's viscosity change as a function of temperature. Higher VI indicates less viscosity change as oil temperature changes - a more desirable property. Conventional 5 cSt mineral oils generally have VIs in the range of 85 to 110. Most synthetic base stocks have VI greater than 120.

- **Pour point and low temperature viscosities.** Many synthetic base stocks have low pour points, -30 to -70°C, and superior low-temperature viscosities. Combination of low pour and superior low-temperature viscosity ensures oil flow to critical engine parts during cold starting, thus, offering better lubrication and protection. Conventional mineral oils typically have pour points in the range of 0 to -20°C. Below these temperatures, wax crystallization and oil gelation can occur, which prevent the flow of lubricant to critical machine parts.
- **Thermal/oxidative stability.** When oil oxidation occurs during service, oil viscosity and acid content increase dramatically, possibly corroding metal parts, generating sludge and reducing efficiency. These changes can also exacerbate wear by preventing adequate oil flow to critical parts. Although oil oxidation can be controlled by adding antioxidants, in long term service and after the depletion of antioxidant, the intrinsic oxidative stability of a base stock is an important factor in preventing oil degradation and ensuring proper lubrication. Many synthetic base stocks are designed to have improved thermal oxidative stability, to respond well to antioxidants and to resist aging processes better than mineral oil.
- **Volatility.** Synthetic base stocks can be made to minimize oil volatility. For example, polyol esters have very low volatility because of their narrow molecular weight distribution, high polarity and thermal stability. Similarly, careful selection and processing of raw materials can influence the finished properties of polyalphaolefins (PAO) base stocks.
- **Other properties,** including friction coefficient, traction coefficient, biodegradability, resistance to radiation, etc. can be optimized for synthetic base stocks as required for their intended applications.

1.3   A Brief Overview of Synthetic Lubricant History

Significant commercial development of synthetic lubricants started in the early 1950's with the increased use of jet engine technology[2]. Jet engines must be lubricated properly in extremely high and low temperature regimes where mineral lubricants could not adequately function. Esters of various chemical structures were synthesized and evaluated. Initially, dibasic esters were used as base stock. Later, polyol esters with superior thermal/oxidative stability, lubricity and volatility were developed to meet even more stringent demands. These polyol esters are still in use today.

Another early application that demanded the use of synthetic lubricants came in the mid-1960s during oil drilling in Alaska where conventional mineral oil lubricants solidified and could not function in the severe Alaskan cold weather[3]. Initially, a synthetic lubricant based on an alkylbenzene base stock of excellent low temperature flow properties was used in the field. This base stock was soon replaced by another base stock with better overall properties, namely polyalphaolefins (PAO).

Research on PAO began at Socony-Mobil in early 1950s[4]. The early researchers recognized the unique viscometric properties that could be attained by the proper selection of starting olefins and reaction conditions in the PAO synthesis. After many years of continuous improvements in optimizing the compositions, processes and formulations, Mobil Corporation introduced a synthetic automotive engine oil, Mobil SHC™ in Europe in 1973, followed by a fuel-saving SAE 5W-20 Mobil 1™ in the US. The product was a commercial success and successive generations of Mobil 1™ continue to be the leading synthetic automotive crankcase lubricant today[5].

Since the early introduction of synthetic lubricants in automotive and industrial applications, many products from numerous companies have followed. The total synthetic lubricant market in 1998 amounted to about 200 million gallons/yr, approximately 2% of the total lubricant volume[5]. However, it is estimated to grow at 5-10% per year, much higher than conventional lubricant (less than 2% per year). Although the volume of synthetic lubricants is relatively small compared to conventional lubricants, the overall economic impact from synthetic lubricants is much larger than just the volume number alone, since synthetic lubricants improve energy efficiency, productivity, reliability and reduce waste, etc.

2.   OVERVIEW OF SYNTHETIC BASE STOCKS

Of the total world wide synthetic base stock volume, over 80% are represented by three classes of materials[6]
-    PAO (45%)
-    Esters, including dibasic ester and polyol esters (25%)
-    Polyalkyleneglycol (PAG) (10%)
Other smaller volume synthetic base stocks include alkylaromatics, such as alkylbenzenes and alkylnaphthalenes, polyisobutylenes, phosphate esters and silicone fluids. Among these synthetic base stocks, with the exception of phosphate esters and silicones, the starting materials are all derived from basic petrochemicals - ethylene, propylene, butenes, higher olefins, benzene, toluene, xylenes, and naphthalenes, as illustrated in Figure 1.

As expected, the major producers of PAO, esters, PIB and alkylaromatics are integrated petroleum companies that supply conventional mineral oil base stocks and petrochemicals as well as various synthetic base stocks. PAG, phosphate esters and silicone fluids are manufactured by chemical companies that produce these fluids on a much larger scale mainly for other applications. Their use as lubricant base stocks is only a fraction of the total market. Table 1 summarizes the major synthetic base stock producers.

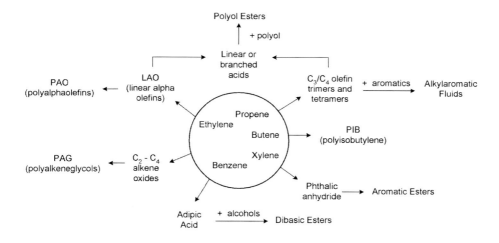

Figure 1. Most synthetic base stocks are derived from petrochemicals

Table 1. Summary of major synthetic base stocks and producers

Synthetic Base Stock	Major Manufacturer	Relative price*
PAO	ExxonMobil Chemical Co., BP, Chevron Phillips Chemical Co., Fortum	4
Dibasic ester	ExxonMobil Chemical Co., Henkel Corp., Hatco Corp., Inolex Chemical Co.	5
Polyol ester	ExxonMobil Chemical Co., Henkel Corp., Hatco Corp., Inolex Chemical Co., Kao Corp.,	7-10
PAG	Dow Chemical Co., BASF	4-10
Alkylaromatic	ExxonMobil Chemical Co., Pilot Chemical Co., Inolex Chem. Co.	4-8
Mineral oil	ExxonMobil, Motiva Enterprise, ChevronTexaco, Valero, BP, Shell, etc.	1

* Estimated relative price vs. Group I mineral oil

3. SYNTHETIC BASE STOCK - CHEMISTRY, PRODUCTION PROCESS, PROPERTIES AND USE

3.1 PAO

PAO with viscosities of 2 to 100 cSt at 100°C are currently produced and marketed commercially[7]. The low viscosity PAO of 4 to 6 cSt account for more than 80% of the total volume. The remaining are mainly medium to high viscosity products of 10 to 100 cSt.

3.1.1 Chemistry for PAO Synthesis

1-Decene is the most commonly used starting olefin for PAO (Figure 2). It is produced as one member of the many linear alpha-olefins (LAO) in an ethylene growth process, which yields C_4 to C_{20} and higher LAO according to the Schulz-Flory distribution[8]. Typically, 1-decene constitutes about 10-25% of the total LAO fraction, depending on the process technology.

To make PAO, the linear 1-decene is further polymerized using Friedel-Crafts catalysts to give C_{20}, C_{30}, C_{40}, C_{50}, and higher olefin oligomers.

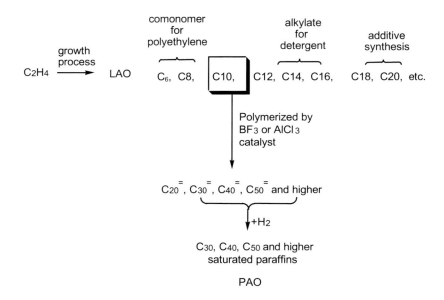

Figure 2. Reaction scheme for converting ethylene into PAO

The degree of polymerization depends on the type of catalyst used and reaction conditions[9]. Generally, BF_3 type catalysts give a lower degree of polymerization. By careful choice of co-catalyst types and reaction conditions, the BF_3 process produces mostly C_{30} to C_{50} oligomers that yield low viscosity base stocks of 4-8 cSt. $AlCl_3$-based catalysts are more suitable for higher viscosity PAO synthesis because they produce oligomers with C_{60}, C_{70} and higher olefin enchainment species. If a C_{20} fraction is produced, it is usually separated and recycled. Fractions containing C_{30} and higher olefin oligomers are then hydrogenated to yield fully saturated paraffinic PAO.

PAO is a class of molecularly engineered base stock with optimized viscosity index, pour point, volatility, oxidative stability and other important lubricant base oil properties. Researchers at ExxonMobil have systematically synthesized polyalphaolefin oligomers of C_{30} to C_{40} by BF_3 catalysis and compared their lubricant properties, as summarized in Table 2.[10]

Table 2. Lubricant base stock property comparison: C_{30}-C_{42} hydrocarbons made from different olefins

Name	Carbon Number	Kinematic Viscosity, cSt, at			Viscosity Index	Pour Point, °C
		100°C	40°C	-40°C		
Propylene decamers	C30	7.3	62.3	>99,000	70	--
Hexene pentamers	C30	3.8	18.1	7,850	96	--
Octene tetramers	C32	4.1	20.0	4,750	106	--
Decene trimers	C30	3.7	15.6	2,070	122	<-55
Undecene trimers	C33	4.4	20.2	3,350	131	<-55
Dodecene trimers	C36	5.1	24.3	13,300	144	-45
Decene tetramers	C40	5.7	29.0	7,475	141	<-55
Octene pentamers	C40	5.6	30.9	10,225	124	--
Tetradecene trimers	C42	6.7	33.8	Solid	157	-20

These data show that the oligomers made from propylene, 1-hexene and 1-octene have relatively low VI and very high viscosity at -40°C. Oligomers from 1-tetradecene have high VI but also have undesirable high pour point and are solid at -40°C. Oligomers from 1-decene have the best combination of high VI, low pour point and -40°C viscosity.

Historically, the market dynamics of LAO supply and demand further drove the trend toward the use of 1-decene as a raw material. Among all the major LAO from the ethylene growth process (Figure 2), C_6 and C_8 LAO are used as co-monomer in the linear low-density polyethylene production; C_{12-16} LAO are used in the manufacture of linear alkylbenzene detergent; C_{18} and C_{20} LAO are used in additives. 1-Decene is not in high demand for other chemical manufacturing and its use as raw material for synthetic base stocks makes a perfect match. When 1-decene supply became tight, other LAO, such as C_8 and C_{12}, have been successfully incorporated with 1-decene as the starting olefins for PAO production. Since 2001, 1-decene supply has increased significantly due to several LAO expansion projects and new production coming on-line around the world[11].

The chemical composition of PAO is very simple. Using 4 cSt PAO as an example, it is made of ~ 85% C_{30} and ~15% C_{40} hydrocarbons. It has a narrow molecular weight distribution compared to typical 4 cSt mineral oils. The gas chromatograms in Figure 3 show that 4 cSt PAO has few low molecular weight components of less than C_{30} that can degrade oil volatility, flash and fire point. Figure 3 also shows that the C_{30} fraction of PAO is not a single compound but a mixture of many isomers. This is because the PAO from BF_3 process contains many isomers, each with different types of branching[12]. This irregular branching may be beneficial to some of PAO's low temperature properties, e.g. pour point.

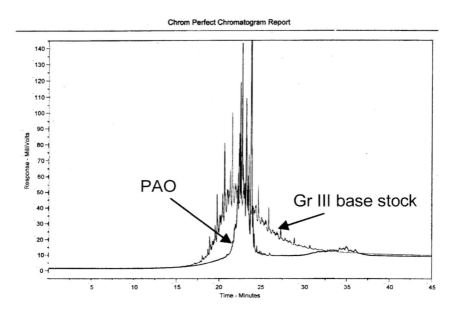

Figure 3. Gas chromatograph comparison of a 4 cSt PAO with a 4 cSt Group III base stock

3.1.2 Manufacturing Process for PAO

Commercial production of PAO using a BF_3 catalyst generally involves a multi-stage, continuous stirred tank reactor (CSTR) process[9]. In early production technology, the catalyst was destroyed with diluted aqueous alkali after polymerization. More recent patents disclosed improved processes using BF_3 catalyst recycle to reduce catalyst usage, minimize process waste and improve process economics[13].

3.1.3 Product Properties

The physical properties of some commercial PAO are summarized in Table 3.[7]

Table 3. General product properties of commercial PAO from ExxonMobil Chemical Company

Fluid type	SHF-20	SHF-41	SHF-61/63	SHF-82/83	SHF-101	SHF-403	SHF-1003
Kinematic Viscosity @100°C, cSt	1.7	4.1	5.8	8.0	10.0	39	100
Kinematic Viscosity @40°C, cSt	5	19	31	48	66	396	1,240
Kinematic Viscosity @-40°C, cSt	262	2,900	7,800	19,000	39,000	--	--
Viscosity Index	--	126	138	139	137	147	170
Pour Point,°C	-66	-66	-57	-48	-48	-36	-30
Flash Point., °C	157	220	246	260	266	281	283
Specific Gravity @15.6°C/15.6°C	0.798	0.820	0.827	0.833	0.835	0.850	0.853

3.1.4 Comparison of PAO with Petroleum-based Mineral Base Stocks

PAO have different chemical compositions compared to mineral oil base stocks. The American Petroleum Institute (API) categorizes lubricant base stocks into five categories, designated Group I to V. The definition of each base stock group is summarized in Table 4.

Table 4. *Definition of API category I to V lubricant base stock*

	Description	% Saturates	% Sulfur	VI
Group I	(Conventional, solvent refined)*	<90	>0.03	80-120
Group II	(Hydroprocessed)*	>/= 90	</= 0.03	80-120
Group III	(Severely hydroprocessed or isomerized wax)*	>/= 90	</= 0.03	>=120+
Group IV	Polyalphaolefins			
Group V	All other base stocks not included in Group I, II, III or IV (e.g. esters, PAG, alkylaromatics, etc.)	--	--	--

* - comments in parentheses are not included in the original API definition

PAO is classified by itself as a Group IV base stock. In addition to the differences listed in Table 3, PAO also contains no cyclic paraffins, naphthenes or aromatics, whereas Group I, II and III base stocks contain different amounts of aromatics ranging from <1% to >40%[14]. With the increasing presence of aromatics and/or naphthenes, oxidative stability and low temperature properties of these fluids are typically degraded. Also, as shown earlier in Figure 3, PAO have discrete carbon numbers with relatively long linear hydrocarbon branches, whereas mineral base stocks contain a continuum of carbon number. As a result, PAO usually have lower volatility.

Table 5 compares the basic properties of low and high viscosity PAO versus Group I to Group III mineral oil base stocks.

Table 5. Typical property comparison of PAO with Group I to III mineral oil

| | Low Viscosity | | | | High Viscosity | |
	Grp I	Grp II	Grp III	PAO	Bright stock	PAO
Kinematic Viscosity @100°C, cSt	3.8	5.4	4.1	4.1	30.5	100
Kinematic Viscosity @40°C, cSt	18	30	19	19	470	1,240
Viscosity Index	92	115	127	126	94	170
Pour Point, °C	-18	-18	-15	-66	-18	-30
Cold Crack Simulator @ -20°C, cP	--	--	750	620	--	--
Noack Volatility, wt%	32	15	14	12	--	--
Aniline point, °C	100	110	118	119	97	>170

- PAO have superior viscometrics properties compared to mineral oil base stocks.

Data in Table 5 show that PAO has higher VI and lower pour point than Group I and II base stocks. Compared to Group III base stocks, PAO has comparable VI, but much lower pour point and improved low-temperature viscosity as measured by Cold Cranking Simulator (CCS) viscosity at -20°C. In an actual engine oil formulation, this lower CCS viscosity observed with PAO results in a wider SAE cross-grade (5W-40) than with Group III base stock (10W-40)[15]. The lower low-temperature viscosity translates into better fuel economy during the engine warm up period.

- PAO has lower volatility.

Data in Table 5 show that PAO has lower volatility than Group I to III base stocks. This lower volatility is the result of the unique chemical compositions of PAO - 100% relatively linear paraffin, little low molecular weight hydrocarbons of less than C_{30} (Figure 3). Low volatility is advantageous for decreased oil consumption and reduced emissions.

- PAO show intrinsic oxidative stability and excellent response to antioxidant additive treatment.

It has been demonstrated that the un-formulated PAO base stock treated with 0.5 wt% antioxidant resists oxidation for more than 2500 minutes in a standard rotary bomb oxidation test (RBOT, D2272 method). In comparison, similarly treated Group II and III base stock started to oxidize much earlier, at less than 800 or 1700 minutes, respectively[16].

This oxidative stability translates into performance advantages in actual engine oil tests (Figure 4).[15]

Figure 4 shows that a fully formulated engine oil with PAO has much lower viscosity increase than with Group III or with Group I/II base stocks in standard length, 64-hour ASTM Sequence IIIE engine test. In an extended-length, 256-hour test, the viscosity increase for PAO-based lubricant is still much less than the maximum increase allowable for this test. In contrast, Group III or Group I/II based engine lubricants become too viscous to measure. Performance advantages in fuel efficiency and oil consumption are also reported.[17]

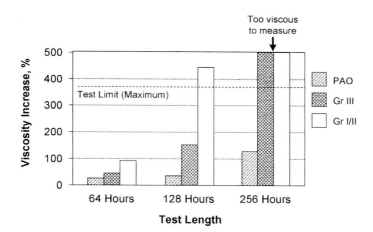

Figure 4. Comparison of viscosity increase in ASTM Sequence IIIE engine test for fully formulated lubricants based on PAO vs. Group III or I/II base stocks[17]

- PAO are available in wide viscosity range.

PAO are available from 2 to 100 cSt at 100°C. The high viscosity PAO maintain excellent VI and low pour point (Table 3 and 5), in a manner that is superior to the highest viscosity mineral oil base stock - bright stock. High viscosity PAO are important when blending with low viscosity fluid to formulate high viscosity grade industrial oils. When used to blend with low viscosity mineral oil, the high viscosity PAO also significantly improves the oxidative stability of the blended base stocks compared to using mineral bright stock[15].

- PAOs have high aniline point, indicating low polarity.

Table 5 shows that low viscosity PAO has a higher aniline point than Group I mineral oils, 119°C vs. 100°C (Table 5). A more pronounced difference is observed for high viscosity fluids (>170°C vs. 97°C). The higher aniline points of PAO mean that they are much less polar than Group I oils. Generally, lubricant additives and oil oxidation by-products are highly polar chemical species. As aniline point has relevance to solvency, additives and oil oxidation by-products are not very soluble in PAO alone. As a result, a polar co-base stock, such as ester or alkylaromatic, is usually added to the formulation to improve the solvency of PAO in a finished lubricant. These co-base stocks can also assist other performance features, such as seal compatibility and improved lubricity.

- PAO possess other important properties, depending on application:
  - Compatibility or miscibility with mineral oil at all concentration levels without phase separation or detrimental effects when cross-contamination occurs
  - Hydrolytic stability

- 10% higher thermal conductivity and heat capacity than comparable mineral oil, allowing equipment to run at lower temperature and improve wear performance[18]
- Lower traction coefficients than conventional fluids, resulting in better energy efficiency for many industrial oil applications[6]
- PAO are non-greasy and non-comedogenic

In summary, PAO have superior VI, pour point, low-temperature viscosity, volatility, and oxidative stability and are available in a wide viscosity range compared to conventional Group I, II or III mineral oils.

3.1.5   Recent Developments – SpectraSyn Ultra as Next Generation PAO

Following the success with PAO, ExxonMobil Chemical Co. recently introduced a new generation of PAO, trade-named SpectraSyn Ultra™. SpectraSyn Ultra™ is produced from the same raw material as PAO, 1-decene, using proprietary catalyst technology[19, 20]. Table 6 summarizes the properties of commercial SpectraSyn Ultra™ products.

Compared to traditional PAOs, SpectraSyn Ultra™ PAO have even higher VI, lower pour point and are available in higher viscosity ranges. This unique class of fluid can be used in automotive engine oil and industrial oil formulations to provide advantages in terms of shear stability, viscometrics properties, thickening power and increased lubricant film thickness.

Table 6. Product properties of next generation PAO - SpectraSyn Ultra™

Product	SpectraSyn Ultra™ 150	SpectraSyn Ultra™ 300	SpectraSyn Ultra™ 1000
Kinematic Viscosity @100°C, cSt	150	300	1,000
Kinematic Viscosity @40°C, cSt	1,500	3,100	10,000
Viscosity Index	218	241	307
Pour Point,°C	-33	-27	-18
Flash Point., °C	>265	>265	>265
Specific Gravity @15.6°C/15.6°C	0.850	0.852	0.855

3.1.6   Applications

PAO is the workhorse base stock for most synthetic lubricants. Low viscosity PAO are used in synthetic automotive crankcase and gear lubricants, industrial oils and greases. High viscosity PAO have found great utility in industrial oils and greases.

Synthetic automotive engine oils command the largest volume among synthetic lubricant products. Taking advantage of the many superior properties of PAO base stocks, performance advantages of synthetic engine oils based on PAO over mineral oil-based engine oils are well-documented in scientific and trade literature[21]. They include:

- Improved engine wear protection
- Extended oil drain interval
- Excellent cold starting performance
- Improved fuel economy
- Reduced oil consumption
- Excellent low-temperature fluidity and pumpability
- High temperature oxidation resistance

Many of these performance advantages are directly attributable to the intrinsically superior properties of PAO, such as high VI, low pour point, low low-temperature viscosity, high oxidative stability, low volatility, etc.

The advantage of using synthetic engine oil is further supported by the fact that many automakers use synthetic lubricant as the "factory fill" lubricant for their high performance cars. For example, in 2003, Mobil 1™ is used as factory-fill lubricant for the Corvette, all Porsches, Mercedes-Benz AMG models, Dodge Viper, Ford Mustang Cobra R and Cadillac XLR[22].

PAO blended with mineral oil are also used in many partial synthetic lubricant formulations. In this case, PAO is used as a blending stock to improve the volatility, high or low-temperature viscosity, oxidative stability, etc. of the mineral oil blend.

Synthetic industrial oils and greases, formulated with PAO, have many specific performance and economic advantages over conventional lubricants[6,21a]. For example, in industrial gear/circulation oils, PAO-based lubricants offer the following documented advantages:

- Energy savings, longer fatigue life and lowered temperatures of operation due to lower traction coefficients
- Wider operating temperature range due to higher VI and better thermal-oxidative stability
- Reduced equipment down-time, reduced maintenance requirements and longer oil life due to the excellent stability of PAO base stock

Because PAO is available in high viscosity grades (up to 100 cSt at 100°C), high ISO grade synthetic industrial oils with improved performance features are more easily formulated. This option is not available for mineral oil-based lubricants.

In compressor oil applications, PAO-based lubricants have advantages due to their better chemical inertness and resistance to chemical attack. Synthetic compressor oils are used in corrosive chemical environments, for example, in sulfuric acid or nitric acid plants. PAO-based lubricants are also used in refrigeration compressor applications due to their excellent low temperature fluidity, lubricity and generally wider operating temperature range.

Other synthetic industrial oil applications with PAO-based lubricants, include gas turbine, wind turbine and food-grade gear lubricants. Synthetic greases based upon PAO are used in industrial equipment, aviation and automotive applications that take advantage of the wide operating temperature range, high degree of stability and other desirable properties and features offered by PAO base stocks.

Recently, PAO is finding its way into personal care products such as shampoos, conditioners and skin lotions because it provides emolliency in addition to good skin feel and is non-greasy and non-comedogenic. It is also used in off-shore drilling fluids because of its good lubricity. New applications for PAO are continuously emerging.

3.2   Dibasic, Phthalate and Polyol Esters - Preparation, Properties and Applications

Lard and vegetable oil, both ester-type compounds derived from natural sources, have been used as lubricants throughout human history. After World War II, thousands of synthetic esters were prepared and evaluated as lubricant base stocks for jet engine lubricants.[2]

3.2.1   General Chemistry and Process

Esters are made by reacting carboxylic acids with alcohols. The elimination of water is shown by the following equation:

The reaction proceeds by heating the mixture to 150°C or higher with or without a catalyst[9]. Catalysts such as p-toluenesulfonic acid or titanium(IV) isopropoxide, are typically used to facilitate reaction rates. The reaction is driven to completion by continuous removal of water from the reaction medium. Sometimes, one component is used in a slight excess to ensure complete conversion. The final product is purified over an adsorbent to remove trace water and acids, both of which are detrimental to base stock quality. Commercially, esters are generally produced by batch processes.

The choice of acid and alcohol determines the ester molecular weights, viscometrics and low temperature properties, volatility, lubricity, as well as the thermal, oxidative and hydrolytic stabilities[23]. The structure-property relationships of ester base stocks are well documented in the literature. Compared to PAO and mineral oil, ester fluids have a higher degree of polarity, contributing to the following unique properties:
-    Superior additive solvency and sludge dispersancy
-    Excellent lubricity
-    Excellent biodegradability
-    Good thermal stability

Three classes of esters are most often used as synthetic base stocks - dibasic ester, polyol ester and aromatic ester. Some basic properties of these esters are summarized in the Table 7.

Table 7. Basic properties of ester base stocks

Acid	Alcohol	Viscosity, cSt 100°C	40°C	VI	Pour Point,°C	Wt% Volati- lity[a]	Wt% Biodegrad ability[b]
Dibasic ester							
Adipate	Iso-$C_{13}H_{27}$	5.4	27	139	-51	4.8	92
Sebacate	Iso-$C_{13}H_{27}$	6.7	36.7	141	-52	3.7	80
Polyol ester							
n-C_8/C_{10}	PE[c]	5.9	30	145	-4	0.9	100
n-C_5/C_7/iso-C_9	PE	5.9	33.7	110	-46	2.2	69
n-C_8/C_{10}	TMP[c]	4.5	20.4	137	-43	2.9	96
Iso-C9	TMP	7.2	51.7	98	-32	6.7	7
n-C9	NPG[c]	2.6	8.6	145	-55	31.2	97
Di- and mono-acids	NPG	7.7	40.9	160	-42	--	98
Aromatic Esters							
Phthalate	Iso-$C_{13}H_{27}$	8.2	80.5	56	-43	2.6	46
Phthalate	Iso-C9	5.3	38.5	50	-44	11.7	53
Trimellitate	Iso-$C_{13}H_{27}$	20.4	305	76	-9	1.6	9
Trimellitate	C7/C9	7.3	48.8	108	-45	0.9	69

(a) Noack Volatility : 250°C, 20 mm-H_2O, and one hour with air purge
(b) by CEC-L-33-A-96 test, % degradable in 21 days
(c) PE: pentaerythritol, TMP: trimethylolpropane, NPG: neopentylglycol

3.2.2 Dibasic Esters

Dibasic esters are made from carboxylic diacids and alcohols. Adipic acid (hexanedioic acid) is the most commonly used diacid (Figure 5). Because it is linear, adipic acid is usually combined with branched alcohols, such as 2-ethylhexanol or isotridecanols ($C_{13}H_{27}OH$) to give esters with balanced VI and low temperature properties (Figure 5). Dibasic ester is most often used as a co-base stock with PAO to improve solvency and seal swell properties of the final lubricants.

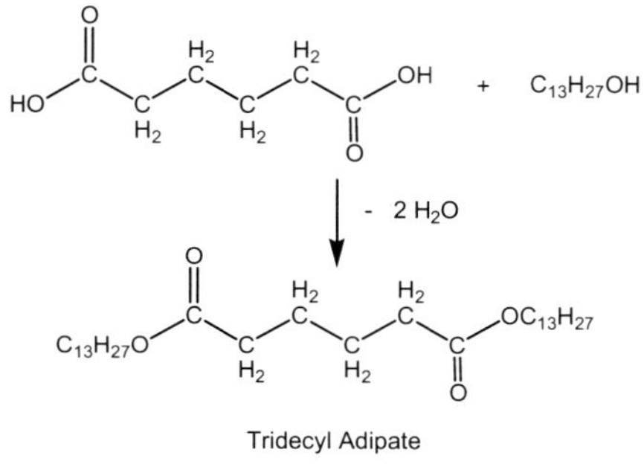

Figure 5. Synthesis of adipate ester

3.2.3   Polyol Esters

The most common polyols used to produce synthetic polyol ester base stocks are pentaerythritol (PE), trimethylolpropane (TMP) and neopentylglycol (NPG), (Figure 6). By carefully choosing the degree of branching and size of the acid functions, polyol esters with excellent viscometric properties - high VI and very low pour points – can be produced (Table 6).

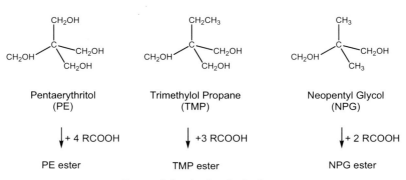

Figure 6. Synthesis of polyol esters

In addition to excellent viscometric properties, polyol esters have the best thermal resistance to cracking. This is because polyols lack β-hydrogen(s) adjacent to the carbonyl oxygen and thus can not undergo the same facile β-H transfer reaction as the dibasic esters (Figure 7). This cracking by β-H transfer leads to two neutral molecules and is a relatively low energy process. Polyol esters can only be cracked by C-O or C-C bond cleavage, leaving two free radicals - a very high-energy process requiring extremely high temperature. Therefore, polyol esters are thermally stable up to 250°.

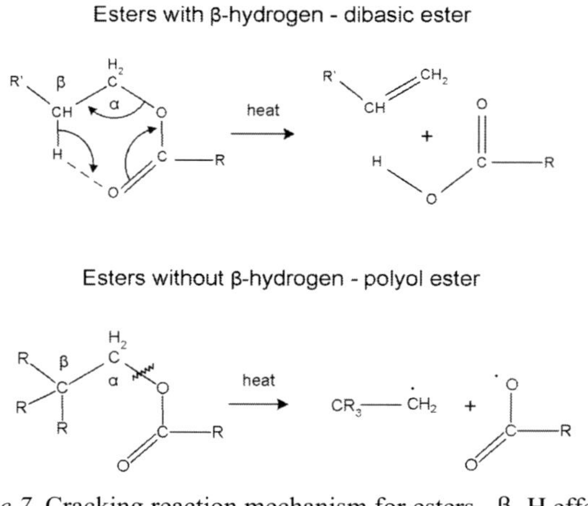

Esters with β-hydrogen - dibasic ester

Esters without β-hydrogen - polyol ester

Figure 7. Cracking reaction mechanism for esters - β–H effect

Among the three polyol ester types, the thermal stability ranking is: PE esters > TMP esters > NPG esters.

3.2.4 Aromatic Esters

Phthalic anhydride or trimellitic anhydride are converted into esters by reactions with alcohols as shown in Figure 8. Phthalic anhydride is produced cheaply and in large volume from oxidation of ortho-xylene. The largest use of phthalate esters is in the plasticizer market. Only a small fraction of its production is consumed by the synthetic lubricants market. Phthalate esters generally have superior hydrolytic stability than adipic esters because the ortho di-ester groups are electronically less available and sterically more hindered[24]. However, they have lower VIs, 50-70, because of their high polarity and the presence of branched alcohol chains. They are used in special industrial oil applications where VI is not a critical parameter. Trimellitate esters are specialty products and relatively expensive. They are of high viscosity and usually are more resistant to oxidation than adipic esters.

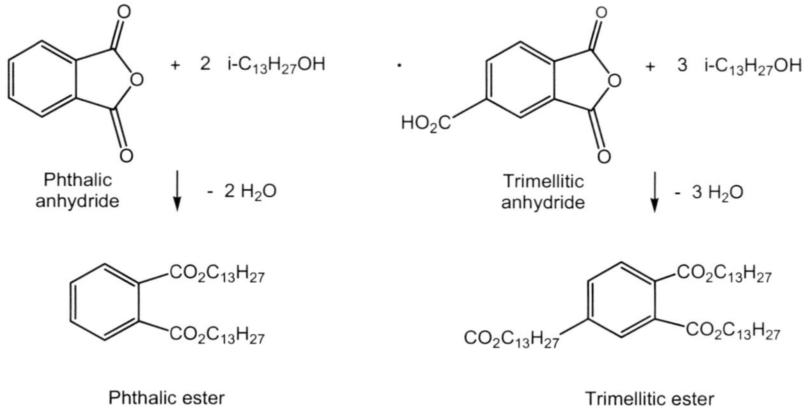

Figure 8. Synthesis of phthalate and trimellitate esters

3.2.5 General Properties and Applications of Ester Fluids

Solvency and dispersancy - Ester fluids are quite polar due to their high oxygen contents. They have high solubility for many commonly used additives. They also have high solubility for the polar acids and sludges generated by oxidation processes during service. This property makes ester based lubricant "clean" compared to hydrocarbon-based lubricants. Typically, low viscosity ester fluids are soluble with non-polar PAO base stocks. These properties make them excellent for use as co-base stocks with PAO in many synthetic automotive and industrial lubricants. Generally, 5 to 25% esters are used with PAO in finished lubricant formulations.

Hydrolytic stability[24] - Hydrolysis of esters to give acids and alcohols is a facile reaction and can proceed at elevated temperatures in the presence of water. Hydrolysis of ester generates acid that can be very corrosive to metal components and can catalyze the base stock decomposition process. Therefore, hydrolytic stability of esters is an important issue. Much work has been carried out to improve the hydrolytic stability by varying the composition of acids and alcohols. Generally, esters made from aromatic acids or from more sterically hindered acids, such as 2-alkyl substituted acids or neo-acids, have improved hydrolytic stabilities. Proper branching of the acids protect the carbonyl ester function from the detrimental attack of water. The presence of impurity, such as trace acid or metal, can catalyze the decomposition and hydrolysis of ester. Compared to PAO or alkylaromatic base stocks, ester hydrolysis is always an issue of concern in many lubrication applications.

Volatility - Ester fluids generally have lower volatility compared to PAO and mineral oil of comparable viscosities. A General volatility ranking for base stocks are as follows:

PE ester > TMP ester > dibasic ester > PAO >> Group I or II mineral oil.

Lubricity - Polar ester fluids show mild boundary film protection at lower temperature. At lower temperature, esters interact with the metal surface via polar interaction, forming a chemisorbed surface film, which can provide better lubrication than the less polar mineral oil or non-polar PAO. When esters decompose, they produce acids and alcohols. Higher molecular weight acids can bind with the metal surfaces to form a film that can offer some degree of wear protection and friction reduction. However, none of these interactions are strong enough to persist when surface or oil temperature rises much above 100°C. At higher temperature, significant wear protection can only be achieved by the use of anti-wear or extreme-pressure (EP) additives. A drawback for the ester high polarity is that esters can compete with metal surface for polar additives, resulting in less efficient usage of anti-wear and EP additives. Therefore, in formulations using esters, it is important to choose the proper additives and concentration levels to obtain the full benefit of the lubricity from both the additives and esters.

Biodegradability - By carefully choosing the molecular compositions, esters of excellent biodegradability can be produced. Generally, esters from more linear acids and alcohols have better biodegradability.

Applications[25] - Esters, both dibasic and polyol esters, are used as co-base stocks with PAO or other hydrocarbon base stocks in synthetic automotive engine lubricants and industrial lubricants. Polyol esters are used in aircraft turbine oils due to their excellent thermal and oxidative stabilities, good lubricity, high VI and excellent low temperature properties (<-40°C)[21a]. Esters are also used in synthetic compressor oils for ozone-friendly refrigeration units. Because of their high biodegradability and low toxicity, esters are often the base oils of choice for many environmentally-aware lubricants or single-pass lubrication applications where ecological impact is critical.

Although ester chemistry has been studied extensively, new esters with unique performance improvements have continuously been reported in the literature[26]. For example, esters with high stability were made from highly branched acids and polyols. Polyol esters formulated with ashless additives can be used as high performance biodegradable hydraulic fluids.

3.3 Polyalkylene Glycols (PAG)

PAG is an important class of industrial chemicals. Its major use is in polyurethane applications. Outside of polyurethane applications, only 20% of the PAG is used in lubricant applications. Compared to PAO or esters, PAG have very high oxygen content and hydroxyl end group(s). These unique chemical features give them high water solubility and excellent lubricity. PAG was first developed as water-based, fire-resistant hydraulic oils during World War II for military use. Other applications have been developed subsequently to take advantage of their unique properties.

3.3.1 Chemistry and Process

PAG are synthesized by oligomerization of alkylene oxides over a base catalyst with an initiator R'OH (Figure 9)[27]. When the initiator is water (R'=H), the final PAG has two hydroxyl end groups. When the initiator is an alcohol (R=alkyl group), one of the end groups is an alkoxy group (RO-). The most commonly used alcohol is n-butanol, although large alcohols have also been used for special applications. Phenol, thiols or thiophenol are also used as initiators.

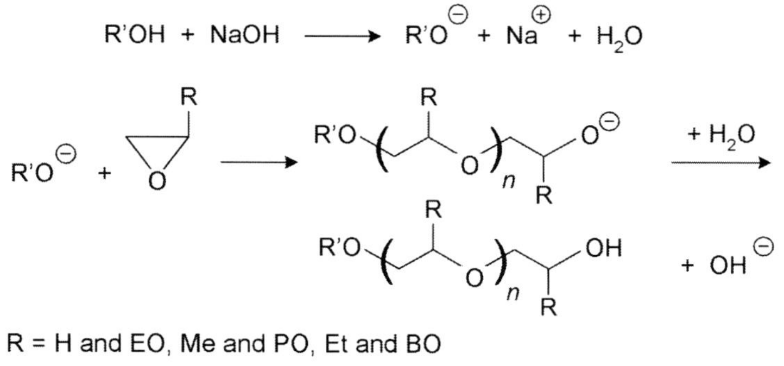

R = H and EO, Me and PO, Et and BO

Figure 9. Reaction scheme for PAG synthesis

Ethylene oxide (EO), propylene oxide (PO), butylene oxides (BO) or combinations of these epoxides are used as starting materials for PAG syntheses. Longer chain alkylene oxides are sometimes added to improve their compatibility with hydrocarbons. PAG with a wide range of viscosities,

VIs, pour points, water solubilities and oil-compatibilities are produced by choosing the proper initiators, monomers, reaction conditions and post treatments. The reaction is highly exothermic (22.6 kcal/mole) and heat removal is important to avoid side-reactions or broadening of the product molecular weight distribution.

3.3.2   Product Properties

Table 8 summarizes the typical lubricant properties of selected PAG produced from EO, PO and BO with several different initiators.[28]

Table 8. Lube properties of PAG fluids from EO, PO and BO with different initiator

	AO Type	End Group	Avg. MW	$KV_{100°C}$ CSt	$KV_{40°C}$ cSt	VI	Pour Point,°C	Density, g/cm³	Solubility in oil	Solubility in water
E300	EO	OH/OH	300	5.9	36	118	-10	1.125	i	s
E600	EO	OH/OH	600	11.0	72	154	22	1.126	I	s
P425	PO	OH/OH	425	4.6	33	26	-45	1.007	--	--
P1200	PO	OH/OH	1200	13.5	91	161	-40	1.007	--	i
PB200	PO	Bu/OH	910	8.3	44	180	-48	0.9831	--	i
EP530	EO/PO	OH/OH	2000	25	168	192	-32	1.017	--	--
EPB100	EO/PO	Bu/OH	--	4.8	101	174	-57	1.0127	--	s
EPB260	EO/PO	Bu/OH	--	11.0	56.1	210	-37	1.0359	--	s
B100-500	BO	OH/OH	500	5.1	44.3	3	-30	0.975	s	s
B100-2000	BO	OH/OH	2000	24.7	234.7	142	-26	0.970	s	i
1500 MW poly BO Mono-ol	BO	Bu/OH	1500	15.8	117.1	153	-30	0.961	s	i

EO-based fluids are typically waxy and have poor low temperature properties. They have high water miscibility and are typically used to formulate water-based lubricants, especially fire-resistant hydraulic oil. PO-based fluids are excellent lubricant base stocks with high VI and low pour point. They have lower solubility in water than EO-based fluids but are not oil miscible.

EO/PO-based fluids have a better combination of VI and low pour points than PO-based products. They are used as base stocks in industrial circulation/gear oils.

BO-based PAG have improved oil solubility and are not water-soluble.

PAG generally have excellent lubricity and low friction coefficients compared to mineral oil as shown in Table 9. These properties result from the facile surface chemisorption of the oxygenate functions or through hydrogen bonding of the terminal OH groups with the metal surface.

Other unique properties for PAG include:

- superior solvency - they dissolve additives, decomposition products and sludges
- non-varnishing and low ash - they leave little or no residue or carbon black upon decomposition

Table 9. Lubricating properties of selected PAG fluids[28]

Fluid Type	Mol. weight	V100°C ,cSt	V40°C cSt	VI	Pour point, °C	Four ball wear scar, mm (a)	Four ball seizure load, kg (a)	Friction coefficient (b)	Soluble in (c)
EO/PO	500	4.6	19	161	-46	0.53	120-140	0.15	water
EO/PO	1300	15	76	218	-42	0.44	180-200	0.11	water
PO	700	6	27	179	-44	0.53	160-180	0.19	oil (d)
PO	1300	14	73	193	-35	0.57	120-140	0.12	none

(a) by DIN 51350 method
(b) determined by oscillation of a steel ball on a steel disc at 30°C under a load of 200 N
(c) determined by mixing equal proportions of water and PAG or oil and PAG.
(d) partially soluble in oi

3.3.3 Application

The major use of PAG is in the industrial oil area[29]:
−   Fire resistant hydraulic fluids. Water-soluble PAG are fire resistant, low toxicity and have excellent lubricity and anti-wear properties.
−   Textile oils. PAG are non-varnishing, non-staining and can be washed away with water.
−   Compressor and refrigeration oils. Low solubility of many industrial gases, such as natural gas and ethylene, makes PAG suitable for gas compressor applications. PAG are compatible with new refrigerants (HFC-143a) and have excellent anti-wear properties.
−   Metal working fluids. PAG are non-varnishing, have excellent lubricity and anti-wear properties.
−   Circulation/bearing/gear oil. Low friction coefficients and traction properties of PAG lead to lower operating temperature and energy consumption. They have good anti-wear properties and are non-varnishing.

3.4   Other Synthetic Base Stocks

Polyisobutylene (PIB) fluids are produced by the oligomerization of isobutylene in a mixed C_4 stream over a BF_3 or $AlCl_3$ catalyst. PIB are seldom used by themselves. They are typically used as blend stocks or as additives to increase lubricant viscosity. Table 10 summarizes the typical properties of selected PIB fluids[30]. The VI and pour points of PIB are comparable to those of conventional mineral oil. PIB usually have a lower flash point and decompose easily into monomer at 200°C and higher. The advantages of PIB are their high compatibility with most synthetic or mineral base stocks and their relatively low cost compared to other synthetic base stocks.

Table 10. Typical physical properties of PIB available from BP Chemical Co.

	H-25	H-50	H-100	H-300	H-1500
Kinematic Viscosity @100°C, cSt	50	100	200	605	3000
Viscosity Index	87	98	121	173	250
Pour Point, °C	-23	-13	-7	3	18
Bromine Number (?)	27	20	16.5	12	8
Flash Point, °C (a)	171	193	232	274	307
Molecular Weight (b)	635	800	910	1300	2200

(a) by Cleveland open cup ASTM D92 method.
(b) by gel permeation chromatography.

Alkylbenzenes and alkylnaphthalenes are produced by the alkylation of benzene or naphthalene with olefins using Friedel-Crafts alkylation catalysts[31]. Their typical properties are summarized in Table 11. One unique feature of these alkylaromatic fluids is their very low pour points. Alkylbenzenes are often mentioned in the patent literature as components for CFC or HCFC refrigeration compressor oil. Alkylnaphthalenes are used in synthetic automotive engine oil, rotary compressor oils, and other industrial oils.

Table 11. Properties of alkylbenzene and alkylnaphthalenes base stocks

Fluid type	Di-alkylbenzenes	Di-alkylbenzenes	Alkylnaphthalenes
Commercial source	V-9050 from Vista Chem. Co.	Zero 150 from Chevron	Synesstic™ 5 from ExxonMobil Chem.
Kinematic Viscosity @100°C, cSt	4.3	4.4	4.7
Kinematic Viscosity @40°C, cSt	22.0	33.5	28.6
VI	100	25	74
Pour Point, °C	-60	-40	-39
Flash Point, °C	215	170	222
Aniline Point, °C	78	--	33

Phosphate esters are produced from phosphorus oxychloride with various alcohols or phenols, or combinations of these hydroxyl compounds[32]. These fluids generally have good thermal and oxidative stabilities and fire-resistancy. However, because of their high polarity, poor VI-pour point balance, facile hydrolysis[33] and inferior elastomer and paint compatibility, their use in general lubrication is limited. The major use for phosphate esters is in fire-resistant hydraulic oils.

4.   CONCLUSION

Synthetic lubricants have significantly raised the performance level of automotive and industrial lubricants with the help of high-quality PAO base stocks and tailored high-performance additive technologies. Equipment builders, industrial users and general consumers have taken advantages of the enhanced performance benefits afforded by synthetic lubricants - reduced maintenance and waste, lower emissions and pollution, higher reliability and efficiency, etc. As a result, in the last ten years, synthetic lubricants have enjoyed yearly growth rates of 5-10%, a range considerably higher than for conventional lubricants[34]. This growth rate has occurred despite the higher initial costs of synthetic products. The higher initial costs have been economically offset by the extended life and performance benefits afforded by synthetic lubricants.

This trend is expected to continue in the finished lubricants market. In the short-term, the growth for some PAO-based synthetic lubricants may slow temporarily due to new competition from hydroprocessed base stocks [35]. However, high-performance synthetic base stocks and finished lubricants should continue to prove their enhanced and well-documented values as further demands on lubricant performance grow. The knowledgeable user, who treats the lubricant as an active machine component and understands the enhanced performance and associated economic benefits, will continue to demand greater efficiency, reduced maintenance, lower emissions and longer service life, etc, offered by high-quality synthetic lubricants. These factors should increase market value and continue market growth for advanced synthetic lubricants. To meet this demand, the leaders of the lubricant industry will need to respond by developing and marketing next-generation, high performance base stocks and products. ExxonMobil's SpectraSyn Ultra[TM] and Mobil 1[TM] with SuperSyn[TM]-Antiwear technology are current examples of this leadership.

5.   ACKNOWLEDGEMENT

The authors thank Hal Murray for his assistance in literature search and Andrew Jackson, Mike Thompson, Charles Foster and Joan Kaminski for their valuable comments about this chapter.

6.   REFERENCES

1. *J. Synth. Lubr.*, **2002**, *18-4*, Publisher's Note.
2. G. J. Bishop, Aviation Turbine Lubricant Development, *J. Synth. Lubr.*, **1987***, 4-1*, 25.
3. Harlacher, E. A.; Krenowics, R.A.; Putnick, C.R. Alkylenzene Based Lubricants. *Prep. 52D26P*, 86th AICHE national meeting, Houston, April 1-5 1979.
4. (a) Garwood, W. E., Synthetic Lubricant, US Patent 2,937,129, 1960.
    (b) Hamilton, L. A. and Seger, F. M. Polymerized Olefins Synthetic Lubricants, US Patent 3,149,178, 1964.
5. (a) *Lubricants World*, "On Track for Growth", June 1999, 19.
    (b) After-Market Business Vol. 100, n3, p. 32, March 2000 "It's Brand vs. Commodity in Choosing Motor Oil"
    (c) Business Wire (23 July 2002), p. 328 "Mobil 1 Chosen as Factory-Fill Motor Oil for Cadillac XLR"

6.   Murphy, W. R.; Blain, D. A.; Galiano-Roth, A. S.; Galvin P. A. Synthetic Basics -
     Benefits of Synthetic Lubricants in Industrial Applications, *J. Synth. Lubr.*, **2002**, *18-4*, 301.
7.   www.exxonmobilsynthetics.com or ExxonMobil Chemical Synthetics, P. O. Box 3272,
     Houston, TX 77253-3272 (281-570-6000)
8.   Lappin, G. R. *Alpha-Olefins Application Handbook*, Lappin, G. R.; Sauer, J. D. (Eds.),
     Marcel Dekker: New York, 1989; 35.
9.   (a) Sacks, M. A private report of Process Economic Program, SRI International, Report
     no. 125, *Synthetic Lubricant Base Stocks*, May 1979
     (b) Bolan, R. E. A private report of Process Economic Program, SRI International, Report
     no. 125A, *Synthetic Lubricant Base Stocks*, Sept. 1989
10. Brennan, J. A. Wide-Temperature Range Synthetic Hydrocarbons Fluids, *Ind. Eng. Chem.*
     Prod. Res. Dev., **1980**, *19*, 2-6
11. (a) "Strong Demand for Synthetic Lubricants Lead to Increased Investment in LAO
     Production", *Ind. Lubr. Tribology*, **2002**, *54(1)*, 32.   (b)
     http://www.the-innovation-group.com/ChemProfiles/Alpha Olefins (Linear).htm
12. Shubkin R.L.; Baylerian, M. S.; Maler, A. R. Olefin Oligomer Synthetic Lubricants :
     Structure and Mechanism of Formation, *Ind. Eng. Chem. Prod. Res. Dev.*, **1980**, *19*, 15-19
13. Hope, K. D.; Ho, T. C.; Archer, D. L.; Bak, R. J.; Collins, J. B.; Burns, D. W. Process for
     Recovering Boron Trifluoride From a Catalyst Complex, US Patent 6,410,812, 2002.
14. Cerny, J.; Pospisil, M.; Sebor, G. Composition and Oxidative Stability of Hydrocracked
     Base Oils and Comparison with a PAO, *J. Synth. Lubr.*, **2001**, *18-3*, 199.
15. *ExxonMobil Chemical Co. Sales Brochure*. ExxonMobil Chemical Synthetics, P. O. Box
     3272, Houston, TX 77253-3272 (281-570-6000)
16. Lubes-n-Greases, March 2002, p. 39, PAO Problem Solver by Chevron Phillips Chemical
     Co.
17. Mattei, L.; Pacor, P.; Piccone, A. Oils With Low Environmental Impact for Modern
     Combustion Engines, *J. Synth. Lubr.*, **1995**, *12-3*, 171.
18. (a) *Synthetic Lubricants and High Performance Functional Fluids*, 2nd ed., Rudnick, L.
     R.; Shubkin, R. L. (Eds.), Marcel Dekker: New York, 1999; 21.
     (b) Lubes-n-Greases, January 2003, p. 39, PAO Problem Solver by Chevron Phillips
     Chemical Co.
19. (a) Wu, M. M. High Viscosity-Index Synthetic Lubricant Compositions, US Patent
     4,827,064, 1989.
     (b) Wu, M. M. High Viscosity-Index Synthetic Lubricant Process, US Patent 4,827,073,
     1989.
20. (c) ExxonMobil SuperSyn[TM] - A New Generation of Synthetic Fluid, Society of
     Tribologists and Lubrication Engineers Annual Meeting, Las Vegas, Nevada, May 26,
     1999
21. (a) Law, D. A.; Lohuis, J. R.; Breau, J. Y.; Harlow, A. J.; Rochette, M. Development and
     Performance Advantages of Industrial, Automotive and Aviation Synthetic Lubricants, *J.*
     Synth. Lubr., **1984**, *1-1*, 6-33.
     (b) Bergstra, R. J.; Baillargeon, D. J.; Deckman, D. E.; Goes, J. A. Advanced Low
     Viscosity Synthetic Passenger Vehicle Engine Oils, *J. Synth. Lubr.*, **1999**, *16-1*, 51.
     (c) Bleimschein, G.; Fotheringhan, J.; Plomer A. On the Road to New Diesel Regs -
     Synthetic Lubes Push on With Fuel to Burn, *Lubes-n-Greases*, November 2002, p. 22
22. www.mobil1.com What auto experts say.
23. (a) Szydywar, J. Ester Base Stocks, *J. Synth. Lubr.*, **1984**, *1-2*, 153.
     (b) Debuan, F.; Hanssle, P. Aliphatic Dicarboxylic Acid Esters for Synthetic Lubricants,
     J. Synth. Lubr., **1985**, *1-4*, 254.
     (c) Zeman, A.; Koch, K.; Bartle, P., Thermal Oxidative Aging of Neopentylpolyol Ester
     Oils: Evaluation of Thermal-Oxidative Stability by Quantitative Determination of
     Volatile Aging Products, *J. Synth. Lubr.*, **1985**, *2-1*, 2-21.

(d) Denis, J. The Relationships between Structure and Rheological Properties of Hydrocarbons and Oxygenated Compounds Used as Base Stocks, *J. Synth. Lubr.*, **1984**, *1-3*, 201.

24. Boyde, S. Hydrolytic Stability of Synthetic Ester Lubricants, *J. Synth. Lubr.*, **2000**, *16-4*, 297.
25. Carnes, K, Ester? Ester Who? *Lubricants World*, October 2002, p. 10
26. (a) Schlosberg, R. H.; Chu, J. W.; Knudsen, G. A.; Suciu, E. N.; Aldrich, H. S. High Stability Esters for Synthetic Lubricant Applications, *Lubr. Eng.*, **2001**, 21.
    (b) Duncan, C.; Reyes-Gavilan J.; Costantini, D.; Oshode, S. Ashless Additives and New Polyol Ester Base Oils Formulated for Use in Biodegradable Hydraulic Fluid Applications, *Lubr. Eng.*, September 2002, p. 18
27. Kussi, S. Chemical, Physical and Technological Properties of Polyethers as Synthetic Lubricants, *J. Synth. Lubr.*, **1985**, *2-1*, 63.
28. The Dow Chemical Co., Midland, Michigan 48674, Dow Polyglycol product brochure. www.dow.com
29. Matlock, P. L.; Brown, W. L.; Clinton, N. A. Polyalkylene Glycols, Chapter 6, In *Synthetic Lubricants and High Performance Functional Fluids*, 2nd Ed. Rudnick, L. R.; Shubkin, R. L. (Eds.), Marcel Dekker: New York, 1999; p. 159.
30. www.bpchemicals.com/polybutene/
31. Wu, M. M. Alkylated Aromatics, Chapter 7, In *Synthetic Lubricants and High Performance Functional Fluids*, 2nd Ed. Rudnick, L. R.; Shubkin, R. L. (Eds.), Marcel Dekker: New York, 1999; p. 195.
32. Marino, M. P.; Placek, D. G., Phosphate Esters, Chapter 4, *Synthetic Lubricants and High Performance Functional Fluids*, 2nd Ed. Rudnick, L. R.; Shubkin, R. L. (Eds.), Marcel Dekker: New York, 1999; p. 103
33. Okazaki, M. E.; Abernathy, S. M. Hydrolysis of Phosphate-Based Aviation Hydraulic Fluids, , *J. Synth. Lubr.*, **1993**, *10-2*, 107.
34. Petroleum Technology Quartery, Vol. 4, #4, Winter/2000, p. 22
35. Slower growth forecast for PAO lubes, Lube Report, Industry News from *Lubes-n-Greases*, Vol. 2 Issue 16, April 17, 2002.

Chapter 18

CHALLENGES IN DETERGENTS AND DISPERSANTS FOR ENGINE OILS

James D. Burrington*, John K. Pudelski, and James P. Roski
The Lubrizol Corporation, 29400 Lakeland Blvd, Wickliffe, OH 44092

1. INTRODUCTION

This chapter will focus on the function and chemistries of today's detergents and dispersants, and how they are being transformed to meet increasing performance and cost demands. A significant trend to address market needs by the combination of additive chemistry with additional technologies will also be presented and some examples discussed.

2. ENGINE OIL ADDITIVE AND FORMULATION

Detergents and dispersants are the dominant performance additives components in engine oil formulations. For example, a "typical" gasoline engine oil contains 5-20% of a performance package, which is the largest component, on average, after base oil. The additive supplier supplies the oil marketer with the additive performance package, the pour point depressant, and the viscosity modifier (sometimes known as the viscosity index improver). In a typical crankcase oil the performance package is between 5 and 20% of the formulation. A typical level is about 10%. The viscosity modifier cannot be blended with the performance package and is supplied separately. The performance package is dominated by dispersant and detergent.

The dispersant and detergent together make up about 55-70% of the performance package. Thus, the chemistry of the total package and finished oil is greatly influenced by these two components. As in this example (Figure 1) of a "typical" finished engine oil, the detergent and dispersant must not only perform their intended chemical functions, but must also provide the proper bulk and rheological properties consistent with the application.

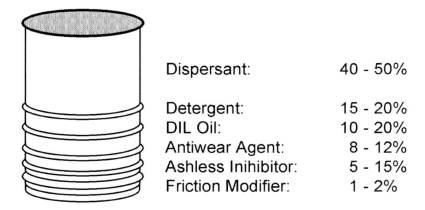

Formulating a Performance Package
for Passenger Car Motor Oils:
Additive Company Perspective

Dispersant:	40 - 50%
Detergent:	15 - 20%
DIL Oil:	10 - 20%
Antiwear Agent:	8 - 12%
Ashless Inihibitor:	5 - 15%
Friction Modifier:	1 - 2%

Figure 1. A typical finished engine oil

2.1   Detergents

The detergent functions to solubilize polar components, inhibit rust & corrosion, and prevent high temperature deposits, in part, through neutralization of acids. They are composed of two components, a substrate or surfactant and a colloidal inorganic phase, generally resulting from the overbasing process. The major variables effecting performance are the substrate, which is generally sulfonate, phenate, or salicylate, the metal, which is generally Ca, Mg or Na, the degree of overbasing or conversion, which is the level of basic phase present relative to the amount of surfactant. The structural features of a detergent responsible for these unique properties is shown in Fig. 2.

The combination of a surfactant molecule with a colloidal inorganic core results in a micelluar-type structure as shown in the figure. This gives both the ability to solubilize polar materials in a continuous matrix of oil, and provides acid neutralization capacity, which is also intimately contacted with the oil in a dispersed amorphous colloidal phase, shown in the figure as $CaCO_3$ for a Ca-detergent. This basic colloidal carbonate neutralizes acids formed during the combustion process, such as nitric acid, sulfuric acid, and hydrochloric acid, which lead to metal corrosion and wear, as well as organic acids, which lead to polymerization, viscosity increase and resin formation. Nitro-hydroxy-carbonyl-compounds also form and, if not neutralized, are the precursors of varnish and sludge.

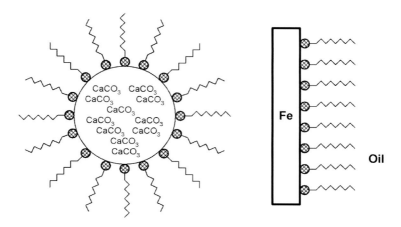

Overbased & surfactant components work in synergy to:

- *Inhibit rust and corrosion*
- *Inhibit oil degradation*
- *Reduce high temperature deposits*
- *Solubilize polar contaminants*

Figure 2. Detergents: Surfactant/base oil synergies

The surfactant component of the detergent can also form a protective layer on metal parts as shown here, resulting in the inhibition of rust and corrosion. Together, the surfactant and the basic components work in synergy to inhibit rust and corrosion and oil degradation, reduce high temperature deposits and solubilize polar components.

This is an idealized representation of a detergent. Calcium carbonate is suspended in oil with a sulfonate or phenate. The excess calcium carbonate provides a base reserve to oils and neutralizes acids that are formed during combustion. Detergents are also effective at keeping surfaces in the engine clean. The metals used to make detergents are typically calcium, magnesium, and sodium. Calcium is the most common.

The need for this acid neutralization capacity is evident from Fig. 3, which shows the continual decrease in total base number (TBN) and increase in total acid number (TAN) for a high- and low-TAN gasoline engine oil. The TBN/TAN equivalence point occurs at around 3000 to 6000 miles. (The metal content in the drain increases after the equivalence point is reached in field testing.) Typically, at this point, acids build up to unacceptable levels, and it is, therefore, desirable to change the oil before the TBN and TAN cross.

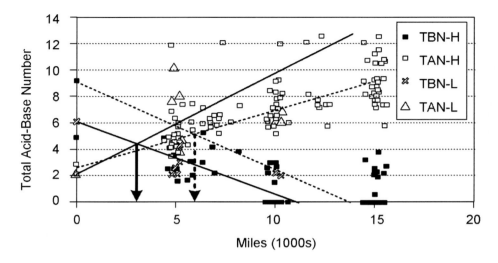

Figure 3. Field testing data demonstrate the need for base capacity (solid lines: TBN, dotted lines: TAN)

These are data from a recent field test showing the decrease in TBN and the increase in TAN with usage. Drain recommendations are often determined by this type of testing. In this example the oil represented by the blue (darker) lines has a lower TBN initially than the oil represented by the pink (lighter) lines. It is not surprising that the cross over point occurs sooner with the lower TBN oil. It is evidence such as this that results in oils with higher TBN being recommended for longer drain intervals.

Another important detergent function is the prevention of high temperature deposit. These deposits, such as the varnish shown in Fig. 4 result, in part, from the acidic precursors formed from high temperature reaction of nitrogen oxides and oxygen with the mixture of fuel and lubricant. The pictures show the effect when these high temperature deposit-forming processes *are* (acceptable) and when they *are not* (unacceptable) sufficiently controlled.

2.2   Dispersants

Dispersants function to suspend soot, thereby mitigating the deleterious effects of large particle agglomerates inside the crankcase. Dispersants include a polymer backbone component, which is predominantly polyisobutylene, or PIB, connected to a polar group, normally an amino group. There are two major classes of dispersants, both of which use PIB and a polar amine group: succinimide dispersants, which use maleic anhydride hook, and Mannich dispersants which use formaldehyde. The major variables affecting performance are the nature of the backbone (composition and structure) and the nature, and relative levels of, the hook and polar group.

Unacceptable   ## Acceptable

Figure 4. Detergents prevent high temperature deposits

In practice, the dispersant, like the detergent, also solubilizes polar contaminants, but in this case is designed with a longer tail (M_n = thousands vs. hundreds for detergents) to provide greater steric stabilization to the dispersed carbon (or other contaminant) particle in the micelular structure (Fig. 5). A polyamine head group is used, which is tailored specifically for strong adsorption of soot particles. The nature of the dispersant interactions is tailored to meet the performance characteristics of the particular engine and application. Like detergents, dispersants have been finely tuned over decades to arrive at an optimal structure and composition of the various parameters.

Thus, the combination of the longer hydrocarbon chain and the polar amine head group provides for: (1) the suspension of soot particles (Fig. 6) and the reduction in the corresponding wear and viscosity increases; (2) the solubilization of other polar contaminants; and (3) the prevention of low temperature deposits.

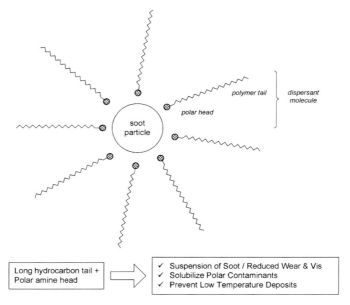

Figure 5. Dispersant chemistry

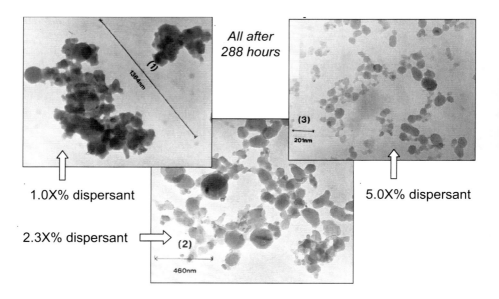

Figure 6. Soot particle growth in a sequence VE test

3. PERFORMANCE CHEMISTRY

The rest of this chapter will discuss the chemistries of these additive systems, the opportunities for performance improvements based on chemistry and incorporation of other technologies with additives. Dispersants and detergents are an important class of performance products made by Lubrizol and other additives suppliers based on alkyl succinimides, succans and phenols. These molecules represent a class of surfactant-type materials which are composed of alkyl chains of varying lengths and polar heads.

The dispersant or detergent molecule can be pictured as a "typical" functionalized molecule with a non-polar tail connected to a polar head via a hook connecting group (Fig.7). Detergents generally employ an alkylated aromatic sulfonate, phenol or salicylate where the hydrocarbon chain is a C_{12}-C_{32}, typically C_{16}, linear or branched alpha olefin or olefin oligomer mixture. These are converted to the corresponding sulfonate, phenate or salicylate salt (usually Ca, Mg or Na) and converted to an "overbased" product with a metal base and CO_2 to incorporate an amorphous carbonate phase which provides base capacity. Again, the short chain and polar head surfactant combination provides the mechanism for solubilization of polars and adsorption/protection of metals. The base capacity neutralizes acid, which can contribute to high temperature deposits.

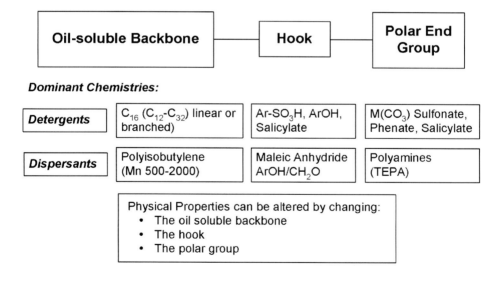

Figure 7. Typical functional molecules

Dispersants, on the other hand, utilize longer alkyl chains, mainly polyisobutylene of 500-2000 Mn. For succinimide-types, these are converted to the succinic anhydride intermediates (or succans) by reaction with maleic anhydride, and subsequently to the final succinimide dispersant by condensation with a polyamine such as TEPA. For Mannich-types, the long-chain alkylphenol is converted with formaldehyde and the amine to the corresponding dispersant. The polyamine end group strongly adsorbs to soot particles as they are formed in the combustion process and the longer tail provides the steric stabilization of the dispersant/soot micelle structure to inhibit agglomeration into larger, more harmful wear particles.

The physical properties of these end products can be tailored by controlling the nature of each component and the relative amounts of hooks and end groups per molecular chain. These have been optimized over several decades, resulting in today's cost effective and high performance products. But despite the large body of knowledge regarding these structure-function relationships, there are only a handful of chemistries commonly in actual commercial use today for production of detergents and dispersants. This is especially true of the "hooks", and "polar end groups", which generally make up only 10% by weight or less of the product, and where the cost effective polyamine succinimides (from maleic anhydride and polyamines) and overbased alkyl sufonates and phenates dominate dispersants and detergents, respectively. While polyisobutylene and aromatic alkylate backbones are most commonly used, there has been a significant emphasis on backbone modifications as a means to greater leverage of chemical and physical properties, for optimal cost and performance, particularly for dispersants.

Table 1. Currently available backbones

Backbones	Sources (Examples)
Conventional PIB	Lubrizol, BP
High vinulidene PIB	BASF, BP, Nippon Petrochemical Co.
Olefnin copolymer (OCP)	Mitsui, ExxonMobil
Poly alpha olefin (PAO)	Mitsui, ExxonMobil

4.   CURRENT DISPERSANT AND DETERGENT POLYMER BACKBONES

While dispersant hydrocarbon backbones are currently dominated by conventional polyisobutylene, many more backbones are on the horizon with the potential to provide improved properties, processing, overall performance per cost, and the ability to optimize properties to respond to specific engine performance characteristics. Some of these (Fig. 8) include high vinylidene PIB, olefin copolymers (OCP) and poly-alpha olefins (PAO). Each of these will be discussed in terms of their structure and reactivity, physical properties and how these translate into strengths and weaknesses in the final application.

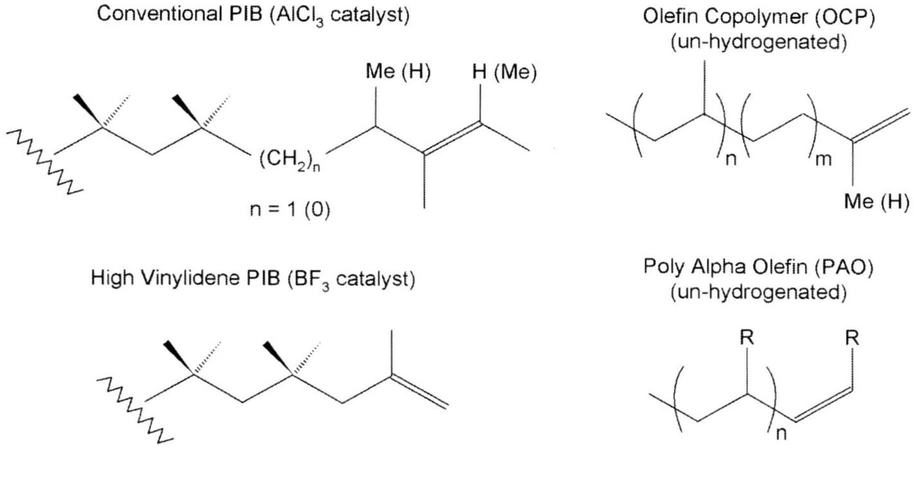

Conventional PIB (AlCl₃ catalyst)

Olefin Copolymer (OCP)
(un-hydrogenated)

Me (H)   H (Me)

(CH₂)ₙ

n = 1 (0)

Me (H)

High Vinylidene PIB (BF₃ catalyst)

Poly Alpha Olefin (PAO)
(un-hydrogenated)

R   R

Figure 8. Polyolefin backbone structures

Conventional PIB, made using AlCl₃ as the isobutylene polymerization catalyst, has a distinctive 5-carbon end grouping with either a tetra- or trisubstituted terminal olefin. High vinylidene PIB, on the other hand, which can be made from a BF₃-based catalyst, contains a predominance (usually >70%) of the more straightforward gem-disubstituted vinylidene group. These PIB's have similar physical and rheological properties dictated by the common gem-di-methyl groups on every other carbon of the backbone, while the reactivities with the olefinic end groups are very different. Olefin co-polymers and poly alpha olefins, on the other hand, have very different rheological properties than either of the PIB's, but are more similar to high vinylidene PIB in chemical reactivity. These properties are discussed in more detail below, first dealing with their relative reactivities.

Shown here are some representative structures of common backbones and their corresponding succinic anhydrides. Conventional PIB, with multiple allylic carbon atoms, is ideally suited to maleination reaction conditions which promote the formation of diene intermediate, as has been proposed for chlorine-assisted maleination. The diene, once formed, undergoes rapid Diels-Alder reaction with maleic anhydride to form the corresponding polyisobutenyl succinic anhydride, or PIBSA. The vinylidene-type polymers, on the other hand, are more suited to the thermal succination process, thought to proceed by an ene reaction as shown. PAO can also be made with a high degree of vinylidene end groups.

The vinylidene-type containing backbones - that is, high vinylidene PIB, OCP's and PAO's, are also the good reactants for the acid catalyzed alkylation of phenol. The resulting long-chain alkyphenols are intermediates for the Mannich-type dispersants.

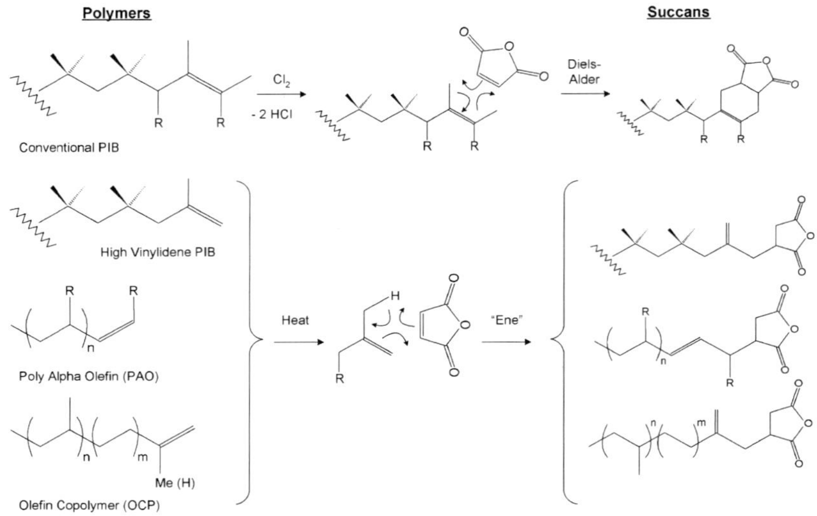

Figure 9. Succination chemistry

5.   FUTURE POLYMER BACKBONES

Besides reactivity, there are also a number of other properties that are desirable in an "ideal" polymer backbone for detergents and dispersants. As we have discussed, high reactivity with maleic anhydride and phenol is of course important.   For example, conversion to dispersants is also key to achieve high conversions at low temperatures, short reaction times, and with minimal excess and decomposition of maleic anhydride.   Good overall viscometrics and low temperature properties are becoming increasingly important, especially in light of high energy efficiency and low emissions formulations. These requirements will also drive changes to other formulation components, making it important to use dispersants and detergents with broad flexibility in formulating. Of course, backbones must be produced at low overall cost with minimal capital outlay by well-developed, commercialized (or at least commercializable) processes.

A summary of the strengths and weaknesses of current backbones against these criteria is presented in Table 2.

Conventional PIB has the advantage of providing very low cost products that utilize tried and true technology and already having large volume capacity in place. However, the process is not amenable to thermal processing technology and the resulting products can limit flexibility in formulating to an optimal elemental composition.

As has been discussed, high vinylidene PIB is more thermally reactive,[3] giving more process and formulation flexibility, but requires more expense and is thinner than conventional PIB. The OCP and PAO's have excellent reactivity and improved low temperature viscometrics, but are even more expensive.

Table 2. Polyolefin backbone alternatives

	Low Overall Cost	Reactive w/MAA (thermal)	Reactive w/Phenol	Low Capital Requirements (mainly succination)	Viscosity Credit / Thickening Power	Low Temperature Performance	Low Treat Rate	Process / Formulation Flexibility	Developed Technology	Commercially Available
Conventional PIB (AlCl3)	+	-	-	+	+	-	0	-	+	+
High Vinylidene PIB (BF3)	-	+	+	-	-	-	0	+	+	+
LZ High Vinylidene PIB (developmental)	+	+	+	-	+	-	0	+	-	-
OCP, PAO	--	++	++	-	-	+	+	+	+	+
Branched Polyethylene (potential)	+	++	++	-	-	+	+	+	-	-

This situation leads to several opportunities for improvement, two of which are: (in the shorter term) to reduce the cost of high vinylidene PIB, and, (in the longer term) for maximal performance, to provide a backbone that has OCP/PAO-type performance at PIB cost. The first opportunity is discussed below.

A catalytic process for producing high vinylidene PIB using a non-BF$_3$ catalyst was discovered and patented at Lubrizol,[1] uses a solid heteropolyacid-based heterogeneous catalyst,[2] and produces PIB form IOB with high >70% terminal olefin content, and thus, has high reactivity with maleic anhydride and phenol. In addition, it has a unique molecular weight distribution, which gives it and the resulting dispersant products unique properties, especially viscometrics. The process is very efficient, with almost no by-products, is amenable to continuous or batch processing and has a simple product recovery stage. At the heart of this process is a heteropolyacid-based catalyst.

Besides reduced overall cost because of the simplified process and materials requirements, it also provides some improvement in thickening power lost from the BF$_3$-produced material, while maintaining the good reactivity properties. However, this process is not yet fully developed, so the total costs are not yet fully certain.

Another alternative backbone, this time moving to a more OCP/PAO-type architecture, is shown in Fig. 10.

● **Branched Ethylene Oligomers**

| Brookhart (UNC/DuPont) - 1995 |

n CH$_2$=CH$_2$ ■ Ni or Pd "single site" catalysts ➡

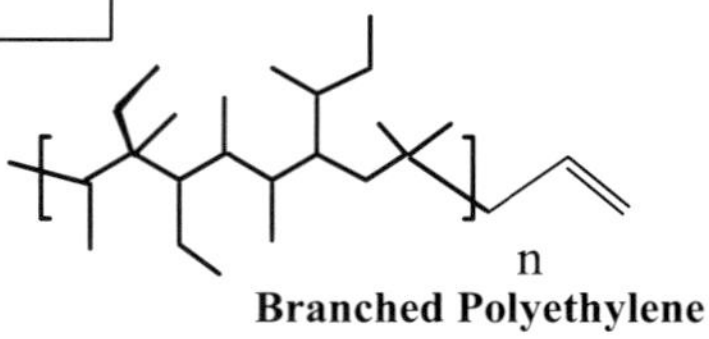

Branched Polyethylene

■ **Many variations of catalysts, MW, and branching level/length**
  ✓ **Grubbs / Caltech**
  ✓ **Turner / Symyx**
  ✓ **Others**

■ **Terminal olefin: thermal succination reactivity**

■ **Potential: PE-like price; PAO-like properties:**
  ✓ **Low deposit detergents**
  ✓ **High performance dispersants**
  ✓ **Alternative to PAO in other additives**

Figure 10. New backbone: branched polyethylene

Branched backbones resulting from polymerization of homopolymerization of ethylene was first reported by Brookhart at UNC/Chapel Hill and co-workers at DuPont in 1995.[4] The discovery generated a flurry of literature reports and patents with many claims around the materials and the catalysts, the majority of which are Ni and Pd systems with diimide ligands. The excitement was based on the possibility of generating branched backbones with broad range of molecular weights solely from ethylene. Since then, many other groups have investigated these so-called "single-site", late transition metal systems for production of branched polyethylenes, including Bob Grubbs at Caltech and workers at Symyx. The latter investigation involved the use of combinatorial-type methods for rapid screening of di and multi-dentate ligands for catalysts for ethylene polymerization activity.

Included in the reported literature of these single-site Ni and Pd catalysts are branched polyethylenes in the dispersant MW ranges and with terminal olefins, which provide thermal succination reactivity to provide PAO-like properties such as low deposits and low temperature performance, at polyethylene-type prices.

These alternate backbones are summarized in the Table 2.

6.   FUTURE TRENDS

Besides these new backbones, there are also many other approaches being pursued to improve detergents and dispersants to better address market needs.

Cost reduction activities by reducing treat rate and process improvements are continually occurring, as are formulating approaches to improved compatibility with low friction/fuel economy systems, improved viscometrics and incorporation of EP/antiwear and low S/P/Cl and ashless components.

In the longer term, customers are continually expecting extended life and durability while demanding lower cost, compatibility with catalyst systems and accountability for everything with which the lube comes into contact during its lifetime. These broader and more demanding challenges will be difficult to address with chemistry and formulation technology alone. Presented below is the growing trend to incorporate other technologies and systems to complement the work that detergents and dispersants do as chemical additives.

These trends include:

- Advanced fluids technology
- Technologies for new product introduction, and
- Performance systems

6.1   Advanced Fluids Technology

The evolution of Advanced Fluids Technology through improvements in understanding of compatibility of key components with each other and with base oils is having a significant impact on our ability to predict performance based on high speed computations. Simple empirical models which may have worked in the past are inadequate for today's and future complex formulations. The resulting solutions provide advanced fluids technology not possible even 2-3 years ago. This is described in more detail below.

As an example, some of the major interactions of base oil with additives to be considered in the design of a lubricant package are shown in Fig. 11. This list isn't complete by any means, but it gives a feel for some of the major issues in base oil quality.

The properties are particularly important to dispersant and detergent surface activity, chemical activity, thermal/oxidative stability and solvent power. Furthermore, maximum performance for one test requires very different and frequently conflicting properties.

Before talking about specific issues, it's worth reminding ourselves about the reasons we use additives at all. With respect to our today's oil industry customers, it is suggested if all oils were perfect, then additives would not be so widely needed. Yet today we see great improvements in base oil quality in combination with additive developments which – together – give us outstanding finished lubricant performance. The conflicting requirements for a dispersant formulation are discussed below.

Major Interactions of Base **Oil** with Additives

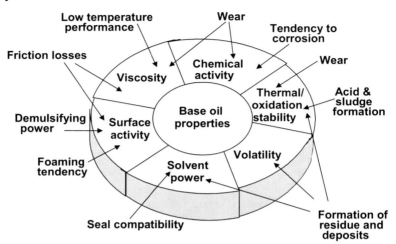

Figure 11. Major interactions of base oil with additives

Four of the performance criteria for dispersants are shown on each axis of Fig. 12; the higher the performance, the larger the area of the figure. The typical dilemma is represented by the dotted lines. Thermal stability and soot control require high conversion normally and high TBN dispersants, which are also the ones which attack seals and give poor VE (sludge) performance. Adjusting the formulation with this conventional wisdom to improve seal compatibility and VE performance results in unacceptable thermal stability. The solid line is a computer-generated formulation, which takes into account many more variables and interactions than would be possible without a very large statistical design and computational power. This design is shown in more detail in Fig. 13.

6.2   Technologies for New Product Introduction

The way in which a thorough understanding of structure-performance relationships is obtained is through statistically designed experiments. This design results in a better balance between the competing requirements by consideration of a large statistically designed matrix of variables, including PIB mol. wt. and mol. wt. cut-off, succination level, amine type, and amine charge. Two new formulations were identified which significantly improved performance (the lower the number the better) over the current baseline. Continued use of these powerful, but computationally demanding statistical models will continue to improve our abilities for formulation to competing requirements.

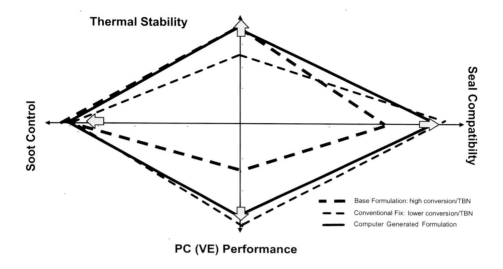

Figure 12. Multiple dispersant requirements

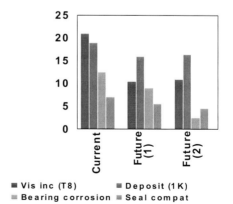

- **Large statistically designed matrix run**
- **Variables included:**
  PBU mol wt
  PBU mol wt "cut"
  Succination level
  Amine type
  Amine charge
- **2 formulations were better in all areas of performance**

Figure 13. Balance between competing requirements

6.3   Performance Systems

*Performance System*s integrate mechanical, chemical and electronic technologies. An example is Lubrizol's water-blended emulsion fuel system called PuriNOx.[5] PuriNOx is a fill-and-go solution for the simultaneous reduction of both particulate and NOx without the need for engine modifications. Typical results are 50% particulate reduction and 30% NOx reduction by simply switching to PuriNOx fuel, with no other modifications.

The system is composed on the additive package, which includes dispersant-type chemistry, water, diesel fuels and season components, and a blending unit, which includes the mechanical and electronic components necessary to provide a stable water-fuel emulsion, the final PuriNOx fuel product. The system is currently being used in several key cities in North America and Europe with emissions problems for both off and on-road applications.

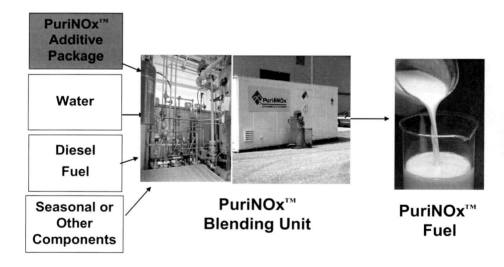

Figure 14. PuriNOx water blended fuel

7.   SUMMARY AND CONCLUSIONS

In summary, dispersants and detergents perform critical functions in lubricants, and the chemistry has been well-optimized over the years for existing PIB backbones. However, new backbones provide the potential for beyond-incremental improvements in performance and cost. Integration of additives chemistry with artificial intelligence, and mechanical and electronic systems will provide the technological basis for future step changes in performance.

8. ACKNOWLEDGEMENTS

The authors thank the following people for their help in preparing this talk: John Pudelski (Co-author, Dispersants), Jim Roski (Co-author, Detergents), John Johnson (Lubrizol High Vinylidene PIB), Saleem Al-Ahmad (Branched Polyethylene), William Chamberlin (Performance Testing Examples), Steve Di Biase (Trends & New Technologies), Richard Biggin, Phillip Shore and Tom Curtis (Advanced Fluids Technology Example), and Casimir Trotman-Dickenson (Detergents).

9. REFERENCES

1. US Pat 5,710,225, **1998**.
2. Burrington, J.D.; Johnson, J.R.; Pudelski, J.K., *Top. Catal*, 2003, *23(1-4)*.
3. US Pat. 5,286,823; EP Pat 145 235; US Pat 5,068,490; RP Pat 322 241; EP Pat 481 297; US Pat 4,152,499.
4. US Pat. 5,880,323, **1999**.
5. Daly, D. T.; Langer, D. A., *Prep. Am. Chem. Soc. Div. Pet. Chem.,* **2000**, *45(1)*, 28-31.

Chapter 19

THE CHEMISTRY OF BITUMEN AND HEAVY OIL PROCESSING

Parviz M. Rahimi and Thomas Gentzis*
National Centre for Upgrading Technology
1 Oil Patch Drive, Suite A202, Devon, Alberta, Canada T9G 1A8
*Current address: CDX Canada Co., 1210, 606-4th Street SW,
  Calgary, Alberta, Canada T2P 1T1

1.   INTRODUCTION

Petroleum consists of a complex hydrocarbon mixture. The physical and chemical compositions of petroleum can change with location, age and depth. Beside carbon and hydrogen, the organic portion of petroleum contains compounds combined with sulphur, oxygen and nitrogen, as well as metals such as nickel, vanadium, iron, and copper.

In recent years, technological advancements in bitumen and heavy oil processing and the stabilization of crude oil prices have made production of synthetic crude oil from these resources attractive and economical. With the goal of processing heavy oil, bitumen, and residua to obtain gasoline and other liquid fuels, an in-depth knowledge of the constituents of these heavy feedstocks is an essential first step for any technological advancement.

It is well established that upgrading and refining of different feedstocks is related to their chemical properties. Since petroleum is composed of complex hydrocarbon mixtures, it is impossible to identify each individual component and its upgrading chemistry. Most investigations, to date, relate the thermal or catalytic behavior of petroleum feedstocks to their fractional composition. The hydrocarbon components of petroleum fall into three classes: [1]

1. *Paraffins* - saturated hydrocarbons, straight or branched chains but no ring structures;
2. *Naphthenes* - saturated hydrocarbons with one or more rings; these hydrocarbons may have one or more paraffinic side chains;

3. *Aromatics* - hydrocarbons with one or more aromatic nuclei; these hydrocarbons may be connected to naphthenic rings and/or paraffinic side chains.

In general, as the MW (molecular weight, or boiling point) of the petroleum fraction increases, there is a decrease in the amount of paraffins and an increase in the amount of naphthenes and aromatics.

This chapter briefly reviews the topics related to bitumen and heavy oil properties and their thermal chemistry; it includes:

• the chemical composition of heavy oil and bitumen in terms of fractional composition — an overview of the methods that have been applied to the separation of petroleum into hydrocarbon types;
• asphaltenes – physical and chemical properties, stability and thermal chemistry;
• chemistry of bitumen and heavy oil upgrading;
• reaction chemistry of bitumen components;
• application of hot stage microscopy;
• stability and compatibility of petroleum.

2.   FRACTIONAL COMPOSITION OF BITUMEN/ HEAVY OIL

Conversion (upgrading) of bitumen and heavy oils to distillate products requires reduction of the MW and boiling point of the components of the feedstocks. The chemistry of this transformation to lighter products is extremely complex, partly because the petroleum feedstocks are complicated mixtures of hydrocarbons, consisting of 10^5 to 10^6 different molecules.[2] Any structural information regarding the chemical nature of these materials would help to understand the chemistry of the process and, hence, it would be possible to improve process yields and product quality. However, because of the complexity of the mixture, the characterization of entire petroleum feedstocks and products is difficult, if not impossible. One way to simplify this molecular variety is to separate the feedstocks and products into different fractions (classes of components) by distillation, solubility/insolubility, and adsorption/desorption techniques.[2] For bitumen and heavy oils, there are a number of methods that have been developed based on solubility and adsorption.[1] The most common standard method used in the petroleum industry for separation of heavy oils into compound classes is SARA (saturates, aromatics, resins, and asphaltenes) analysis.[3] Typical SARA analyses and properties for Athabasca and Cold Lake bitumens, achieved using a modified SARA method, are shown in Table 1. For comparison, SARA analysis of Athabasca bitumen by the standard ASTM method is also shown in this table. The discrepancy in the results between the standard and modified ASTM methods is a result of the aromatics being eluted with a mixture of 50/50vol% toluene/pentane in the modified method instead of 100 vol% pentane in the standard method. A different SARA composition was reported by Clark and Pruden when heptane was used to precipitate asphaltenes.[4] Compositional analyses, including SARA, for a number of Chinese and Middle Eastern vacuum resids were reported by Liu et al.[5] These authors also used modified SARA, which was developed in their laboratory (Table 2). In general, regardless of the method used, the properties of the feedstocks as determined by the SARA method give some indication of the "processability" or problems that may occur during upgrading of these relatively heavy materials. The knowledge of the chemical composition of the feedstocks, as will be shown later, plays a major role in predicting their behaviour in terms of phase separation, coke formation, molecular interactions, and the cause of catalyst deactivation.

Table 1. Properties and SARA fractionation results for Athabasca and Cold Lake bitumens

	Athabasca Modified ASTM 2007		Athabasca ASTM2007	Cold Lake Modified ASTM 2007		Cold Lake ASTM 2007
API Gravity			8.05		10.71	
	MW			MW		
Saturates (wt%)	381	17.27	16.9	378	20.74	21.52
Aromatics (wt%)	408	39.7	18.3	424	39.2	23.17
Resin (wt%)	947	25.75	44.8	825	24.81	39.36
Asphaltene (wt%)	2005	17.28	17.18	1599	15.25	15.95
Carbon (wt%)		83.34			83.62	
Hydrogen (wt%)		10.26			10.5	
Sulfur (wt%)		4.64			4.56	
Oxygen (wt%)		1.08			0.86	
Nitrogen (wt%)		0.53			0.45	
Residue (wt%)		0.15			0.01	

Table 2. SARA composition of various crudes[5]

Sample	Saturates wt%	Aromatics wt%	Resin wt%	C7 Asphaltene wt%
Daqing	34.8	35.5	29.5	0.0
Dagang VR	27.8	28.7	43.4	0.0
Guado VR	14.6	33.0	47.7	4.7
Shengli VR	16.1	30.6	51.1	2.2
Huabei VR	37.6	31.4	29.9	1.1
Liaohe VR	20.8	31.8	41.6	5.7
Oman VR	26.3	40.6	31.2	2.0
S-A-L VR	16.5	49.5	26.8	7.3

Since the trend with petroleum, including bitumens, is towards heavier feedstocks, greater emphasis will be placed here on the composition of material with high boiling point and molecular weight (MW). In a series of papers, Boduszynski[6,7] and Altgelt and Boduszynski[8,9] published data on the composition of heavy petroleum fractions up to the AEBP (atmospheric equivalent boiling point) of 760°C. The fractions were obtained by a combination of DISTACT distillation and sequential elution fractionation (SEF). Data from this work showed that heavy crude oils and petroleum residues had a wide range of MW distributions that extended to relatively small molecules. Quantitative analysis demonstrated that the MWs of most heavy petroleum components did not exceed 2000 g/mol. The hydrogen deficiency (aromaticity) of the petroleum fractions increased with increasing boiling point. The carbon content of petroleum feedstocks and their products varied over a narrow range (80-85wt%) whereas the difference between the hydrogen content of the feed and products varied over a much wider range (5-14wt%). Wiehe[10] suggested that the use of hydrogen concentration rather than H/C atomic ratio was more instructive for assessing fuel materials and processes. The work by Boduszynski[6,7] also showed that the distribution of sulphur, nitrogen and oxygen constituents of petroleum increased with increasing AEBP. It was further demonstrated that S and N associated with metals also occur in the same molecular structure.[6]

Recently, Chung et al.[11] applied SFE (Supercritical Fluid Extraction) to reveal properties of Athabasca bitumen resid. Using this technique, it was possible to fractionate Athabasca bitumen vacuum bottoms (VB) into narrow cut fractions (based on boiling point), the properties of which are shown in Table 3. Approximately 60 wt% of the residue was composed of relatively small molecules (500-1,500 g/mol). The remaining 40 wt% contained larger molecules, including asphaltenes, with an average MW of 1,500-4,200 g/mol. As the MW of the fractions increased, the fractions became more refractory (higher contents of S, N, metals and MCR (Micro Carbon Residue)). The fractions also became more deficient in hydrogen (lower H/C ratio). The use of this fractionation technique resulted in the removal of all asphaltenes and the concentration of the refractory material in the highest boiling point fraction. In comparison to oil sands bitumen, conventional crude oils are of better quality in terms of asphaltenes content. The crudes consist of small amounts of asphaltenes, a moderate amount of resins and a significant amount of oils.[12]

Table 3. Characteristics of Athabasca bitumen vacuum bottom fractions obtained by SFE technique[11]

	Fraction No.									
	1	2A	2B	3	4	5	6	7	8	9	Pitch[b]
Pressure MPa	4-5	5-5.5	5.5-6	6-7	7-8	8-9	9-10	10-11	11-12	>12	
Wt% f itch p	12.7	9.8	7.6	10.6	6.5	4.4	3.3	2.6	2.1	40.4	100
Density, g/mL at 20°C	0.975	0.993	1.006	1.023	1.043	1.054	1.065	1.068	1.074	N/A[c]	1.087
Molecular eighty a D	506	755	711	799	825	948	1134	1209	1517	4185	1191
Sulfur, t% w	4.0	4.5	5.0	5.4	6.0	6.2	6.5	6.8	6.8	7.6	6.5
Nitrogen, ppm	3080	4100	4330	5070	6160	6810	7370	7530	7900	10500	4600
Carbon, t% w	84.5	83.5	83.5	84	83	84	83	83	82.5	78.5	82.7
Hydrogen, t%w	11.5	11.15	10.95	10.55	10.25	10.05	9.8	9.7	9.5	8	9
C/H atomci)	0.612	0.624	0.635	0.664	0.675	0.697	0.706	0.713	0.724	0.818	0.766
Aromatic carbon[a], a f	0.26	0.29	0.25	0.33	0.36	0.40	0.37	0.43	0.43	0.49	0.41
Nickel, pm p	12.8	21.3	30.1	44.8	71.1	89.7	123	138	162	339	148
MCR, wt%	5.6	7.9	10.8	14.3	18.2	21.5	24.7	26.5	28.7	78.9	26.7
Saturates, t% w	26.8	164	9.7	4.1	1.4	0.7	0.6	0.1	0	0	6.3
Aromatics, t% w	57.2	62.4	65.7	66.7	63.9	53.3	45.4	45.9	40.8	2	33
Resin, t% w	15.9	21.2	24.6	29.2	34.8	46	54	53.8	59.2	9.4	29.4
Asphaltenes, t% w	0	0	0	0	0	0	0	0	0	88.03	31.4

[a] C13-NMR.   [b] Pitch = 524°C + fraction   [c] N/A - not applicable

In recent years, the increased interest in refining of heavy crudes and processing VB from oil sands has clearly shown the need for a deeper knowledge of the composition and the chemical structure of these materials. At the National Centre for Upgrading Technology (NCUT) in Devon, Alberta, Canada, a series of fundamental studies was undertaken in order to understand the chemistry of bitumen and heavy oil VB and relate their chemical compositions to processing chemistry. The chemical properties of a number of heavy oils and bitumen VB are shown in Table 4. The data show that although the Forties VB contained a higher percent of pitch compared with the other resids, it also contained less asphaltenes, MCR, metal content, and heteroatoms. In order to gain more insight into the chemical properties of these vacuum residues, the maltenes (pentane solubles) of Athabasca, CL, and Forties VB were subjected to separation using a method similar to SARA analysis. VB maltenes were separated into saturates (M_1), mono/di-aromatics (M_2), polyaromatics (M_3) and polars (M_4). The data for Athabasca bitumen were first reported by Dawson et al.[13] and are compared with other feedstocks in Table 5.

Different laboratories may use different methods for the evaluation of feedstock quality. Method variation makes it difficult to compare the data among laboratories. For example, when comparing the properties of different feedstocks – for instance their SARA analysis – the feedstocks should have been distilled to the same nominal cut point. The data in Table 5 show that

Table 4. Properties of Feedstocks

Analyses	Forties VB	CLVB	Lloyd VB/CLVB
Density (15°C, g/mL)	1.039	1.039	1.03
Oils (wt%)	91.76	75.6	-
Asphaltenes (wt%)	8.15	24.4	20.9
Aromaticity, ^{13}C NMR	30	35	33 (31)
Pitch Content (+525°C, wt%)	93.7	75.4	74.1
MCR (wt%)	16.3	20.2	18 (18)
Molecular weight (g/mol)	948	1071	687 (800)
Viscosity (cST)			
100°C	509.1	2748	80.8 (at 60°C)
120°C	185.9	1010	-
135°C	100.8	377.8	-
Elemental (wt%)			
Carbon	87.1	82.66	84.44
Hydrogen	10.6	9.82	9.69
Nitrogen	0.43	0.51	0.58
Sulphur	1.24	5.42	5.43
Oxygen	<0.5	0.8	
Ash	-	0.04	0.1
H/C	1.46	1.42	1.37
Metals (ppmw)			
Vanadium	29	269	207
Nickel	13	104	96
Iron	3	175	-

Table 5. Components of vacuum resids (wt% on heavy oil/bitumen)

	Pitch (+525°C)	Maltenes (wt%)	Asphaltenes (wt%)	Saturates (M1)	Mono-diaromatics (M2)	Polyaromatics (M3)	Polars (M4)
Athabasca VB	96.4	59.9	40.1	3.36	3.18	38.1	15.36
Cold Lake VB	75.4	75.6	24.4	11.12	4.76	41.52	18.22
Forties VB	93.7	90.9	9.1	14.5	8.83	51.11	16.35

Athabasca bitumen VB and Forties VB have similar pitch contents (fractions boiling above 525°C) and that the concentrations of subcomponents of maltenes are also comparable. However, CLVB (Cold Lake vacuum bottoms) has much lower pitch content and the product composition cannot directly be compared with the other two feedstocks.

Another widely used technique for the separation of heavy oils into subcomponents is IEC (Ion Exchange Chromatography). In this method, the petroleum samples are separated into acid, base and neutral fractions. Walton[14] reported this technique in 1992, and it has also been used extensively by Green et al.[15] in studying the relationships between the composition of different feedstocks with product slate and composition in catalytic cracking. The properties of the Hamaca resid from Venezuela and its fractions, using the technique developed by Green, were reported by Rahimi et al.[16] and are shown in Table 6.

Table 6. Properties of Hamaca resid (+510°C) and its fractions

	Amphoteres	Bases	Acids	Aromatics	Saturates	Hamaca resid	Losses
Yield (wt%)	30	9.8	8.9	41.1	5.7		4.5
MW, VPO	1832	1048	996	600*	620*	NA[a]	
TGA 600°C	43.4	19.2	24.9	5.5	0	NA[a]	
Residue, wt%							
C	83	84.5	83.6	84.6	85.2	85.9	NA
H	8.2	9.6	10.4	11.5	15	9.7	NA
N	2.1	1.7	1.8	ND	ND	0.9	NA
O	2.9	1.6	2.5	1.6	<0.1	2.2	NA
S	2.8	2.8	2.3	2.5	<0.05	4.0	NA

[a] NA, not applicable. ND, not detectable. * values by desorption chemical ionization MS.

3. HETEROATOM-CONTAINING COMPOUNDS

Bitumens and heavy oils present a challenge for upgrading, partly because of their high levels of metals, N, S and O. There have been numerous publications related to organometallic compounds in heavy oils and bitumen, and the effects that they have during thermal and catalytic processing.[11,17-22] The concentrations of heteroatom-containing molecules such as sulphur, nitrogen and oxygen may be relatively small, but their influence during upgrading can be significant. Although heteroatoms are mostly concentrated in the heavier fractions, they are present throughout the range of boiling points. The presence of these molecules creates considerable process constraints during catalytic upgrading, causing catalyst poisoning and deactivation. In the finished products, these heteroatom-containing compounds also may cause problems, including lack of stability on storage and discolouration. Moreover, because of environmental issues future transportation fuels will contain no or significantly less heteroatoms compared with fuels today. Sulphur compounds in heavy oils, bitumen and transportation fuels have been studied by a number of investigators.[1,23-28] It has been shown in the asphaltenes fraction of petroleum that some of the sulphur in the sulphidic form connects two-ring structures (bridge) that can be easily cleaved under HDS conditions (300-345°C) and hydrogen pressures of 500-1,000psi.[1,29] Most S compounds are relatively easy to remove during hydrotreating. However, there are some that are very resistant and create problems during HDS reactions. There are excellent reviews by Whitehurst et al.[30], Toshiaki et al.[31] and Te et al.[32] regarding the HDS of polyaromatic S compounds. In general, it has been shown that the sulphur in ring compounds such as thiophene and benzothiophene can be removed rather easily during hydrotreating. Alkylation of the parent dibenzothiophene, especially at the 4 and 6 positions, reduces the catalyst activity for S removal because of steric hindrance.

Other molecules containing nitrogen and oxygen have a strong inhibition effect on desulphurization reactions.[33-35] H_2S, which is produced during HDS reactions, is also known to act as an inhibitor.

Because of the current limitations in HDS technology, Whitehurst et al, suggested alternative approaches listed below:[30]

1. Development of higher activity catalysts.
2. Altering the desulphurization reaction pathways for hindered sulphur compounds by using zeolite-containing catalysts.
3. HDS reactions that take place in more than one stage: modifying the feed in the first stage, such as isomerization of alkyl groups in hindered alkyldibenzothiophenes, to produce less sterically hindering positions for sulphur removal.
4. Developing a novel reactor design: knowledge of the composition and reactivity of sulphur compounds in different ranges has led to the design of a novel reactor. In this design, the desulphurization of lighter sulphur-containing compounds occurs in the top part of the reactor (co-current with the hydrogen stream). The higher boiling-point sulphur compounds react more efficiently, in the presence of catalyst, in a countercurrent mode.[36]
5. Developing a process other than hydrotreating: this involves the selective adsorption of sulphur compounds on materials such as activated carbon[37] and/or selective oxidation of the sulphur compounds followed by extraction[38-41].

There has been a significant increase in the number of studies to understand hydrotreating reactions for the development of HDN catalysts. Nitrogen compounds are known to be catalyst poisoning in hydrotreating processes[42] and are involved in the formation of gum, causing severe stability problems in the finished petroleum products[43].

Nitrogen content in petroleum and bitumen is much lower than sulphur content (0.1-0.9wt%), although some crude oils may contain up to 2wt%.[1] The amount of N increases as a function of boiling point, in a similar fashion to S. It has been shown that N concentration increases significantly around 350°C and continues to rise.[23] Nitrogen compounds can be classified into a) basic, including pyridine and its derivatives; b) neutral, including alkylindoles and alkylacridines; and c) acidic, including indoles, carbazoles, amides, and nonmetallic porphyrins.[1] Holmes,[44] demonstrated that most of the nitrogen in oil sands bitumen is tied up in pyridinic structures, including quinolines and acridines. Some molecules contain more than one N compound and others contain other elements such as oxygen and S in addition to nitrogen.[45] It should be pointed out that porphyrins are also considered nitrogen-containing molecules and their concentration is relatively high in the heavy distillates and asphaltenes fractions.

Although the chemical structures of some common N compounds are similar to their S counterparts, they behave differently under hydrotreating conditions. Based on the resonance energy of two-ring (Benzothiophene 56 kcal/mole vs. Indole 43 kcal/mole) structures there is no reason to believe that N removal would be more difficult than S removal.[42] However, from the available published data on N compounds, it can be concluded that the removal of nitrogen is indeed more difficult when compared with the removal of sulphur. Basic N compounds with a relatively low MW can be extracted with dilute mineral acids. Nitrogen compounds such as pyridine, quinoline, and isoquinoline also can be extracted from petroleum distillates using dilute mineral acids. However, the carbazole, indole, and pyrrole types of N compounds cannot be extracted with these acids.[1] In addition, N compounds can easily adsorb on the catalyst surface and inhibit other hydrotreating reactions. Removal of N requires prehydrogenation followed by hydrogenolysis of strong C-N bonds. In the transportation of fuels, traces of N compounds can have a significant effect on the stability of those fuels.

The oxygen content of petroleum is relatively small (<1wt%) and is concentrated in the heavier fractions (>350°C boiling point). The chemical functionalities of the oxygen-containing molecules include the following: hydroxyl groups (phenols), carboxyl groups (carboxylic acids and esters), carbonyl groups (ketones), and cyclic and acyclic ethers.[9] Phenols, carboxylic acids and naphthenic acids have been identified in a number of crude oils.[47] Ketones, esters, ethers, and anhydrides are difficult to identify because most of them occur in the higher molecular weight, nonvolatile residua.[1] Various analytical techniques, including LC and HPLC, have been developed to identify and quantify acidic compounds. Recently, using Ion Cyclotron Resonance Mass Spectrometry, Marshall et al.[48] identified over 3000 acids in heavy petroleum. However, in day-to-day refinery operation, a standard method (titration) is used for the measurement of the *total acid number*. In some crude oils, the concentration of acids reaches approximately 1 wt%. The presence of these acids is known to cause problems with corrosion in pipelines during transportation. It has been suggested that naphthenic acids react with iron salts to form iron naphthenates.[49] Upon decomposition of iron naphthenate, FeO is formed that reacts with S compounds such as H_2S, thiols, and disulphides to produce iron sulphide (Figure 1). Iron sulphide is involved in the fouling of refinery equipment. Formation of gum and sludge during storage of fuels can be attributed, in part, to the presence of phenolic compounds.[50]

$$(R\text{-}CH_2\text{-}COO)_2Fe \xrightarrow{\text{heat}} FeCO_3 + R\text{-}CH_2\text{-}CO\text{-}CH_2\text{-}R$$

$$FeCO_3 \xrightarrow{\text{heat}} FeO + CO_2$$

$$FeO + H_2S \longrightarrow FeS + H_2O$$

Figure 1. Mechanism of formation of FeS [49]

4. PROPERTIES OF ASPHALTENES (SOLUBILITY, MOLECULAR WEIGHT, AGGREGATION)

Petroleum consists of four hydrocarbon-types (saturates, aromatics, resins, and asphaltenes) that may be defined in terms of solubility, polarity, and MW.[1] Of these structural types, asphaltenes have markedly adverse effects on the processability of petroleum and play a significant role in the physical properties of heavy oils and bitumen. Because of these effects, in this chapter, asphaltenes will be discussed in detail in terms of their properties, composition, and thermal chemistry during upgrading, as well as their influence on instability/incompatibility during the production, transportation and upgrading of petroleum.

Asphaltenes are probably the most studied fraction of petroleum and bitumen. By definition, they are a solubility class: a fraction of petroleum that is not soluble in paraffinic solvents but soluble in aromatic solvents. In general, they are believed to contain large polynuclear aromatic ring systems ranging between 6 and 20 rings. The condensed aromatic structures bear alkyl side chains varying in size between 4 and 20 carbons.[51-53] Asphaltenes are also known to self-associate (aggregate) in solutions. The difficulty in measuring the MW of asphaltenes has been related to this phenomenon. Asphaltenes MWs as high as 300,000 g/mol (using the ultracentrifuge technique) have been reported.[53] Using the VPO technique, a MW of 80,000 g/mol for the same asphaltenes was reported.[53] Even lower MWs (1,000-5,000 g/mol) were reported in the same study for the same material using VPO under different conditions.

In terms of processability, aggregation of asphaltenes can hinder the conversion of heavy residues to lighter products and can enhance coke formation. Using surface tension measurements and rheological data, Sheu et al.[54-55] and Storm et al.[56] have shown that asphaltenes exhibit properties similar to colloids. These colloids exist in the heavy oil matrix in a micelle form.[57] Since the mole fraction of resins is higher than asphaltenes in petroleum, micelles are richer in resins. Asphaltenes also exhibit properties of colloidal systems such as the 'critical micelle concentration' (CMC) at which aggregates begin to form.

According to the asphaltenes-resin model, resins provide steric stabilization against precipitation of asphaltenes in petroleum fractions. The precipitation and phase separation of asphaltenes upon the addition of a nonpolar solvent can be rationalized in terms of a reduction in the solubility parameter or the polarity of the hydrocarbon medium.[58-60] The most direct evidence for the presence of asphaltenes aggregates in oil has been demonstrated by means of SANS (Small Angle Neutron Scattering). Ravey et al.[61] and Overfield et al.[62] further demonstrated that the physical dimensions and shape of the aggregates are a function of the solvent used. Ravey et al.[61] demonstrated that asphaltenes from Middle Eastern crude oils form sheet-like aggregates in tetrahydrofuran. Sheu et al.[54] showed that Ratawi asphaltenes in toluene and pyridine solutions form spherical aggregates having diameters between 60 and 66 Å. Watson and Barteau[62] observed STM (Scanning Tunneling Microscope) images of Ratawi asphaltenes aggregates separated from pyridine solutions, which showed that the asphaltenes aggregates formed orderly flat arrays covering hundreds of angstroms of the surface. The self-aggregation of asphaltenes plays an important role in the chemistry of asphaltenes conversion during thermal treatment of petroleum residues[63] and will be discussed in detail later in this chapter. Neves et al.[64] using light scattering and electrophoresis techniques, determined the aggregate size and charges present in asphaltenes from a Brazilian crude oil. When n-heptane was added to a mixture of asphaltenes in toluene, depending on concentration, the size of aggregate changed significantly varying between 125-186 nm in high concentration solutions to 238-398 nm in the low concentration solutions. It was also shown that asphaltenes possess a positive charge.

Asphaltenes from Athabasca bitumen were first separated using n-pentane by Pasternak and Clark in 1951.[65] In most refinery practices, the solvent of choice is n-heptane and asphaltenes are defined as materials soluble in toluene or a solvent having a solubility parameter in the 17.5-21.6 Mpa$^{1/2}$ range. As the carbon number of the extracting solvent increases, the amount of asphaltenes that precipitate decreases. Fundamentally, it is important to note that during asphaltene precipitation by any solvent, smaller asphaltene molecules, as well as some maltene materials, co-precipitate because of chemisorption. In order to obtain "pure asphaltenes," Strausz et al.[66] suggested that after standard asphaltenes extraction, the material should be further extracted by acetone for one week. Following this procedure would result in asphaltenes that do not contain any foreign material. The formation of aggregates in Athabasca bitumen has been studied in detail by Murgich and Strausz.[67] It has been shown that the molecular shapes of asphaltenes and resins play a significant role in the sizes and lifetimes of the aggregates.

4.1   Chemical Structure of Asphaltenes

Recent studies by Strausz et al.[68-70], Strausz et al. and Peng et al.[71-72], and Murgich et al.[73] revealed the close structural similarities of asphaltenes from different sources. Asphaltenes are thought to be molecular units consisting of small- to mid-size alkyl and naphthenoaromatic hydrocarbons. Some units contain S and to a lesser extent N. The molecular units are linked together by C-C, C-S, and C-O linkages. A molecular representation of a petroleum asphaltenes model has been given by Artok et al.[74] This model was based on extensive analytical data using ^{1}H and ^{13}C NMR, GPC, pyrolysis gas chromatography/mass spectrometry, and MALDI TOF (Matrix-assisted laser desorption/ ionization time-of-flight) mass spectrometry.

Chemical and degradation methods were employed by Peng et al.[72] to study the structure of Athabasca asphaltenes. In the oxidation reaction of asphaltenes, the RICO (Ruthenium Ions Catalyzed Oxidation) reaction permits the selective oxidation of aromatics while leaving saturated hydrocarbons intact. A two-dimensional structure for Athabasca asphaltenes having the general formula of $C_{412}H_{509}S_{17}O_9N_7$, a H/C ratio of 1.23 and MW of 6,239 g/mol was proposed by Strausz et al. Recently, Leon et al.[75] investigated the structural characterization and self-association of asphaltenes having different origins. These authors argued that problems during crude oil production could be related to the asphaltene properties. They showed that n-heptane asphaltenes from two problematic crude oils had higher aromaticity, lower H/C atomic ratio, and significantly lower CMC (in different solvents) compared with the properties of asphaltenes derived from non-problematic crude oils. Although the data from this work could explain operational problems during crude oil production, the average structural model proposed for asphaltenes contains highly fused aromatics. This model is significantly different from the asphaltenes model proposed by Strausz et al.[66], which contains a smaller number of fused aromatic rings. In another study, Shirokoff et al.[76] investigated the structure of Saudi crude asphaltenes using compositional analysis as well as XRD (X-Ray Diffraction). Based on these analyses, it was concluded that the n-pentane asphaltenes derived in these crudes contained condensed aromatic sheets with a tendency to stack. The condensed aromatics had naphthenic and alkyl chains on their periphery. Yen et al.[77] postulated a similar structure with less dense alkyl regions on the periphery of asphaltene particles.

The effect of asphaltenes on the physical properties of heavy oils and bitumen has been studied extensively.[78-82] It has been demonstrated that the viscosity of petroleum is significantly influenced by the presence and concentration of asphaltenes. Storm et al.[81] demonstrated that when the relative viscosity of heavy oils was plotted versus asphaltenes concentration in both toluene (at room temperature) and vacuum residue (at 93°C), a straight line resulted. Thus, it was concluded that toluene is as good a solvent for asphaltenes as for vacuum resid. However, the amount of solvation is temperature dependent. By analyzing the temperature dependency of solvation, Storm et al. showed that the forces holding asphaltenes in the resid are very weak. Moreover, the fact that the solvation constant is the same for toluene at 25°C as in a vacuum resid at 93°C implies that the forces between asphaltene colloidal particles and toluene are weaker.

In a similar study aimed at shedding light on the aggregation of asphaltenes, Rao and Serrano[78] studied the physical interactions of asphaltenes in heavy oil. The viscosities of Arab resids containing different amounts of asphaltenes were measured in toluene at 27°C. It was concluded that aggregation of asphaltenes in heavy oils is stepwise and causes high viscosity and an apparent increase in MW. At low asphaltene concentrations, smaller aggregates were formed that could be dissociated to monomers at the processing temperature. However, at high asphaltene concentrations, the aggregates could not be dissociated and formed asphaltene clusters. Aggregate formation can cause process upset and limit process yields. Since formation of aggregates is stepwise and reversible, dissociation of asphaltene clusters to monomers may be accomplished by diluting the resid with an appropriate solvent, thus improving the process efficiency.

The dependency of feedstock viscosity on asphaltenes concentration has significant implications, because reducing viscosity could make pipeline transportation of heavy oil less dependent on diluent. Removal of even 100wt% asphaltenes from Athabasca bitumen does not reduce viscosity enough to meet pipeline specifications in Alberta (viscosity of 350 cSt at operating temperature, °API of 19, and BSW of <0.5 vol%). However, removal of approximately 30 wt% of asphaltenes has been shown to reduce diluent requirement by almost 30%. The benefit of partial removal of asphaltenes on thermal processing will be discussed later.

4.2   Thermal Chemistry of Asphaltenes

Asphaltenes are considered "bad actors" in refinery upgrading processes, because they cause coke and sludge formation, and in catalytic processes because they cause severe catalyst deactivation. It is widely accepted that petroleum is colloidal in nature and that asphaltenes exist in the petroleum in a micelle form. The formation of micelles is believed to be primarily due to the interaction between asphaltene species or asphaltene-resin fractions. The nature of the intermolecular or intramolecular forces that cause the formation of asphaltene micelles is not clear at present. It has been suggested that a number of forces may be involved including Van der Waals attraction, dipole-dipole interaction, hydrogen bonding, electron-transfer or charge transfer between aromatics (π-π bonding), and porphyrin interaction.[83-84] Wiehe [84] stated that the primary interaction between asphaltene molecules is the Van der Waals attraction between large areas of flat polynuclear aromatics. He used the solubility parameter to measure the attractive interaction between petroleum molecules, which is inversely related to the hydrogen content of the fractions. Using the generated phase diagram, it was shown that asphaltene molecules are relatively insoluble because of their high molecular weight and low H/C atomic ratio (high aromaticity). Wiehe also proposed a hybrid model for petroleum materials. According to this model, the asphaltenes are held in a delicate balance that can be easily upset by the addition of saturates or by the removal of resins and small aromatics.[85] During thermal treatment of petroleum feedstocks, the asphaltene micelles break down to form smaller aggregates. Further heating can result in the breakup of the protective resin layer and, finally, at about 300°C, the cores become "bare" resulting in precipitation of asphaltenes.

From a processing point of view, the microphase behaviour of asphaltenes plays an important role during catalytic and thermal upgrading of heavy oils and bitumens.[86-87] Storm et al. demonstrated that during hydroconversion of vacuum residues, the amount of sediment formation (cyclohexane insolubles) is strongly related to the amount of heptane insolubles in the residues.[86] When the heptane asphaltenes were removed from the residue, no sediment was formed. The authors further showed that a specific fraction of the residue, namely the pentane insoluble-heptane soluble fraction (which is relatively rich in hydrogen), plays an important role in reducing coke formation. They reasoned that the hydrogen in this fraction is used to cap the radicals generated in the asphaltenic-rich phase during conversion, hence retarding the formation of less soluble molecules.

Other properties of the feedstocks that could be correlated to the sediment formation are the degree of condensation of polynuclear aromatics and the degree of alkyl-substitution of polynuclear aromatics. In a later publication, Storm et al.[87] suggested the involvement of "macrochemistry" in the formation of sediment during hydroprocessing at lower temperatures. These particles (sediments) are then converted to macroscopic particles at higher temperatures. According to this model, grouping of certain molecules in the resid results in the formation of micelles with a dimension of 0.004:m. Rheological and SAXS (Small Angel X-ray Scattering) results showed that these small particles could be transformed into larger two-dimensional particles with dimensions of 0.02-0.03:m. As discussed earlier, these microstructure particles form well below temperatures that are characteristic of chemical reactions. Also, during low severity catalytic hydrocracking, polymerization between the larger two-dimensional particles does not occur and only a semisolid, which is soluble in the reaction media, is formed. However, at higher reaction temperatures, polymerization of the particles takes place and results in the formation of insoluble coke. It was thought that if the flocculation of larger particles could be interrupted below reaction temperatures, coke formation could be reduced. To reduce flocculation, Storm et al.[87] used 1 wt% of a polymeric additive (functionalized poly propyleneoxide diole with PCl_3) in a series of hydrocracking experiments with VB from Arabian medium and heavy crude oils. It was shown that the use of the additive, which was soluble in the heavy asphaltenic phase, interrupted the coalescence of asphaltic micelles required for coke formation. Thus, in these experiments, pitch conversion, desulphurization, and demetallation were improved and the formation of sediment was reduced.

In the pendant-core model proposed by Wiehe[88], every molecule in petroleum consists of two building blocks: an aromatic core, which is coke-producing, and a pendant block, which is attached to the core and can be cracked to produce volatile liquids. Using this model, the building blocks of each hydrocarbon type, including asphaltenes, were constructed. As the polarity or the boiling point of petroleum fractions increases, the core part of the fraction or its aromaticity also increases. As the resins and asphaltene fractions contain more cores, these fractions contribute significantly more coke than the other petroleum fractions during thermal processes. It has also been suggested that coke formation is the result of separation of a second liquid phase formed from partially converted asphaltene cores. The liquid-liquid phase separation is evidenced by the presence of spherical liquid crystalline coke called mesophase. The stacking of the polynuclear aromatic structure, present in asphaltenic material, forms the mesophase. During reside hydrocracking, mesophase lacks a significant degree of ordering (fast solidification) as a result of the high reactivity of the asphaltene cores. Wiehe[2] has shown in the thermal conversion of Cold Lake (CL) vacuum resid that there is a delay in coke formation (an induction period) and that the onset of coke formation is triggered by phase separation.

As indicated earlier, there is a delicate balance between the concentration of resins and asphaltenes in petroleum fractions. Any interruption in the ratio of asphaltenes:resins can cause operational problems because of asphaltene precipitation (coke formation) and plant shutdown. It is therefore crucial, if coke formation is to be minimized, to monitor the ratio of asphaltenes:resins during the thermal processing of bitumens and heavy oils. Clarke and Pruden[89] developed a heat transfer analysis technique that can detect the onset of asphaltene precipitation. They showed that precipitated asphaltenes could be re-peptized using polynuclear aromatic compounds such as phenanthrene. It has been demonstrated in the 5000 bbl/d CANMET hydrocracking unit operated by Petro-Canada at its Montreal refinery, that the recycling of heavy resids during the hydrocracking of heavy oils improves the plant's operability, resulting in higher conversions.[90] In Texas City, the H-Oil process for the catalytic conversion of resid blends has shown that the presence of highly aromatic byproduct streams is most effective in minimizing asphaltenes precipitation and solids formation.[91] The review by Mansoori[92] on asphaltenes deposition and control during production suggested that the addition of resins (peptizing agents) in proper amounts might prevent or control the heavy organic deposition problem.

Considering the "average chemical structure" of asphaltenes as proposed by Strausz et al.[66], understanding the chemistry of asphaltenes conversion could be extremely complicated. The literature suggests that asphaltenes, apart from producing coke, are also converted to lower molecular weight components that are later converted to liquid and gaseous products. Calemma et al.[93] studied the pyrolysis kinetics of four different asphaltenes using TGA/FTIR (Thermo Gravimetric Analyzer/Fourier Transform Infrared Spectroscopy). They concluded that for asphaltenes containing more sulphur, the activation energy was lower by about 2 Kcal/mole. The data were interpreted on the basis of weaker bond strength in C-S and S-S bonds, which are approximately 10 Kcal/mole lower than in C-C bonds. The data also showed that the activation energy increased at higher conversion. The results were explained as follows: as the reaction proceeds, the structures in the asphaltenes become more aromatic (dehydrogenation) and the alkyl groups attached to the rings become shorter. These reactions make the subsequent decomposition more difficult, which translates to higher activation energy for decomposition reactions.[94] Rahimi et al.[95] have shown that approximately 50 wt% of Ç asphaltenes from Athabasca bitumen can be converted to maltenes at a relatively moderate severity.

During the thermal cracking of vacuum residues, maltenes play an important role in the conversion of asphaltenes. Wiehe[60] has shown that in the hydrocracking of CL VB (Cold Lake Vacuum Bottoms), the presence of maltenes in the resid increases (prolongs) the coke induction period significantly. The results were interpreted based on the effectiveness of maltenes as hydrogen donors to cap free radicals produced by the thermal cracking of asphaltenes.

At this stage, a general overview of the chemistry of upgrading is necessary prior to discussing the conversion of bitumen and heavy oil constituents at elevated temperatures.

5.   CHEMISTRY OF UPGRADING

It is well documented that the conversion of resid to lighter products, whether or not in the presence of hydrogen and/or a catalyst, is largely thermally driven.[91,96] The hypothetical molecular structure of bitumen and

asphaltenes consists mostly of C-C, C-H, C=C (in the aromatic rings) and to a lesser degree C-S, C-O, C-N, S-H, and O-H. The metal impurities are mostly attached to nitrogen in porphyrin and non-porphyrin structures. Since most of the chemical reactions during bitumen upgrading are thermally driven, there is no selectivity in bond cleavage. Under non-selective thermal reaction conditions, the weakest bonds break first. The bond dissociation energies of the most common bonds are shown in Table 7. According to this table, C-S (sulphide) has the lowest bond dissociation energy and will break first at a relatively moderate severity. At low to moderate severities in typical visbreaking conditions (380°C-410°C), 10-20wt% pitch conversion can be achieved without major coke formation. Under these conditions, the changes to the molecular structure of bitumen are relatively small since most of the C-C bonds remain intact. A significant MW reduction must take place before bitumen molecules are converted to distillates. The following chemical reactions are known to occur during this transformation to distillates:
1. homolytic cleavage of C-C bonds;
2. side chain fragmentation (cleavage);
3. ring growth;
4. hydrogen shuttling;
5. hydrogenation of aromatics/dehydrogenation of cycloparaffins;
6. ring opening;
7. heteroatom and metals removal.

Table 7. Bond dissociation energies

Bonds	Kcal/mole
H-H	103
C-C	83-85
C-H	96-99
N-H	93
S-H	82
O-H	110-111
C=C	146-151
C-N	69-75
C-S	66
Ar-CH$_2$-CH$_2$-Ar	71
Ar-H	111

The most important reaction in upgrading that leads to a significant molecular weight reduction and produces distillate fractions is probably cleavage of the C-C bonds. The bond dissociation energies for C-C bond cleavage can vary depending on the type of molecules. It has been suggested that the reaction mechanism for the cleavage of C-C bonds during upgrading is free radical in nature, and proceeds through a free radical chain mechanism.[97]

The reaction kinetic (homolysis) of a hypothetical molecule (M) proceeding through a free radical chain mechanism is shown in Figure 2. The overall reaction rate is related to the rate of the initiation step. If one assumes that the initiation step involves the homolytic cleavage of the C-C bond, then one can calculate the half-life ($t_{1/2}$) for the reaction. For example, the $t_{1/2}$ for the cleavage of the C-C bond in PhCH$_2$-CH (CH$_3$)$_2$ is 2.4 hours at 540°C and 46.3 days at 440°C. The fact that thermal hydrocracking of bitumens and heavy oils can be accomplished at a much lower temperature and a relatively high conversion indicates that the cleavage of C-C bonds is not a rate-determining step. The initiation step may involve cleavage of the C-S bond, or the breakage of the C-C bonds must be accomplished by a mechanism other than homolytic cleavage.

Initiation $\quad$ M $\quad\xrightarrow{k_i}\quad$ 2 R$^{\bullet}$

Propagation $\quad$ R$^{\bullet}$ + M $\xrightarrow{k_p}$ RH + M$^{\bullet}$

$\qquad\qquad$ M$^{\bullet}$ $\xrightarrow{k_{p2}}$ R'$^{\bullet}$ + O

Termination $\quad$ R$^{\bullet}$ + R$^{\bullet}$ $\xrightarrow{k_t}$ R-R

$\qquad\qquad$ R$^{\bullet}$ + M$^{\bullet}$ $\xrightarrow{k_t}$ R-M

$\qquad\qquad$ M$^{\bullet}$ + M$^{\bullet}$ $\xrightarrow{k_t}$ M-M

$$\text{Rate} = k_p\left(\frac{k_i}{2k_t}\right)^{1/2}[M]^{2/3}$$

Figure 2. Radical chain mechanism for homolysis of a hypothetical molecule M

There are a number of mechanisms proposed for the initiation step of the cleavage of strong C-C bonds. These reactions are shown in Figures 3 to 5. In the radical hydrogen transfer mechanism proposed by McMillen et al.[98], there is a direct transfer of hydrogen from a radical to another molecule. Another mode of transferring H atoms includes RRD (Reverse Radical Disproportionation), proposed by Stein et al.[99] Besides transferring an H atom from solvents, a free H atom can also be produced by β-elimination from various radicals or by the reaction of a stabilized radical with molecular hydrogen[100] (Figure 5).

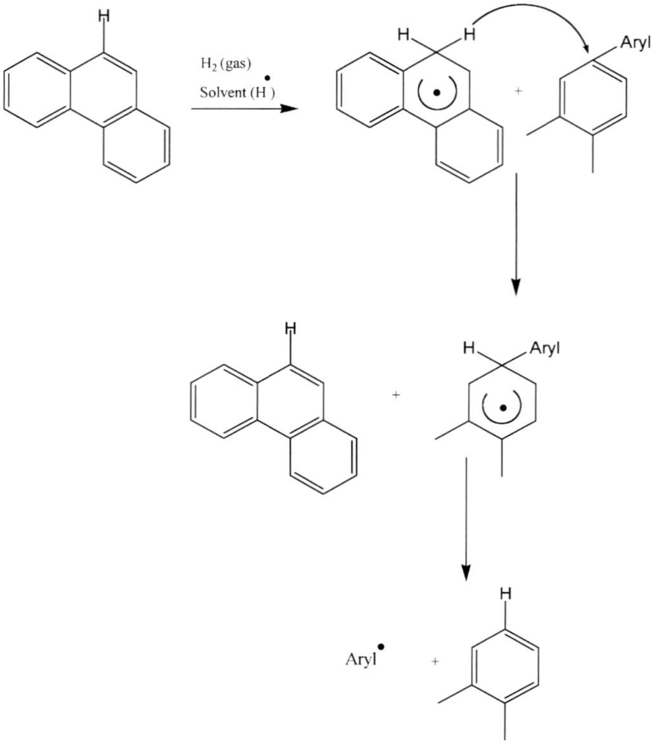

Figure 3. Radical hydrogen transfer for the cleavage of C-C bond[98]

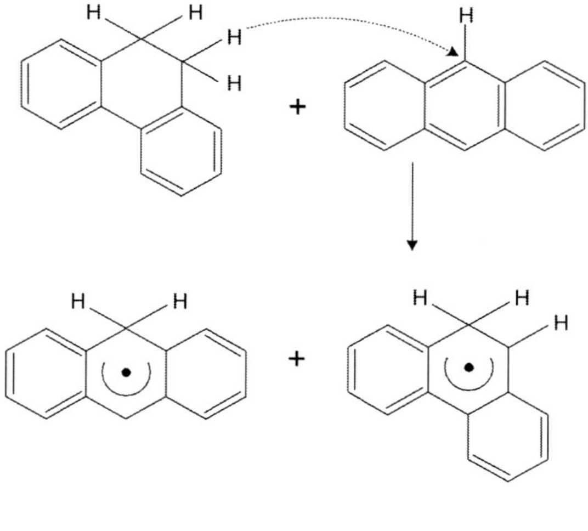

Figure 4. Reverse radical disproportionation mechanism for H-transfer [99]

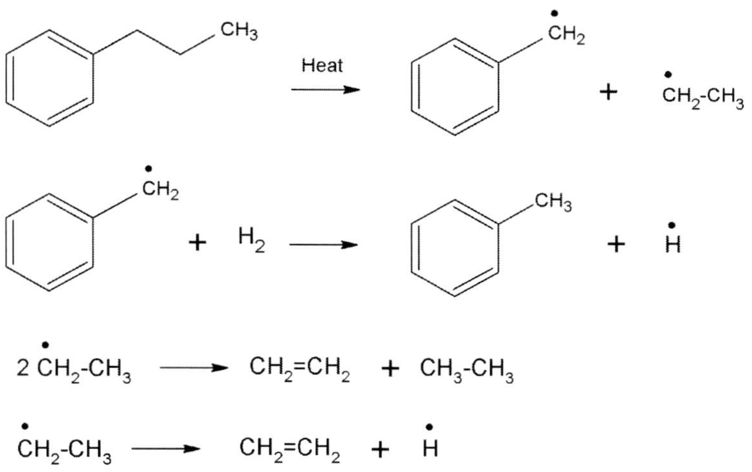

$$2\ \overset{\bullet}{C}H_2\text{-}CH_3 \longrightarrow CH_2\text{=}CH_2 + CH_3\text{-}CH_3$$

$$\overset{\bullet}{C}H_2\text{-}CH_3 \longrightarrow CH_2\text{=}CH_2 + \overset{\bullet}{H}$$

Figure 5. Formation of H by β elimination [100]

An alternative reaction mechanism to unfavorable homolytic C-C bond cleavage is the electron transfer mechanism shown in Figure 6. In this reaction, an electron is transferred from an aromatic core of a molecule to a metal (Ni, V, or Fe) to produce a radical cation. An electron transfer mechanism has been proposed for the reaction of the model in the presence of carbon black, and with tetralin as the solvent.[101] In this model, selective C-C bond cleavage (at the position of poly condensed aromatic moiety) is achieved at 320°C in the presence of carbon black where no thermal reaction is known to occur. The authors rationalized their observations based on the aforementioned electron transfer mechanism. The selective C-C bond cleavage occurs by an electron transfer from the condensed aromatic ring to the carbon black surface, which has become positively charged at a reaction temperature of 320°C. This reaction mechanism has been recently disputed by Penn et al.[102] They argue the C-C bond cleavage is caused by a radical hydrogen transfer mechanism in which the H atom adds to the ipso position of the polyaromatic moiety. It is also possible that this electron transfer may occur during bitumen and heavy oil upgrading since there is a significant concentration of transition metals capable of accepting electrons from highly condensed poly aromatics.

$$M^{3+} + PhCH_2\text{-}CH(CH_3)_2 \longrightarrow M^{2-} + \overset{+\ \bullet}{PhCH_2}\text{-}CH(CH_3)_2$$

$$\overset{+\ \bullet}{PhCH_2}\text{-}CH(CH_3)_2 \longrightarrow \overset{+}{PhCH2} + \overset{\bullet}{C}H(CH_3)_2$$

Figure 6. Cleavage of C-C bond via electron transfer mechanism

Whatever the reaction mechanism, it has been demonstrated in heavy oil upgrading that the conversion to lighter products is mainly thermally driven, whether or not the reaction is carried out in the presence of hydrogen and/or a metal catalyst.[103-106]

5.1   Reaction of Feedstock Components - Simplification of Upgrading Chemistry

The chemistry of resid upgrading is extremely complicated.[107-110] This is in part due to the complexity of the chemical nature of the feedstocks. In order to understand the chemistry of upgrading, it would be helpful to reduce this complexity prior to reaction, by separating the feedstocks (bitumen and heavy oils) into well-known components such as SARA – saturates, aromatics, resins and asphaltenes – which are useful tools in understanding bitumen chemistry.

Speight[111] investigated the chemistry of the thermal cracking of asphaltenes from Athabasca bitumen and deasphalted oil using a destructive distillation technique. A comparison of the analytical data from the feedstocks and products indicated that considerable changes occurred during cracking. The H/C atomic ratio data showed that simultaneous hydrogenation and dehydrogenation reactions took place. The results also indicated that DAO was more thermally labile than asphaltenes. Approximately 83 wt% of the DAO was converted into resins (maltenes), light oil and gases. In contrast, only 52 wt% of asphaltenes was converted into cracked products. The presence of n-paraffins in the light oil fractions indicated dealkylation (side chain fragmentation) of alkylaromatic compounds. A separate investigation of the thermal reaction of Athabasca bitumen at 440°C for 30 minutes using a micro-autoclave showed that approximately 46 wt% of pentane solubles (maltenes) was formed.[112] Further analysis revealed that the maltenes consisted of 17 wt% saturates, 5 wt% mono-/diaromatics, 20.5 wt% polyaromatics and 3.5 wt% polar compounds.

In a recent study by Rahimi et al.[113], partial deasphalting of Athabasca bitumen resulted in bitumen with an improved quality in terms of lower viscosity, MCR, and metals content. The better quality of the deasphalted oil was reflected in its coking behaviour. At laboratory conditions comparable to a delayed coking operation, deasphalted bitumen produced similar liquid yield but lower coke yield compared with the coking of the whole bitumen. Also, the liquid product resulting from the coking experiments of the partially deasphalted bitumen feed had less olefins. This product would be more stable and would need less hydrogen during upgrading and refining.

Klein et al.[114] investigated the pyrolysis kinetics of resids, isolated asphaltenes and maltenes from Hondo, Arabian heavy, Arabian light, and Maya oils. At 400°C and 425°C, isolated asphaltenes reacted selectively to form maltenes. At higher temperatures (450°C), asphaltenes reacted predominantly to form coke. Furthermore, pyrolysis of the maltenes indicated that asphaltenes and coke were formed in the following order:

maltenes $\rightarrow$ asphaltenes $\rightarrow$ coke

Karacan and Kok recently studied the pyrolysis of two crude oils and their SARA fractions.[115] Differential scanning calorimetry and thermogravimetry techniques were used to evaluate the pyrolysis behaviour of the feedstocks. The results indicated that the pyrolysis mechanisms depend on the nature of the constituents. Thermogravimetric data showed that asphaltenes are the main contributors to coke formation and that resins are a second contributor. The weight loss for the SARA components was additive. The authors argued that each fraction in a whole crude oil follows its own reaction pathway and there is no interaction or synergy between the components.

The chemistry of upgrading is expected to become significantly more complex when vacuum bottoms are processed. Dawson et al.[116] investigated the thermal behaviour of Athabasca bitumen VB and SARA fractions at temperatures between 420°C and 460°C. Detailed product analyses revealed that saturates and mono- and diaromatics are relatively unreactive, whereas polyaromatics and resins are converted to smaller molecules including saturates. The aromatic fraction constituted the major components of the vacuum residues. Approximately 54 wt% of CLVB consisted of polyaromatic hydrocarbons.[117] The upgrading chemistry of this fraction was investigated in a batch autoclave at different severity conditions (420°C-440°C, 30 min, 13.9 MPa H_2). The product analyses showed that at all severities the aromatic fraction (M_3) not only decomposed to form smaller molecules but also polymerized to form larger molecules and a small amount of coke (Table 8). The reaction sequence can be summarized as :

aromatics $\rightarrow$ resins $\rightarrow$ asphaltenes $\rightarrow$ coke

Table 8. Thermal hydrocracking of polyaromatics derived from Cold Lake vacuum bottoms

Products (wt%)	420°C	440°C	450°C
Gases	5.6	14.3	11
Asphaltene	6.0	9.5	12
Coke	0.7	0.5	0.3
Maltenes			
Saturates	10.3	15	21.7
Mono-diaromatics	8.4	9.8	9.1
Polyaromatics	57.8	33.1	41.6
Polars	11.2	17.7	4.6

Furthermore, the analysis of polyaromatic fractions following the reaction (M_3 products) showed that these molecules had relatively lower MW, shorter chains and were more aromatic. The results of this work confirm that side chain fragmentation and hydrogenation/dehydrogenation reactions are major routes in the thermal cracking of heavy oils and bitumen (see Tables 9-10).

Table 9. NMR analysis of M3 fractions from different sources

Source	Aliphatic carbon types (wt %)				
	Alpha CH$_3$	Naphthenic (Beta)	>C$_6$ Chains	Paraffinic CH	Others
Feed	8.5	4.9	6.1	9.4	24.2
420°C	6.8	2.8	4.4	9.3	18.1
440°C	6.6	1.7	1.1	5.1	13.9
440°C (cat)	7.6	3.1	4.3	8.5	19.6
450°C	7.1	1.5	0.9	5.2	14.3

Assignment of Aliphatic Carbons
in a Hypothetical Molecule
1 - Alpha CH3
2 - Naphthenic CH2
3 - CH2 in >C6 Chain
4 - Paraffinic CH
5 - Others

Table 10. NMR analysis of M3 fractions from different sources

Source	Aromatic carbon types (wt%)						Aromaticity	Cluster size*
	Q1	Q2	Q1+Q2	Ha	Hb	Ha+Hb		
Feed	9.0	8.5	17.5	5.8	6.3	12.1	0.36	14
420°C	11.1	10.1	21.2	11.1	8.7	19.8	0.50	12
440°C	11.4	20.2	31.6	15.7	8.1	23.8	0.66	18
440°C (cat)	10.4	10.7	21.1	10.8	9.4	20.3	0.49	12
450°C	14.3	13.1	27.4	17.2	9.3	26.5	0.65	11

χ_b = / (cluster size) given in reference (7)

$$X_b = Q_2 / (Q_1 + Q_2 + H_a + H_b)$$

In an attempt to correlate the thermal cracking behavior of heavy oils to their properties, Liu et al.[118] studied the thermal cracking of 40 heavy oil fractions obtained by supercritical extraction from six Chinese light crude oils and oils from Oman and Saudi Arabia. The thermal cracking experiments were performed at 410°C, 0.1 MPa N$_2$ for 1 hour. A non-linear regression fit indicated that the thermal cracking of the fractions could be correlated with the H/C, S (wt%), N (wt%) and molecular weight. A similar correlation was obtained with SARA analysis, S, and MW. In this study, the coke yields (toluene insolubles) correlated with the concentration of asphaltenes/resins plus aromatics of the cracked residues. These results were rationalized in terms of coke formation, not only from asphaltenes but also as a result of the phase separation of the colloidal system in the residues.

6. APPLICATION OF HOT STAGE MICROSCOPY IN THE INVESTIGATION OF THE THERMAL CHEMISTRY OF HEAVY OIL AND BITUMEN

Investigating the effects of process variables on coke formation is usually achieved by autoclave and pilot plant experiments, which are time consuming and expensive. Another tool that offers the advantage of real-time observation of the thermal reaction is hot-stage microscopy. This section reviews the usefulness of hot-stage microscopy for a better understanding of bitumen chemistry.[119]

The coke induction period (the time that is required for coke precursors or mesophase to start forming at a specific temperature) during thermal processing of heavy oils and bitumen is an important measurement for understanding operational problems. These problems may include coke formation and fouling in heat exchangers and fractionators leading to an unscheduled plant shutdown. The induction period coincides with the moment at which the asphaltenes in the reaction system reach their maximum concentration.[120] At the National Centre for Upgrading Technology's laboratory, hot-stage microscopy techniques were used to investigate those parameters that influence the thermal chemistry pathways during upgrading of petroleum feedstocks. Such studies could lead to possible solutions aimed at reducing coke and maximizing product yields. The parameters studied included the following: feed composition (i.e., SARA, acid, base); boiling point range (343ºC-675ºC); degree of asphaltenes removal (0-18 wt%, C_5 asphaltenes); coke suppressing agent (H-donors); and solid additives (clays) that inhibit coalescence of coke precursors.

6.1 Effect of Feedstock Composition

The thermal chemistry of heavy oils and bitumen is extremely complicated because of wide variations in chemical compositions. The most refractory components in petroleum feedstocks are asphaltenes, which contribute the most to coke formation during thermal cracking. Next to asphaltenes, resins and large aromatics also contribute to coke. To investigate the effect of these three heavy oil components on the mesophase induction period, Athabasca bitumen fractions containing varying amounts of asphaltenes (obtained by supercritical fluid extraction) and Venezuelan heavy oil fractions varying in polarity (obtained by Ion Exchange Chromatography) were selected.

One of the Athabasca bitumen fractions investigated under a hydrogen atmosphere consisted of 88 wt% asphaltenes. This fraction exhibited a very short induction period (48 minutes at room temperature), as expected for a highly-asphaltenic feed. These results are consistent with the findings of Wiehe[120], who demonstrated, using an autoclave, that Cold Lake asphaltenes (neat) formed coke immediately with no induction period. Using hot-stage microscopy it is also possible to follow the coalescence of mesophase particles in real-time and observe the changes in the apparent viscosity and, finally, the solidification process (Figure 7a). The second feedstock examined was a fraction of Athabasca bitumen containing 45.4 wt% resins, 54.0 wt% aromatics and no asphaltenes. The coke induction period was significantly longer (~72 minutes) compared with the fraction containing asphaltenes. This fraction developed mesophase of various sizes during thermal treatment, which later coalesced to form bulk mesophase and even-flow domains (Figure 7b).

In another study, in order to examine the relationship between reactivity and composition of a Venezuelan heavy oil, the Hamaca resid (510°C+) was separated into fractions including amphoteric, acidic, basic, neutral and aromatic.[121] Results showed that the amphoteric fraction exhibited the shortest induction period for coke formation (50 minutes), followed by the basic and the acidic fractions. Amphoterics contain polynuclear aromatic systems having 5-6 rings per system [122]; as such, they are the most viscous and showed the fastest solidification. The basic fraction, which consists of 4 aromatic rings per system, showed a longer induction period (58 minutes). The acidic fraction, with only 1-3 aromatic rings per system, had an induction period of 68 minutes. The neutral fraction, which contains non-basic nitrogen and oxygen species, formed small mesophase spheres (Figure 7c) and the induction period was 82 minutes. The aromatic fraction had the longest induction period of 93 minutes and developed large mesophase (Figure 7d). The total resid containing all of the above components had an induction period of 61 minutes, indicating the synergy or interaction among the components during thermal reaction.

6.2   Effect of Boiling Point

Recently, major synthetic oil producers in Western Canada have switched from atmospheric bottoms to vacuum resids for processing bitumen. This raises the question of what impact this change might have on the coke induction period during thermal processing of these materials. To address this question, Athabasca bitumen (+343°C) was fractionated using Distact distillation into four distillates and resids.[123] The selected boiling point cuts were 525°C, 575°C, 625°C, and 675°C. The coke induction periods of bitumen and its resid fractions were measured under hydrogen and nitrogen atmospheres. The results indicated no major differences in the coke induction periods between bitumen (68 minutes) and the four fractions (68, 66, 77 and 61 minutes, respectively). This may have important process implications in that processing higher boiling fractions does not necessarily shorten the coke induction period.

Fig 7a: Semicoke formation from asphaltenes   Fig 7b: Semicoke formation from resins

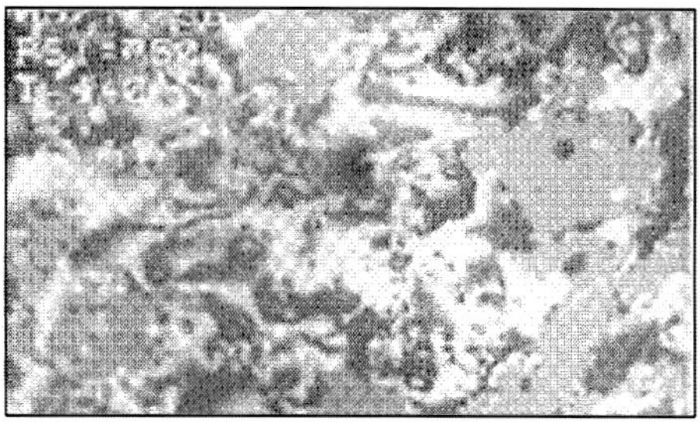

Fig 7c: Mesophase from neutral fraction 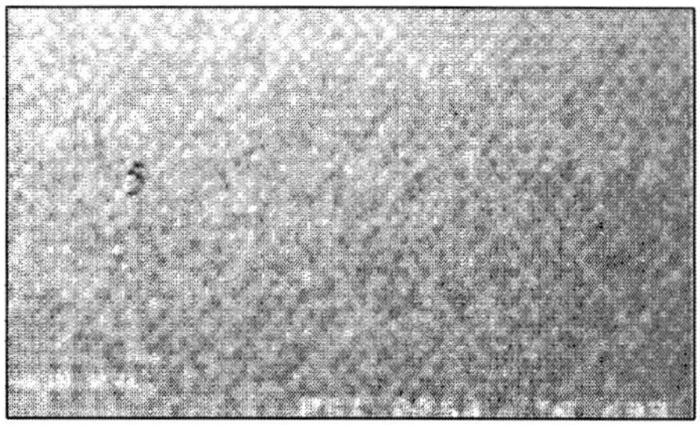   Fig 7d:Mesophase from aromatic fraction

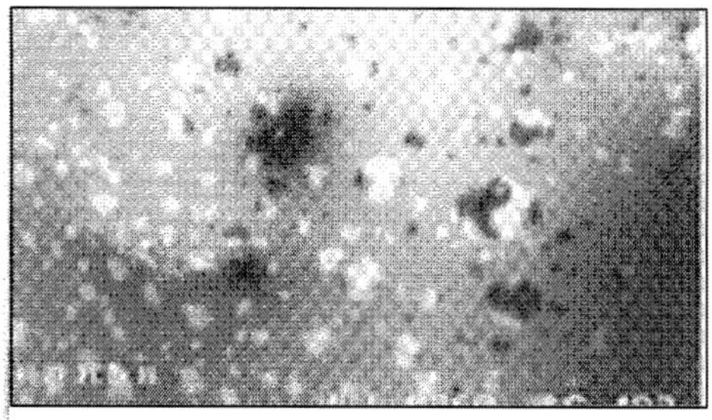

Fig 7e: Mesophase in presence of clay 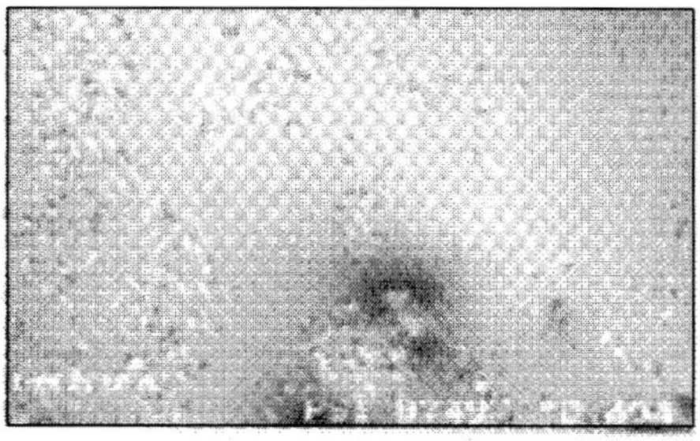   Fig 7f: Mesophase in presence of H-donor

40 microns

Figure 7. Effect of different variables on mesophase formation

6.3   Effect of Additives

Coke formation during thermal treatment of bitumen proceeds via the formation of mesophase spheres that coalesce to form larger mesophase, which eventually deposits as coke on the surfaces of equipment. If the coalescence process can be slowed down or prevented, the size of mesophase would be smaller and, consequently, be carried out of the process lines and vessels without fouling the equipment. To investigate the effectiveness of clay minerals as additives that interfere with the growth of mesophase, three clays (kaolinite, illite and montmorillonite) were added to Athabasca bitumen at concentrations of 2 wt% and 5 wt%.[124] Although the presence of clays did not result in a delay of the coke induction period, the presence of kaolinite reduced the size of mesophase and decreased the mesophase coalescence under nitrogen (Figure 7e). Thus, it was deduced that clays may reduce or prevent fouling on the walls of furnaces, exchangers and reaction vessels during bitumen upgrading.

Liquid additives, such as hydrogen donors, have also been shown to reduce coke formation during heavy oil hydroprocessing.[125] It has been shown that H-donors have the ability to scavenge free radicals, to act as antioxidants, and to inhibit coke formation while improving asphaltenes conversion. In order to examine the effectiveness of hydrogen donors on the induction period during upgrading, Athabasca bitumen vacuum bottoms was mixed with 5 wt% of an H-donor derived from a petroleum stream.[126] The results showed that the H-donor prolonged the coke induction period by as much as 20 min. It was also clear from the visual observation under microscope that the presence of additives reduced the rate of mesophase formation under nitrogen gas (Figure 7f).

6.4   Effect of Deasphalting

Asphaltenes are known to be the most refractory components in heavy oils and bitumen and can be converted to coke during thermal cracking. Therefore, it would be beneficial to selectively remove, if possible, the "worst" asphaltenes using solvents with different solubility parameters (polarity). To achieve the above objective, Athabasca bitumen was treated with mixtures of pentane and toluene ranging from P/T =100, 90/10, 85/15, 75/25 and 65/35. As the pentane to toluene (P/T) ratio decreased the amount of asphaltenes remaining in the partially deasphalted oil (PDAO) increased and their quality deteriorated, as indicated by progressively higher MCR content. To investigate the effect of asphaltenes removal on the coke induction period, the PDAOs were subjected to hot-stage microscopy studies.[127] It was shown that, as the P/T ratio decreased (solubility parameters increased), the coke induction period became shorter, ranging from 72 minutes at P/T=100 to 60 minutes at P/T=65/35. This study shows that deasphalting enhances the bitumen quality and results in a lower coking propensity.

7.   STABILITY AND COMPATIBILITY

In nature, the components of petroleum, including bitumen and heavy oils, are in a fine balance. Any changes to this balance, as a result of physical or chemical treatments, can result in instability followed by asphaltenes precipitation, phase separation and sediment formation. In fact, the changes to the properties of petroleum occur from the time of its production and throughout transportation and processing. Asphaltenes precipitation can also occur when two petroleum streams are incompatible. Using the colloidal hybrid model for petroleum discussed earlier by Wiehe,[84] it is relatively easy to follow the changes in the structure of petroleum as a result of physical and chemical treatment. Either removing or converting the resins layer that protects (peptizes) asphaltenes, results in instability of asphaltenes and, finally, insolubility in the media and precipitation. To prevent further system deterioration, one must bring asphaltenes back into the solution (re-peptize), for example by the addition of small amounts of dispersants.

There are no standard tests for measuring the onset of asphaltenes precipitation. Among the techniques and analytical methods frequently used to measure sediment and asphaltenes onset for the adjustment of different process parameters in the refineries are:
1. spot test (ASTM-D-4740-95);
2. total sediment (ASTM-4870-96);
3. solubility parameters, optical microscope;[128]
4. light scattering (PORLA);[129]
5. peptization value (P-value);[130]
6. colloidal instability index (CII);[131-132]
7. coking index[133].

In recent years there has been a significant effort by different groups[129, 134-135] to automate the measurement of asphaltenes precipitation. This will, supposedly, create more reliable and consistent data.

In this section the stability and compatibility of petroleum will be discussed in terms of physical treatment such as distillation, deasphalting and diluent addition for pipeline transportation, and in terms of chemical treatment such as upgrading.

7.1   Physical Treatment

7.1.1   Effect of Distillation

Any physical treatment that may disturb the balance existing between components of oil may cause instability in the system and, finally, asphaltenes precipitation. One such physical treatment is distillation. Conversion of Athabasca bitumen to synthetic crude oil at both Syncrude Canada and Suncor Energy Inc., has changed from processing full-range bitumen (atmospheric bottoms, 343°C) to processing vacuum tower bottoms (524°C). Due to improvements in distillation technology, it is feasible to go to even deeper cut points without any cracking by using a short path distillation unit.

At the National Centre for Upgrading Technology in Canada, full-range Athabasca bitumen was distilled into four fractions using a DISTACT unit. The results on SARA analysis of the original feed and the fractions are shown in Table 11, and reveal that as the boiling point increased, the saturates, aromatics and resins decreased, whereas the asphaltenes content (C_5) increased significantly. As a result, the ratio of resins/asphaltenes decreased from 1.49 in the full-range bitumen to 0.30 in the fraction with a nominal boiling point of 675°C. As shown in Table 12, the stability as defined by the ratio of solubility number and peptization value (p-value) point to a less stable feedstock as the boiling point increases (although all fractions are stable and have values well above 1, which is considered to be the borderline between stable and unstable material). The fouling tendency as measured by the colloidal instability index (CII) showed that the feedstocks with a CII greater than about 0.6 would have a greater chance of forming deposits or coke when subjected to thermal treatment.[132] From the data in Table 12, it could be concluded that, although all the distillate fractions from Athabasca bitumen are stable, distillation into deeper cuts removed the asphaltenes' protective layers, making them more prone to separation/deposition and, subsequently, to coke formation.

Table 11. Effect of distillation cut point on SARA analysis of Athabasca bitumen

Cut point, °C	Saturates, wt%	Aromatics, wt%	Resins, wt%	Asphaltenes,wt%	Resins/asph
343	17.3	39.7	25.8	17.3	1.49
525*	5.9	31.8	20.8	41.5	0.50
575	5.3	30.9	19.7	44.1	0.45
625	4.8	23.3	17.1	54.7	0.31
675	3.5	20.4	17.5	58.5	0.30

* From DISTACT distillation

Table 12. Effect of cut point on satiability and solubility parameters of Athabasca bitumen

Cut point, °C	Resins/asph	S_{BN}/I_N	P-value	CII**
343	1.49	3.5	3.6	0.53
525*	0.50	3.0	3.1	0.90
575	0.45	2.4	2.7	0.98
625	0.31	2.5	2.4	1.5
675	0.30	2.4	2.5	1.6

* From DISTACT distillation
** Colloidal Instability Index = (S+Asph)/(A+P)

7.1.2   Effect of Addition of Diluent

Heavy oils and bitumen are very viscous (>100,000 cP) and as such cannot be transported by pipeline. The Canadian pipeline specification for viscosity is 350 cSt at operating temperature. At the present time, the viscosity specification is met by the addition of approximately 25 vol% diluent, usually natural gas condensate. Depending on the characteristics of the diluent, i.e., if it is paraffinic, it may cause asphaltenes precipitation during pipelining.

It is well known that the presence of asphaltenes in heavy oils and bitumen plays an important role in its rheological properties and is responsible for the observed high viscosity of these materials.[81] A recent study by Rahimi et al. showed that the viscosity of Athabasca bitumen varies significantly with the asphaltenes concentration.[127] Although the removal of all the asphaltenes from Athabasca bitumen (17wt% of C_5 asphaltenes) did not result in a product that met the pipeline specification for viscosity, the amount of diluent required to meet that specification was significantly reduced. It was further shown that a large excess of a paraffinic diluent (80 vol%) was required before asphaltenes precipitation occurred.[130] In the case of partial deasphalting for the purpose of producing cleaner feedstock, the largest and the most refractory asphaltenes could be removed by the addition of a small amount of paraffinic diluent. However, if the diluent in the bitumen is not removed immediately, the remaining asphaltenes might precipitate.

7.1.3   Thermal/Chemical Treatment

As discussed earlier, the asphaltene molecules are peptized by resins. In the deasphalting process, which is usually performed at low temperatures, the disruption between maltenes and asphaltenes is done deliberately and results in the precipitation of the latter. However, during thermal processes, the nature and the chemical characteristics of both maltenes and asphaltenes can change significantly. Because of the nature of the chemical reactions – such as side chain fragmentation – the paraffinic content of the reaction medium will increase. The reaction medium will not keep large molecules in solution and phase separation will occur. The reverse may also occur — wax (long paraffin chain molecules) separates because of a drop in temperature or an increase in the aromaticity of the liquid medium.[136]

There is another type of solid-like (coke-like) material formed during visbreaking and hydrocracking of vacuum residues. This type of solid is usually referred to as sediment or sludge. Sediment may form from both the inorganic or from the organic constituents of petroleum. Formation of sediment and sludge limits process conversion because of its accumulation in downstream equipment.[86] High severity (high conversion) processes promote condensation and polymerization reactions. When the solvent power of the liquid phase is not sufficient to keep the coke precursors in solution, coke is formed. Even during catalytic hydrocracking, such as in the H-Oil process, the formation of deposits cannot be avoided. The degree and the amount of deposit depend on the severity of the process. It has been shown that during H-Oil operation for moderate resid conversion (60%), there are no problems associated with sediment formation.[137] However, conversions above 60% are accompanied by increased fouling and sedimentation problems in the operating units. The fouling can be 1) the result of asphaltenes precipitation (phase separation) because of the incompatibility of the effluent stream; 2) the result of the formation of polyaromatics (PAH) from naphthenic hydrocarbons by dehydrogenation reactions; and 3) phase separation because of the existence of supercritical conditions in the hydroprocessing equipment.[138] However, the authors of this study believe that reactor fouling is mostly related to the rejection of Ni and V sulphides to the catalyst surface. Moreover, the extract from the deposit formed in the reactor was more aromatic than the feed to the H-Oil unit, indicating dehydrogenation under relatively severe conditions (440°C). From the analyses of the deposits obtained from the vacuum distillation tower, it was concluded that asphaltenes deposition is the major contributor to vacuum tower fouling in the H-Oil process.[137]

Incompatibility is believed to be the major cause of fouling during crude oil refining and hydrotreating when using a fixed-bed reactor.[138,139] Mixing crude oils may cause asphaltenes precipitation (incompatibility), which results in rapid fouling of the preheat exchanger and coking furnace tubes. Wiehe[138] has developed a compatibility model based on solubility parameters of crude oils. This model demonstrates that not only is the ratio of the blend important for obtaining a compatible mixture, but also the correct order of mixing crude oils will determine if the mixture is compatible. The incompatibility of oils has also been shown to result in the plugging of hydrotreaters. In Wiehe's study [139], the foulant accumulated in the top few inches of the bed and consisted of carbonaceous material with little inorganic matter (ash). In order to apply the oil compatibility model to oils that contain no asphaltenes, Wiehe developed new tests and was able to diagnose and resolve the plugging problem.

Polymerization and retrograde reactions lead to coke formation during thermal treatment of heavy oils and bitumen. Tests have been developed to diagnose the initial stage of the problem and prevent fouling. Depending on the thermal process used, there are different tests available to determine and control the sediment formed. The Shell hot filtration test is a common test used in hydrocracking to determine the stability of the operating process without possible shutdown. The amount of solids (n-heptane insoluble) formed is measured at intervals, and should not exceed a certain percentage (0.15-0.5 wt%) so that the stability of the operation can be maintained.

In visbreaking, which is a relatively mild thermal cracking process, the amount of asphaltenes in the visbroken residue (+350°C) increases and the ability of the medium to disperse asphaltenes decreases. The flocculation ratio of different concentrations of the resid is first determined using various concentrations of a binary solvent (n-heptane and xylene). A plot of the flocculation ratio versus the dilution ratio is then constructed. Extrapolation of the flocculation ratio to zero on the X-axis produces a value that is called the P-value (peptization value). This value must be higher than 1.1 (P-valve of the tar 350°C+ > 1.1). The stability of visbroken products from Athabasca bitumen obtained at different severities was reported recently by Rahimi et al.[130] During thermal reaction, the asphaltene aromatic cores, which were stabilized by aliphatic side chains and the presence of resins, were exposed by the cracking off of the side chains and by the conversion of resins. Thus, as the severity of the thermal process increased, the asphaltenes became increasingly less soluble (increased insolubility number). Thermal reactions also produce light hydrocarbons by breaking off the side chains from aromatic rings. This can result in a decrease in the solvency of the media (decreased solubility blending number). However, for the very aromatic Athabasca bitumen the solubility blending number remained nearly constant with increasing severity, showing that the light aromatic and light saturated products compensated for each other. Therefore, partially thermally cracked feedstocks are not only unstable because of the presence of olefins and diolefins, but also because the solubility of asphaltenes has significantly been reduced. The addition of hydroaromatic compounds (H-donors) significantly improves the quality and the stability of the visbroken tars.[140]

The stability of the reaction products in reactors can significantly be improved by controlling the resin:asphaltenes ratio. Benham and Pruden[90] demonstrated in the CANMET Hydrocracking Process® that controlling the ratio of polar aromatics to asphaltenes is the key to achieving a better unit operability and to obtaining higher pitch conversion and lower coke yield. By recycling the heavy gas oil fraction that is rich in polar aromatics, asphaltenes could be kept in peptized form leading to high pitch conversion with low coke yield.

8.   REFERENCES

1. Speight, J.G. *The Chemistry and Technology of Petroleum,* 3rd ed., Dekker: New York, 1998; 215 pp.
2. Wiehe, I.A. Tutorial on Resid Conversion and Coking, Proc. 2nd Intl. Conf. on refinery processing, AIChE 1999 Spring National Meeting, Houston, TX, March 14-18, 499-505.
3. *ASTM. 1995. Annual Book of Standards.* American Society for Testing and Materials: Philadelphia, PA; Method D-2007.
4. Clarke, P.; Pruden, B. Asphaltene precipitation from Cold Lake and Athabasca bitumen, *Pet. Sci. Technol.,* **1998**, *16(3&4)*, 287-305.

5. Liu, C.; Zhu, C.; Jin, L.; Shen, R.; Liang, W. Step by step modeling for thermal reactivities and chemical compositions of vacuum residues and their SFEF asphalt, *Fuel Process. Technol.*, **1999**, *59*, 51-67.
6. Boduszynski, M.M. Composition of heavy petroleum. 1. Molecular weight, hydrogen deficiency, and heteroatom concentration as a function of atmospheric equivalent boiling point up to 100°F (760°C), *Energy Fuels* **1987**, *1*, 2-11.
7. Boduszynski, M.M. Composition of heavy petroleum. 2. Molecular characterization, *Energy Fuels* **1988**, *2*, 597-613.
8. Altgelt, K.H.; Boduszynski, M.M. Composition of heavy petroleum. 3. An improved boiling point-molecular weight relation, *Energy Fuels* **1992**, *6*, 68-77.
9. Boduszynski, M. M.; Altgelt, K.H. Composition of heavy petroleum. 4. Significance of the extended atmospheric equivalent boiling point (AEBP) scale, *Energy Fuels* **1992**, *6*, 72-76.
10. Wiehe, I.A. The pendant-core building block of petroleum residua, *Energy Fuels* **1994**, *8*, 536-544.
11. Chung, K. H.; Xu, C.; Hu, Y.; Wang, R. Supercritical fluid extraction reveals resid properties, *Oil Gas J.,* **1997**, *95*, 66-69.
12. Koots, J.A.; Speight, J.G. Relation of petroleum resins to asphaltenes, *Fuel*, **1975**, *54*, 179-184.
13. Dawson, D.W.; Chornet, E.; Tiwari, P.; Heitz, M. Hydrocracking of individual components isolated from Athabasca bitumen vacuum resid, *Prep. Div. Petrol. Chem.,* American Chemical Society, Dallas national meeting, 1989, *34 (2)*, 384-394.
14. Walton, H.F. Ion exchange chromatography, In *Chromatography,* 5th ed., *Fundamentals and applications of chromatography and related migration methods*, E. Heftmann (Ed.), Elsevier: Amsterdam, 1996; Chapter 5, pp. A227-A265.
15. Green, J.B.; Zagula, E.J.; Reynolds, J.W.; Wandke, H.H.; Young, L.L.; Chew, H. Relating feedstocks composition to product slate and composition in catalytic cracking. 1. Bench scale experiments with liquid chromatographic fractions from Wilmington, CA, >650°F resid, *Energy Fuels* **1994**, *8*, 856-867.
16. Rahimi, P.M.; Gentzis, T.; Cottʊ, E. Investigation of the thermal behavior and interaction of Venezuelan heavy oil fractions obtained by ion exchange chromatography, *Energy Fuels* **1999**, *13*, 694-701.
17. Pearson, C.D.; Green, J.B. Comparison of processing characteristics of Mayan and Wilmington heavy residues. 2. Characterization of vanadium and nickel complexes in acid-base-neutral fractions, *Fuel* **1989**, *68*, 456.
18. Speight, J.G., *Fuel science and technology handbook,* Marcel Dekker: New York, 1990; Chapter 3, p.80.
19. Quann, R.J.; Ware, R.A. Catalytic hydrodemetallation of petroleum, *Adv. Chem. Eng.,* **1988**, *14*, 96-259.
20. Franceskin, P.J.; Gonzalez-Jiminez, M.G.; Darosa, F.; Adams, O.; Katan, L. "First observation of an iron porphyrin in heavy crude oil," *Hyperfine Interact.* **1986**, *28*, 825.
21. Biggs, J.C.; Brown, R.J.; Fetzer, W.R. Elemental profiles of hydrocarbon materials by size exclusion chromatography/inductive coupled plasma atomic emission spectroscopy, *Energy Fuels* **1987**, *1*, 257-262.
22. Ware, R.A.; Wei, J. "Catalytic hydrodemetallation of nickel porphyrins," *J. Catal.* **1985**, *93*, 100-121.
23. Altgelt, K.H.; Boduszynski, M.M. *Composition and Analysis of Heavy Petroleum Fractions,* Marcel Dekker: New York, NY, 1994; 495p.
24. Strausz, O.P.; Lown, E.M.; Payzant, J.D. Nature and geochemistry of sulphur-containing compounds in Alberta petroleums, In *Geochemistry of Sulphur in Fossil Fuels, ACS Symposium series 429.* W.L. Orr; C.M. White (Eds.) American Chemical Society: Washington, D.C., 1990; Chapter 22, 366-395.
25. Payzant, J.D.; Montgomery, D.S.; Strausz, O.P. Novel terpenoid sulphoxides and sulphides in petroleum, *Tetrahedron Lett.,* **1983**, *24*, 651.

26. Cyr, T.D.; Payzant, J.D.; Montgomery, D.S.; Strausz, O.P. A homologous series of novel hopane sulphides in petroleum, *Org. Geochem.,* **1986**, *9*, 139-143.
27. Sarowha, S.L.S.; Dogra, P.V.; Ramasvami, V.; Singh, I.D. Compositional and structural parameters of saturate fraction of Gujarat crude mix residue, *Erdol Kohle,* **1988**, *41*, 124-125.
28. Waldo, G.S.; Carlson, R.M.K.; Maldowan, J.M.; Peters, K.E.; Penner-Hahn, J.E. Sulphur speciation in heavy petroleum: Information from x-ray absorption near-edge structure, *Geochim. Cosmochim. Acta,* **1991**, *55*, 801-814.
29. Shaw, J.E. Molecular weight reduction of petroleum asphaltenes by reaction with methyl iodide-sodium iodide, *Fuel* **1989**, *68*, 1218-1220.
30. Whitehurst, D.D.; Isoda, T.; Mochida, I. Present state of the art and future challenges in hydrodesulphurization of polyaromatic sulphur compounds, *Adv. Catal.,* **1998**, *42*, 345-471.
31. Kabe, T.; Ishihara, A.; Qian, W. *Hydrodesulfurization and hydrodenitrogenation.* Kodansha Ltd.: Tokyo, 1999.
32. Te, M.; Fairbridge, C.; Ring, Z. Various approaches in kinetic modeling of real feedstock hydrodesulfurization, *Pet. Sci. Technol.,* **2002**, in Press
33. Kabe, T.; Ishihara, A.; Nomura, M.; Itoh, T.; Qi, P. Effects of solvents in deep desulfurization of benzothiophene and dibenzothiophene, *Chem. Lett.,* **1991**, *12*, 2233-2236.
34. Nagai, M.; Kabe, T. Selectivity of molybdenum catalyst in hydrodesulfurization, hydrodenitrogenation, and hydrodeozygenation: Effect of additives on dibenzothiophene hydrodesulfurization, *J. Catal.,* **1983**, *81*, 440-449.
35. Whitehurst, D.D.; Mitchell, T.O.; Farcasiu, M.; Dickert, J.J. Exploratory studies in catalytic coal liquefaction, *Report on EPRI Project,* **1979**, 779-18.
36. Sucharek, A.J. How to make low sulfur, low aromatics, high cetane diesel fuel – Synsat technology, *Prep. Div. Pet. Chem.*, American Chemical Society, **1996**, *41*, 583.
37. Savage, D. W.; Kaul, B.K. Deep desulfurization of distillate fuels, U.S. Patent: 5454933, 1995.
38. Bonde, S.E.; Gore, W.; Dolbear, G.E.; Skov, E.R. Selective oxidation and extraction of sulfur-containing compounds to economically achieve ultra-low proposed diesel fuel sulfur requirements, *Prep. Div. Pet. Chem.,* American Chemical Society, 219th National Meeting, San Francisco, CA, March 26-31, 2000; 364-366.
39. Collins, F.M.; Lucy, A.R.; Sharp, C. Oxidative desulphurization of oils via hydrogen peroxide and heteropolyanion catalysts, *J. Mol. Catal. A: Chemical,* **1997**, *117*, 397-403.
40. Dolbear, G.E.; Skov, E.R. Selective oxidation as a route to petroleum desulfurization, *Prep. Div. Pet. Chem.,* American Chemical Society, 219th National Meeting, San Francisco, CA, March 26-31, 2000; 375-378.
41. Zannikos, F.; Lois, E.; Stournas, S. Desulfurization of petroleum fractions by oxidation and solvent extraction, *Fuel Process. Technol.,* **1995**, *42*, 35045.
42. Ho, T.C. "Hydrodenitrogenation catalysis," *Catal. Rev.-Sci. Eng.,* **1988**, *30 (1)*, 117-160.
43. Mushrush, G.W.; Speight, J.G. Petroleum products: Instability and Incompatibility, *Applied Energy Technology Series,* Taylor & Francis, 1995, 183,pp.
44. Holmes, S.A., Nitrogen functional groups in Utah tar sand bitumen and product oils, *AOSTRA J. Res.,* **1986**, *2*, 167-175.
45. McKay, J.F.; Weber, J.H.; Latham, D.R. Characterization of nitrogen bases in high-boiling petroleum distillates, *Anal. Chem.,* **1976**, *48*, 891-898.
46. Satterfield, C.N.; Cocchetto, J.F. Pyridine hydrodenitrogenation: An equilibrium limitation on the formation of pyridine intermediate, *AIChE J.,* **1975**, *21(6)*, 1107-1111.
47. Seifert, W.K.; Teeter, R.M. Identification of polycyclic aromatic and heterocyclic crude oil carboxylic acids, *Anal. Chem.,* **1970**, *42*, 750-758.
48. Qian, K.; Robbins, W.K.; Hughey, C.A.; Cooper, H.J.; Rodgers, R.P.; Marshall, A.G. Resolution and identification of elemental compositions for more than 3000 crude acids in

heavy petroleum by negative-ion microelectrospray high-field fourier transformed ion cyclotron resonance mass spectrometry, *Energy Fuels* **2001**, *15*, 1505-1511.

49. Panchal, C.B. Fouling induced by dissolved metals, *Proc. Intl Conf. on Petroleum Phase Behavior and Fouling*, AIChE Spring National Meeting, Houston, TX, March 14-18, 1999, 367-372.

50. Motahashi, K.; Nakazono, K.; Oki, M. Storage stability of light cycle oil: studies for the root substance on insoluble sediment formation, *Proc. 5th Intl. Conf. on Stability and Handling of Liquid Fuels,* Rotterdam, The Netherlands, October 3-7, 1994, 829-844.

51. Hasan, M.-U.; Ali, M. F. Structural characterization of Saudi Arabian extra light and light crudes by ^{1}H and ^{13}C NMR spectroscopy, *Fuel* **1989**, *68*, 801.

52. Mojelsky, T.W.; Ignasiak, T.M.; Frakman, Z.; McIntyre, D.D.; Lown, E.M.; Montgomery, D.S.; Strausz, O.P. Structural features of Alberta oil sands bitumen and heavy oil asphaltenes, *Energy Fuels* **1992**, *6*, 83.

53. Payzant, J.D.; Lown, E.M.; Strausz, O.P. Structural units of Athabasca asphaltenes: the aromatics with linear carbon framework, *Energy Fuels* **1991**, *5*, 445.

54. Sheu, E.Y.; Storm, D.A.; DeTar, M.M. Asphaltenes in polar solvents, *J. Non-Cryst. Solids,* **1991**, *341*, 131-133.

55. Sheu, E.Y.; DeTar, M.M.; Storm, D.A.; DeCanio, S.J. Aggregation and kinetics of asphaltenes in organic solvents, *Fuel* **1992**, *71*, 299-302.

56. Storm, D.A.; Barresi, R, J.; DeCanio, S.J. Colloidal nature of vacuum residue, *Fuel* **1991**, *70*, 779-782.

57. Speight, J.G. *Petroleum Chemistry and Refining,* Applied Technology Series, Taylor & Francis, 1998; 116.

58. Speight, J.G. *The Chemistry and Technology of Petroleum,* 2nd ed., Marcel Dekker: New York, NY, 1991.

59. Wiehe, I.A. Solvent-resid phase diagram for tracking resid conversion, *Ind. Eng. Chem. Res.,* **1992**, *31*, 530-536.

60. Wiehe, I.A. A phase-separation kinetic model for coke formation, *Ind. Eng. Chem. Res.,* **1993**, *32*, 2447-2454.

61. Ravey, J.C.; Decouret, G.; Espinat, D. Asphaltene macrostructure by small angle neutron scattering, *Fuel* **1988**, *67*, 1560-1567.

62. Watson, B.A.; Barteau, M.A. Imaging of petroleum asphaltenes using scanning tunneling microscopy, *Ind. Eng. Chem. Res.,* **1994**, *33*, 2358-2363.

63. Overfield, R.E.; Sheu, E.Y.; Shina, S.K.; Liang, K.S. SANS study of asphaltene aggregation, *Fuel Sci. Technol. Int.,* **1989**, *7*, 611-624.

64. Neves, G.B.M.; dos Anjos de Sousa, M.; Travalloni-Louvisse, A.M.; Luca, E.F.; González, G. Characterization of asphaltene particles by light scattering and electrophoresis, *Pet. Sci. Technol.,* **2001**, *19 (1&2),* 35-43

65. Pasternack, D.S.; Clark, K.A. The components of the bitumen in Athabasca bitumous sand and their signification in the hot water separation process, *Alberta Research Council Report,* **1951**, No. 58, 1-14.

66. Strausz, O.P.; Mojelsky, T.W.; Faraji, F.; Lown, E.M. Additional structural details on Athabasca asphaltene and their ramification, *Energy Fuels* **1999**, *13*, 207-227.

67. Murgich, J.; Strausz, O.P. Molecular mechanics of aggregates of asphaltenes and resins of Athabasca oil, *Pet. Sci. Technol.,* **2001**, *19 (1&2),* 231-243

68. Strausz, O.P.; Mojelsky, T.W.; Lown, E.M. The molecular structure of asphaltenes; an unfolding story, *Fuel* **1992**, *71*, 1355-1363.

69. Peng, P.; Morales-Izquierdo, A.; Hogg, A.; Strausz, O.P. Molecular structure of Athabasca asphaltenes; Sulfide, ether, and ester linkages, *Energy Fuels,* **1997**, *11*, 1171-1187.

70. Strausz, O.P.; Mojelsky, T.W.; Lown, E.M.; Kowalewski, I.; Behar, F. Structural features of Boscan and Duri asphaltenes, *Energy Fuels* **1999**, *13*, 228-247.

71. Peng, P.; Morales-Izquierdo, A.; Lown, E.M.; Strausz, O.P. Chemical structure and biomarker content of Jinghan asphaltenes and kerogens, *Energy Fuels* **1999**, *13*, 248-265.

72. Peng, P.; Fu, J.; Sheng, G.; Morales-Izquierdo, A.; Lown, E.M.; Strausz, O.P. Ruthenium-ions-catalyzed oxidation of an immature asphaltene: Structural features and biomarker distribution, *Energy Fuels* **1999**, *13*, 266-286.
73. Murgich, J.; Abanero, J.A.; Strausz, O.P. Molecular recognition in aggregates formed by asphaltene and resin molecules from the Athabasca oil sand, *Energy Fuels,* **1999**, *13*, 278-286.
74. Artok, A.; Su, Y.; Hirose, Y.; Hosokawa, M.; Murata, S.; Nomura, M. Structure and reactivity of petroleum-derived asphaltene, *Energy Fuels* **1999**, *13*, 287-296.
75. Leon, O.; Rogel, E.; Espidel, J. Structural characterization and self-association of asphaltenes of different origins, *Proc. Intl. Conf. on Petroleum Phase Behavior and Fouling,* 3rd International Symposium on Thermodynamics of Heavy Oils and Asphaltenes, AIChE Spring National Meeting, Houston, TX, March 14-18, 1999; 37-43.
76. Shirokoff, J.W.; Siddiqui, M.N.; Ali, M.F. Characterization of the structure of Saudi crude asphaltenes by X-ray diffraction, *Energy Fuels* **1997**, *11*, 561-565.
77. Yen, T.F.; Saraceno. Investigation of the nature of free radicals in petroleum asphaltenes and related substances by electron spin resonance, *J. Anal. Chem*, **1962**, *34*, 694.
78. Rao, B.M.L.; Serrano, J.E. Viscosity study of aggregation interactions in heavy oil, *Fuel Sci. Technol. Int.,* **1986**, *4*, 483-500.
79. Mehrotra, A.K. A model for the viscosity of bitumen/bitumen fractions-diluent blends, *J. Can. Pet. Technol.,* **1992**, *31(9)*, 28-32.
80. Schramm, L.L.; Kwak, J.C.T. The rheological properties of an Athabasca bitumen and some bituminous mixtures and dispersions, *J. Can. Pet. Technol.,* **1988**, *27*, 26-35.
81. Storm, D.A.; Sheu, E.Y.; DeTar, M.M.; Barresi, R.J. A comparison of the macrostructure of Ratawi asphaltenes in toluene and vacuum residue, *Energy Fuels* **1994**, *8*, 567-569.
82. Storm, D.A.; Barresi, R.J.; Sheu, E.Y. Rheological study of Ratawi vacuum residue in the 298-673 K temperature range, *Energy Fuels* **1995**, *9*, 168-176.
83. Li, S.; Liu, C.; Que, G.; Liang, W.; Zhu, Y. A study of the interactions responsible for colloidal structures in petroleum residua, *Fuel* **1997**, *14*, 1459-1463.
84. Wiehe, I. Tutorial on the phase behavior of asphaltenes and heavy oils, Third international symposium on the thermodynamics of heavy oils and asphaltenes, AIChE Spring National Meeting, Houston, TX, March 14-18, 1999, 3-8.
85. Panchal, C.B. Petroleum fouling mechanism, predictions and mitigation, AIChE short course, AIChE Spring National Meeting, Houston, TX, March 14-18, 1999.
86. Storm, D.A.; Decanio, S.J.; Edwards, J.C.; R.J. Sheu, Sediment formation during heavy oil upgrading, *Pet. Sci. Technol.,* **1997**, 15, 77-102.
87. Storm, D.A.; Barresi, R.J.; Sheu, E.Y.; Bhattacharya, A.K.; DeRosa, T.F. Microphase behavior of asphaltic micelles during catalytic and thermal upgrading, *Energy Fuels* **1998**, *12*, 120-128.
88. Wiehe, I.A., The pendant-core building block model of petroleum residua, *Energy Fuels* **1994**, *8*, 536-544.
89. Clarke, P.F.; Pruden, B.B. Asphaltene precipitation: Detection using heat transfer analysis, and inhibition using chemical additives, *Fuel* **1997**, *76*, 607-614.
90. Benham, N.K.; Pruden, B.B., CANMET residuum hydrocracking: Advances through control of polar aromatics, *NPRA Annual Meeting,* San Antonio, TX, March 17-19, 1996, offprint.
91. Beaton, W.I.; Bertolacini, R.J., Resid hydroprocessing at Amoco, *Catal. Rev.-Sci. Eng.,* **1991**, *33*, 281-417.
92. Mansoori, G.A. A review of advances in heavy organics deposition control, Internet: http://www.uic.edu-mansoori/hod html
93. Calemma, V.; Montanari, L.; Nali, M.; Anelli, M. Structural characteristics of asphaltenes and related pyrolysis kinetics, *Prep. Div. Pet. Chem.,* American Chemical Society National Meeting, Washington D.C., August 21-26, 1994; 452-455.
94. Poutsma, M.L. Free-radical thermolysis and hydrogenmolysis of model hydrocarbons relevant to processing of coal, *Energy Fuels* **1990**, *4*, 113-131.

95. Rahimi, P.M.; Dettman, H.D.; Dawson, W.H.; Nowlan, V.; Del Bianco, A.; Chornet, E. Coke formation from asphaltenes – solvent and concentration effects, CONRAD workshop on bitumen upgrading chemistry, June 7-8, 1995, Calgary, Alberta.
96. Heck, R.H.; Diguiseppi, F.T. Kinetic effects in resid hydrocracking, *Energy Fuels* **1994**, *8*, 557-560.
97. Gray, M. R., *Upgrading Petroleum Residues and Heavy Oils,* Marcel Dekker: New York, NY, 1994.
98. McMillen, D.F.; Malhotra, R.; Nigenda, S.E. The case for induced bond scission during coal pyrolysis, *Fuel* **1989**, *68*, 380-386.
99. Stein, S.E.; Griffith, L.L.; Billmers, R.; Chen, R.H. Hydrogen transfer between anthracene structures, *J. Phys. Chem.,* **1986**, *90*, 517.
100. Vernon, L.W. Free radical chemistry of coal liquefaction: Role of molecular hydrogen, *Fuel* **1980**, *59*, 102.
101. Farcasiu, M.; Petrosius, S.C. Heterogeneous catalysis: Mechanism of selective cleavage of strong carbon-carbon bonds, *Prep. Div. Fuel Chem.,* American Chemical Society National Meeting, Washington D.C., August 20-25, 1994; 723-725.
102. Penn, J.H.; Wang, J. Radical cation bond cleavage pathways for naphthyl-containing model compounds, *Energy Fuels* **1994**, *8*, 421-425.
103. Miki, Y.; Yamada, S.; Oba, M.; Sugimoto, Y. Role of catalyst in hydrocracking of heavy oil, *J. Catal.* **1983**, *83*, 371-383.
104. Le Page, J.F.; Davidson, M. *IFP publication,* **1986**, *41*, 131
105. Heck, R.H.; Rankel, L.A.; DiGuiseppi, F. T. Conversion of petroleum resid from Maya crude: Effects of H-donors, hydrogen pressure and catalyst, *Fuel Process. Technol.,* **1992**, *30*, 69-81.
106. Gray, M.R.; Khorasheh, F.; Wanke, S.E.; Achia, U.; Krzywicki, A.; Sanford, E.C.; Sy, O.K.Y.; Ternan, M. Rate of catalyst in hydrocracking of residue from Athabasca bitumen, *Energy Fuels* **1992**, *6*, 478-485.
107. Sanford, E.C. Mechanism of coke prevention by hydrogen during residuum hydrocracking, *Prep. Div. Pet. Chem.,* American Chemical Society, National Meeting, Denver, CO, March 28, April 2, 1993; 413-416.
108. Sanford, E.C. Influence of hydrogen and catalyst on distillate yields and the removal of heteroatoms, aromatics, and CCR during cracking of Athabasca bitumen residuum over a wide range of conversions, *Energy Fuels* **1994**, *8*, 1276-1288.
109. Srinivasan, N.S.; McKnight, C.A. Mechanism of coke formation from hydrocracked Athabasca residuum, *Fuel* **1994**, *73*, 1511-1517.
110. Nagaishi, H.; Chan, E.W.; Sanford, E.C.; Gray, M.R. Kinetics of high-conversion hydrocracking of bitumen, *Energy Fuels* **1997**, *11*, 402-410.
111. Speight, J.G. Thermal cracking of Athabasca bitumen, Athabasca asphaltenes, and Athabasca deasphalted heavy oil, *Fuel* **1970**, *49*, 134-145.
112. Rahimi, P.; Dettman, H.; Gentzis, T.; Chung, K.; Nowlan, V. Upgrading chemistry of Athabasca bitumen fractions derived by super critical fluid extraction, Presented at the 47th CSChE conference, Edmonton, Alberta, Canada, October 5-8, 1997.
113. Rahimi, P.M.; Parker, R.J.; Hawkins, R.; Gentzis, T.; Tsaprailis, H. Processability of partially deasphated Athabasca bitumen, *Prep. Div. Pet. Chem.,* American Chemical Society, National Meeting, San Diego, CA., April 1-5, 2001, 74-77.
114. Yasar, M.; Trauth, D.M.; Klein, M.T. Asphaltene and resid pyrolysis 2: The effect of reaction environment on pathways and selectivities, *Prep. Div. Fuel Chem.,* American Chemical Society, National Meeting, Washington, D.C., August 23-28, 1992; 1878-1885.
115. Karacan, O.; Kok, M.V. Pyrolysis analysis of crude oils and their fractions, *Energy Fuels* **1997**,*11*, 385-391.
116. Dawson, W.H.; Chornet, E.; Tiwari, P.; Heitz, M. Hydrocracking of individual components isolated from Athabasca bitumen vacuum resid, *Prep. Div. of Pet. Chem.,* American Chemical Society, National Meeting, Dallas, TX, April 9-14, 1989; *34*, 384-394.

117. Rahimi, P.; Dettman, H.; Nowlan, V.; DelBianco, A. Molecular transformation during heavy oil upgrading, *Prep. Div. Pet. Chem.*, American Chemical Society, National Meeting, San Francisco, CA, April 13-17, 1997; 23-26.

118. Liu, C.; Zhu, C.; Jin, L.; Shen, R.; Liang, W. Step by step modeling for thermal reactivities and chemical compositions of vacuum residues and their SFEF asphalts, *Fuel Process. Technol.*, **1999**, *59*, 51-67.

119. Rahimi, P.; Gentzis T.; A Delbianco, Application of Hot-Stage Microscopy in the Investigation of the Thermal Chemistry of Heavy Oils and Bitumen – an Overview, Symp. on Kinetics Mechanism of Petroleum Processing, 222nd ACS National Meeting, Chicago, Illinois, USA, August 26-30, 2001.

120. Wiehe, IA. A phase separation kinetic model for coke formations, *Ind. Eng. Chem. Res.*, **1993**, *32*, 2447.

121. Rahimi, P.; Gentzis, T.; Cotté E. Investigation of the thermal behavior and interaction of Venezuelan heavy oil fractions obtained by ion-exchange chromatography, *Energy Fuels*, **1999**, *13*, 694.

122. Speight, J.G. *The Chemistry and Technology of Petroleum*, 3rd Ed., Marcel Dekker: New York, 1999.

123. Rahimi, P.; Gentzis, T.; Taylor, E.; Carson, D.; Nowlan, V.; Cotté, E. The impact of cut point on the processability of Athabasca bitumen, *Fuel*, 2001, forthcoming.

124. Rahimi, P.; Gentzis, T.; Fairbridge, C. Interaction of clay additives with mesophase formed during thermal treatment of solid-free Athabasca bitumen fraction, *Energy Fuels*, **1999**, *13*, 817.

125. Kubo, J.; Higashi, H.; Ohmoto, Y.; Aroa, H. Heavy oil hydroprocessing with the addition of hydrogen-donating hydrocarbons derived from petroleum, *Energy Fuels*, **1996**, *10*, 474.

126. Rahimi, P.; Gentzis, T.; Kubo, J.; Fairbridge, C.; Khulbe, C. Coking propensity of Athabasca bitumen vacuum bottoms in the presence of H-donors – formation and dissolution of mesophase from a hydrotreated petroleum stream (H-donor), *Fuel Proc. Technol.*, **1999**, *60*, 157.

127. Rahimi, P.; Gentzis, T.; Ciofani, T. Coking characteristics of partially deasphalted oils, Proc. AIChE Spring Nat. Mtg., Session 71 – Advances in Coking, Atlanta, GA, 428, 2000.

128. Wiehe, I.A.; Kennedy, R.I. The oil compatibility model and crude oil incompatibility, *Energy Fuels*, **2000**, *14*, 56.

129. Vilhunen, J.; Quignard, A.; Pilvi, O,; Waldvogel, J. Experience in use of automatic heavy fuel oil stability analyzer", *Proc. 7th Intl. Conf. of Stability and Handling of Liquid Fuels*, October 13-17, 1997, Vancouver, B.C.; 985-987.

130. Rahimi, P.M.; Parker, R.J.; Wiehe, I.A. Stability of visbroken products obtained from Athabasca bitumen for pipeline transportation, Symp. on Heavy Oil Resid Compatibility and Stability, 221st ACS National Meeting, San Diego, CA, USA, April 1-5, 2001.

131. Gaestel, C.; Smadja, R.; Lamminan, K.A. Contribution a la Connaissance des proprietes des bitumen routiers. *Rev. Gen. Routes Aerodromes*, **1971**, *85*, 466.

132. Asomaning, S.A.; Watkinson, A.P. Petroleum stability and heteroatoms species effects on fouling heat exchangers by asphaltenes. *Heat Transfer Eng.*, **2000**, *12*, 10-16.

133. Schabron, J.F.; Pauli, A.T.; Rovani Jr., J.F.; Miknis, F.P. Predicting coke formation tendencies," *Fuel*, **2001**, *80*, 1435-1446.

134. Van den Berg, F. Developments in fuel oil blending, Proc. 7th Intl. Conf. on Stability and Handling Fuels, September 22-26, 2000, Graz, Austria; 165-172.

135. Pauli, A.T. Asphalt compatibility testing using the automated Heithaus titration test, *Prep. Div. Fuel Chem.*, American Chemical Society, **1996**, *41*, 1276-1281.

136. Speight, J.G. *Petroleum Chemistry and Refining,* Taylor & Francis, 1998.

137. Bannayan, M. A.. Lemke, H.K.. Stephenson, W. K. Fouling mechanism and effect of process conditions on deposit formation in H-Oil equipment, Catalysis in petroleum refining and petrochemical industries, 1995. *Proc. 2nd Intl. Conf. on Catalysts in Petroleum Refining and Petrochemicals Industries,* Kuwait, April 22-26, 1995, Absi-Halabi et al. Editors, Elsevier Science publisher. 1996; 273-281.

138. Wiehe, I. A. "Prevention of fouling by incompatible crudes with the oil compatibility model," *Proc. Intl. Conf. Petroleum Phase Behavior and Fouling,* AIChE Spring National Meeting, Houston, TX, March 14-18, 1999; 354-358.
139. Wiehe, I. A. "Mitigation of the plugging of hydrotreater with the oil compatibility model," *Proc. Intl. Conf. Petroleum Phase Behavior and Fouling,* AIChE Spring National Meeting, Houston, TX, March 14-18, 1999; 405-411.
140. DelBianco, A.; Garuti, G.; Pirovano, C.; Russo, R. "Thermal cracking of petroleum residues. 3. Technical and economical aspects of hydrogen donor visbreaking," *Fuel,* **1995**, *74*, 756-760.

Chapter 20

MECHANISTIC KINETIC MODELING OF HEAVY PARAFFIN HYDROCRACKING

Michael T. Klein* and Gang Hou
Department of Chemical Engineering, University of Delaware, Newark, DE 19716
*Present Address: Department of Chemical and Biochemical Engineering
School of Engineering, Rutgers University, Piscataway, NJ 08854

1.   INTRODUCTION

Catalytic hydrocracking is a flexible process for the conversion of heavy, hydrogen-deficient oils into lighter and more-valuable products. Part of its flexibility is its ability to handle a wide range of feeds, from heavy aromatics to paraffinic crudes and cycle stocks. The ability to extend this flexibility to "customized" product slates depends on the ability to control and manipulate the process chemistry, which is, in turn, enhanced by a rigorous representation of the process chemistry. This motivates the present interest in a heavy paraffin hydrocracking kinetics model.

Traditional hydrocarbon conversion process models have implemented lumped kinetics schemes, where the molecules are aggregated into lumps defined by global properties, such as boiling point or solubility. Molecular information is obscured due to the multi-component nature of each lump. However, increasing environmental concerns and the desire for better control and manipulation of the process chemistry have focused attention on the molecular composition of both the feedstocks and their refined products. Modeling approaches that account for the molecular fundamentals underlying reaction of complex feeds and the subsequent prediction of molecular properties require an unprecedented level of molecular detail.

This is because modern analytical measurements indicate the existence of the order of 10^5 ($O(10^5)$) unique molecules in petroleum feedstocks. In modeling terms, each species corresponds to one differential equation in a deterministic modeling approach. Therefore not only the solution but also the formulation of the implied model is formidable. This motivated the development of computer algorithms not only to solve but also to formulate the model.

These algorithms have now been organized into a system of kinetic modeling tools – the *Kinetic Modeler's Toolbox (KMT)* - that build and solve the kinetic models on the computer automatically.[1] Monte Carlo techniques are used to model the structure and composition of complex feeds. Graph theory techniques are then utilized to generate the reaction network.[2-4] Reaction family concepts and Quantitative Structure-Reactivity Correlations (QSRC) are used to organize and estimate rate constants. The computer-generated reaction network, with associated rate expressions, is then converted to a set of differential equations, which can be solved within an optimization framework to determine the rate parameters in the model. This automated modeling process enables the modeler to focus on the fundamental chemistry and speed up the model development significantly.

This chapter describes the application of these tools to the development of a mechanistic kinetic model for the catalytic hydrocracking of heavy paraffins. The basic approach and overview synopsis are presented first. This is followed by a detailed description of the steps involved in model formulation, optimization, and use.

2.   APPROACH AND OVERVIEW

Molecular-level modeling can be at either the pathways or the mechanistic level. Pathways-level models include only observable molecules, whereas mechanistic models include reactive intermediates (e.g., carbenium ions) as well. Compared with pathways-level modeling, mechanistic modeling involves a large number of species, reactions, and associated rate constants in the governing network because of the explicit accounting of all reaction intermediates. This can render mechanistic modeling very tedious. However, mechanistic models comprise more fundamental rate constant information and thus can be better extrapolated to various operating conditions and feeds.

This paper thus describes the construction of various mechanistic models for paraffins ranging from C_{16} to C_{80}. All the models incorporate mechanistic acidic chemistry and pathways-level metal chemistry for the prototypical bifunctional hydrocracking catalyst, which has both a metal function for hydrogenation/dehydrogenation and an acid function for isomerization and cracking.

Much of the complexity of these models is better considered bookkeeping than fundamental. That is, there are only roughly 10 different types of reactions or "reaction operators" underlying the process chemistry, which can comprise thousands of literal reactions because of the statistical or combinatorial explosion of matching these reaction operators with all of the feed and product molecules susceptible to each. A related simplification is that the large demand for rate parameters can be handled by organizing the reactions involving similar transition states into one reaction family, the kinetics of which are constrained to follow a Quantitative Structure-Reactivity Correlation (QSRC) or Linear Free Energy Relationship (LFER).

The present modeling approach exploited these notions. The molecules in the paraffin hydrocracking reaction mixture were thus grouped into a few species types (paraffins, olefins, ions, and inhibitors such as NH_3), which, in turn, reacted through a limited number of reaction families on the metal (dehydrogenation and hydrogenation) and the acid sites (protonation, hydride-shift, methyl-shift, protonated cyclopropane (PCP) isomerization, β-scission, and deprotonation). As a result, a small number of formal reaction operations could be used to generate hundreds of reactions.

Figure 1 shows the mechanistic reactions for paraffin hydrocracking via the dual site (both metal site and acid site) mechanism. Each reaction family had a single associated reaction matrix, which was used to generate all possible mechanistic reactions.[2] The reaction matrices are summarized in Table 1.

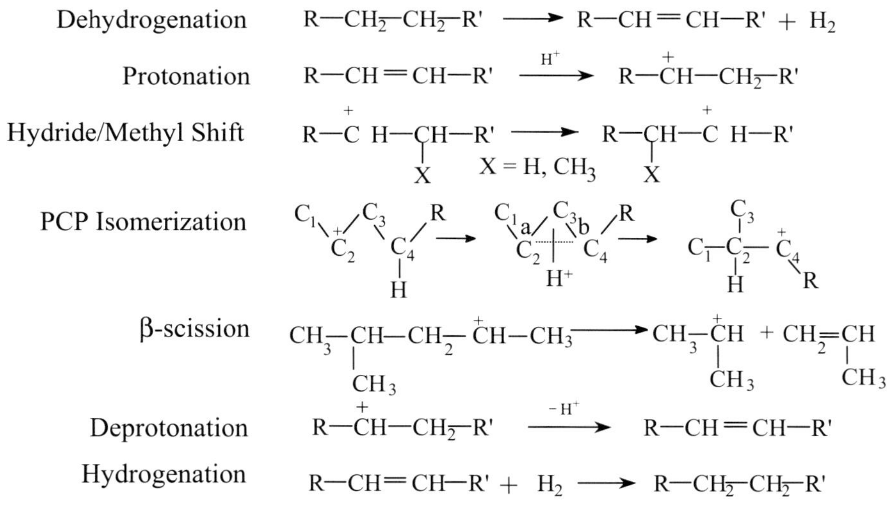

Dehydrogenation $R{-}CH_2{-}CH_2{-}R' \longrightarrow R{-}CH{=}CH{-}R' + H_2$

Protonation $R{-}CH{=}CH{-}R' \xrightarrow{H^+} R{-}\overset{+}{C}H{-}CH_2{-}R'$

Hydride/Methyl Shift $R{-}\overset{+}{C}H{-}\underset{X}{\overset{|}{C}H}{-}R' \longrightarrow R{-}\underset{X}{\overset{|}{C}H}{-}\overset{+}{C}H{-}R' \quad X = H, CH_3$

PCP Isomerization (reaction scheme)

β-scission $CH_3{-}\underset{CH_3}{\overset{|}{C}H}{-}CH_2{-}\overset{+}{C}H{-}CH_3 \longrightarrow CH_3{-}\underset{CH_3}{\overset{|}{\overset{+}{C}H}} + CH_2{=}\underset{CH_3}{\overset{|}{CH}}$

Deprotonation $R{-}\overset{+}{C}H{-}CH_2{-}R' \xrightarrow{-H^+} R{-}CH{=}CH{-}R'$

Hydrogenation $R{-}CH{=}CH{-}R' + H_2 \longrightarrow R{-}CH_2{-}CH_2{-}R'$

Figure 1. Paraffin Hydrocracking Reaction Families at the Mechanistic Level

The rate constant information was organized with a QSRC/LFER for each reaction family, as shown in Eq. 1.

$$E^* = E_0^* + \alpha \Delta H_{rxn} \qquad (1)$$

Eq. 1 is called the Polanyi relationship, which relates the activation energy of each reaction to its heat of reaction. A single frequency factor was used for each reaction family. The use of QSRC/LFER for heterogeneous mechanistic models has been demonstrated by Korre [5] and Russell [6] for hydrocracking, Watson et al. [7,8] and Dumesic et al. [9] for catalytic cracking, and Mochida and Yoneda [10-12] for dealkylation and isomerization.

Table 1. Reaction Matrices for Paraffin Hydrocracking Reactions at the Mechanistic Level.

Reaction Family	Reactant Matrix			
Dehydrogenation Test: The string C - C is required	C C H H	0 1 -1 0 1 0 0 -1 -1 0 0 1 0 -1 1 0		
Hydrogenation Test: The string C = C is required	C C H H	0 1 -1 0 1 0 0 -1 -1 0 0 1 0 -1 1 0		
Protonation Test: The string C = C is required.	C C H+	0 -1 1 -1 0 0 1 0 0		
Deprotonation Test: The string C+ - C is required.	C+ C H	0 0 1 0 0 -1 1 -1 0		
Hydride Shift and Methyl Shift Test: The string C+ - C - X is required in the carbenium ion (X = H or C)	C+ C X	0 0 1 0 0 -1 1 -1 0		
PCP Isomerization Test: The string C+ - C - C is required in the carbenium ion	C+ C C H	0 0 1 0 0 0 -1 1 1 -1 0 -1 0 1 -1 0		
β-Scission Test: The string C+ - C - C is required in the carbenium ion	C+ C C	0 1 0 1 0 -1 0 -1 0		

The remainder of this paper describes the division of the paraffin hydrocracking reactions into mechanistic families with a unique reaction matrix operator for each reaction family. The reaction rules and QSRCs used are then discussed for each reaction family. The technical specifications and the iteration process to find the optimum subset of the mechanistic model will also be discussed.

3.   MODEL DEVELOPMENT

This section first discusses the reaction mechanism for paraffin hydrocracking and the thus-derived modeling specifications for each reaction family. This is followed by a discussion of the automated model building algorithm and the QSRC/LFERs used to organize the rate parameters. Finally, the thus-developed C_{16} paraffin mechanistic hydrocracking model diagnostics are presented.

3.1   Reaction Mechanism

The mechanism of paraffin hydrocracking over bifunctional catalysts is, essentially, the carbenium ion chemistry of acid cracking coupled with metal-centered dehydrogenation/hydrogenation reactions. The presence of excess hydrogen and the hydrogenation component of the catalyst result in hydrogenated products and inhibition of some of the secondary reactions and coke formation.

These mechanistic features were elucidated in detail in the 1960s. Based on the pioneering work of Mills et al.[13] and Weisz[14], a carbenium ion mechanism was proposed, similar to catalytic cracking plus additional hydrogenation and skeletal isomerization. More recent studies of paraffin hydrocracking over noble metal-loaded, zeolite based catalysts have concluded that the reaction mechanism is similar to that proposed earlier for amorphous, bifunctional hydrocracking catalysts.[15-17]

As shown in Figure 1, the elementary steps used to model the mechanism of paraffin hydrocracking over a bifunctional catalyst are as follows:
1.   Dehydrogenation of paraffins to olefins on metal sites,
2.   Protonation of olefins to carbenium ions on acid sites,
3.   Carbenium ion hydride shift on acid sites,
4.   Carbenium ion methyl shift on acid sites,
5.   Protonated cyclopropane (PCP) intermediate mediated branching of carbenium ion on acid sites,
6.   Carbenium ion cracking through β-scission on acid sites,
7.   Deprotonation of carbenium ions to olefins on acid sites, and
8.   Hydrogenation of olefins to paraffins on metal sites.

The KMT model building algorithm contained a reaction matrix for each of the foregoing reaction steps. The model species were also classified into paraffins, olefins, carbenium ions and H^+ ion categories. Algorithms were deployed to identify the degree of branching of each species (1, 2, 3, and more) and the type of carbenium ions (primary, secondary, and tertiary).

The model building process was guided by a set of user-supplied reaction rules that pruned the model from the essentially infinite set of feasible reactions to the kinetically significant subset. These rules enjoy two ultimate foundations: kinetics and analytical chemistry. Kinetics was used to "rule out" feasible but kinetically insignificant reactions, such as reactions leading to the formation of primary carbenium ions. A set of organizational rules was deployed to maintain a reasonable correspondence between the level of detail in the model and in available analytical measurements. The specifics for each reaction family are described in turn.

3.2   Reaction Families

3.2.1   Dehydrogenation/Hydrogenation

In a typical hydrocracking process, the dominant reactions on the metal sites of the catalyst are mainly dehydrogenation and hydrogenation. Very little hydrogenolysis occurs. The metal component of the catalyst dehydrogenates the paraffin reactants to produce reactive olefin intermediates, hydrogenates the olefins, and also prevents catalyst deactivation by hydrogenating coke precursors.

The mechanism of the dehydrogenation reaction involves stripping of two hydrogen atoms by the metal component of the catalyst. The molecular topology test for the dehydrogenation reaction is a search for a C - C string in the molecule.

Figure 2 shows the number of olefins as a function of the carbon number. It can be seen that the number of possible olefins increase almost exponentially with the carbon number, and even one paraffin can form thousands of olefins. Thus the inclusion of all possible olefins and their reactions would generate an enormously large model. Thus, in order to develop a reasonably sized mechanistic model, certain rules were used for this reaction, as summarized in Table 2. All n-paraffins were allowed to undergo dehydrogenation reactions at all sites, whereas all iso-paraffins were allowed to undergo dehydrogenations only at the C - C bonds β to the branch. This rule was based on the relative rate of reactions of these olefins on the acid site.

3.2.2   Protonation/Deprotonation

Protonation transforms an olefin into a carbenium ion. This reaction is much faster than other acid site reactions, and is close to equilibrium under commercial operating conditions. The protonation reaction involves attack of H^+ at a C=C bond. Only three atoms change connectivities during this reaction, as shown in the reaction matrix of Table 1.

Mechanistic insights from the literature were used to synthesize the tests and rules. The deprotonation reaction, which converts carbenium ions to olefins, involves breakage of a C-H bond to give H^+ and an olefin. The deprotonation test required a connected C+ and C atom. The rules are summarized in Table 2. Primary carbenium ions were not allowed to form from protonation due to their thermochemistry.

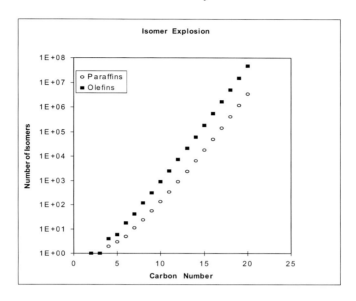

Figure 2. Number of Paraffins and Olefins as a Function of Carbon Number

Table 2. Reaction Rules for Mechanistic Paraffin hydrocracking Reaction Families

Reaction Family	Reactant Rules
Dehydrogenation / Hydrogenation	Dehydrogenation is allowed everywhere on n-paraffins but only β to branch on iso-paraffins. Formation of di-olefins is not allowed.
Protonation / Deprotonation	Primary carbenium ions are not allowed to form.
Hydride Shift and Methyl Shift	Primary carbenium ions are not allowed to form. Only migration to stable or branched ions is allowed. Number of reactions allowed is a function of branch number.
PCP Isomerization	Primary and methyl carbenium ions are not allowed to form. PCP-isomerization increases either the number of branches or the length of the side chains. PCP-isomerization to form vicinal branches is not allowed. Only methyl and ethyl branches are allowed; A maximum of three branches is allowed. Number of reactions allowed is a function of branch and carbon number.
β-Scission	Methyl and primary carbenium ions are not allowed to form.

3.2.3   Hydride and Methyl Shift

Hydride and methyl shifts are responsible for changes in the position of carbenium ions. The net effect is generally to create a stable ion, e.g., tertiary ion, from a less stable ion, e.g. secondary ion. The methyl shift can also change the location of a branch position, which creates isomers.

The rate of hydride shift is considered to be much faster than alkyl shift due to the ease of moving H^+ as compared to the alkyl group. The hydride shift reaction test required a C^+ - C - H string in the molecule; that for the methyl shift reaction required a C^+ - C - (CH_3) string.

The rules are summarized in Table 2. The reactions were allowed for all ions. The number of reactions allowed was constrained as a function of number of branches on the ions. This provided the proper spectrum of isomers and also kept the number of species and reactions manageable and in alignment with available analytical chemistry.

3.2.4   PCP Isomerization

The branching isomerization reaction shown in Figure 1 is postulated to proceed via a protonated cyclopropane (PCP) intermediate with the charge delocalized over the ring.[18] The test for this reaction required a C^+ - C - C string in the molecule. The rate of this reaction is slower than that of hydride/methyl shift. This reaction was further categorized into two types, isomA and isomB, depending on the identity of the bond to be broken in the three-membered ring intermediate.

The rules for this reaction had a dramatic effect on the size of the generated model. The final set of rules used for the model building is summarized in Table 2. As was the case for the hydride shift/methyl shift, the isomerization reaction was allowed for all paraffins and iso-paraffins and the number of reactions was constrained as a function of the number of carbons and branches on the ions to provide the proper spectrum of isomers and an alignment with analytical chemistry.

3.2.5   β-Scission

The β-scission reaction is one of the key carbon number reducing reactions for iso-paraffins. The rate of this reaction is dependent on the acidity of the catalyst. β-scission can lead to the formation of tertiary and secondary carbenium ions, but no primary ions were allowed to form. Several β-scission mechanisms have been suggested for the cracking of branched secondary and tertiary carbenium ions, as summarized in Table 3.[19]

Table 3. β-scission mechanisms for carbenium ion conversion over bifunctional hydrocracking catalyst

Type	Min C#	Ions Involved	Rearrangement
A	8	Tert → Tert	
B1	7	Sec → Tert	
B2	7	Tert → Sec	
C	6	Sec → Sec	

Type A β-scissions convert a tertiary carbenium ion to another tertiary carbenium ion. This is the fastest and the most likely to occur. The reaction rates decrease in the order of A, B1, B2 and C. Each type of reaction requires a minimum number of carbon atoms in the molecule and a certain of branching in order to occur. The proposed β-scission mechanisms suggest that paraffins may undergo several isomerizations until a configuration is attained that is favorable to β-scission.

The test for this reaction requires a C^+ - C - C string. The rules are summarized in Table 2. Unstable species, such as methyl and primary carbenium ions, were not allowed to form from this reaction.

3.2.6   Inhibition Reaction

Nitrogen inhibition effects were accounted for via the dynamic reduction of acid sites by the protonation and deprotonation of the Lewis base on the acid sites. For example, the following ammonia inhibition reaction was included in the reaction network.

$$NH_3 + H^+ \leftrightarrow NH_4^+ \qquad (2)$$

This inhibitor protonation/deprotonation will compete with hydrocarbon protonation/deprotonation and thus reduce the available number of acid sites for hydrocarbons. This is expressed in the site balance of Eq. 3, which depicts the conservation of the total ion concentration.

$$H_0^+ = H^+ + \sum_{i=1}^{N} R^+ + NH_4^+ \qquad (3)$$

3.3   Automated Model Building

The reaction matrices of Table 1, the reaction rules of Table 2, the QSRC/LFER correlations of Eq. 1, and the automated model-building algorithm of Figure 3 were used to construct various kinetic model versions for mixtures containing molecules from C_8 to C_{24}.

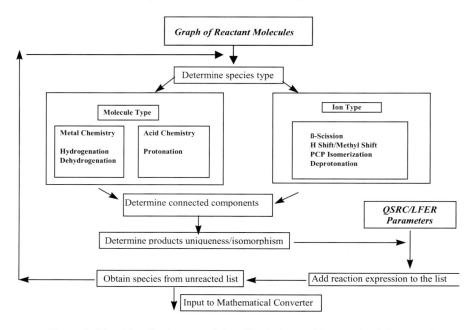

Figure 3. Algorithm for Automated Paraffin Hydrocracking Mechanistic Modeling

Figure 4 depicts a representative result that is presented here to show the types of output that can be obtained. This illustrates the mechanistic level sequences for a large paraffin molecule undergoing various reactions, including dehydrogenation, protonation, H/Me-shift, various isomerizations, the formation of the configuration to carry out the β-scission into a smaller ion and olefin, and then finally deprotonation and hydrogenation to smaller paraffin molecules. It shows that the reactions leading to the formation and subsequent cracking of a tertiary carbenium ion are preferred in the complete reaction network. These insights accrued from building relatively small models guided the extension of model building process to the heavier paraffin hydrocracking model up to as high as C_{80}, as discussed in detail below.

The model building algorithm shown in Figure 3 classified the molecular graphs of the reactants into the families of molecules and ions. The molecules were further filtered to undergo specific metal and acid chemistries. For example, paraffins were only allowed to undergo dehydrogenation reactions on the metal sites, whereas olefins were allowed to hydrogenate on the metal site and protonate on the acid sites. The rate parameters derived from the QSRC/LFER for each fundamental reaction family are considered fundamental because of the elementary step nature of the model.

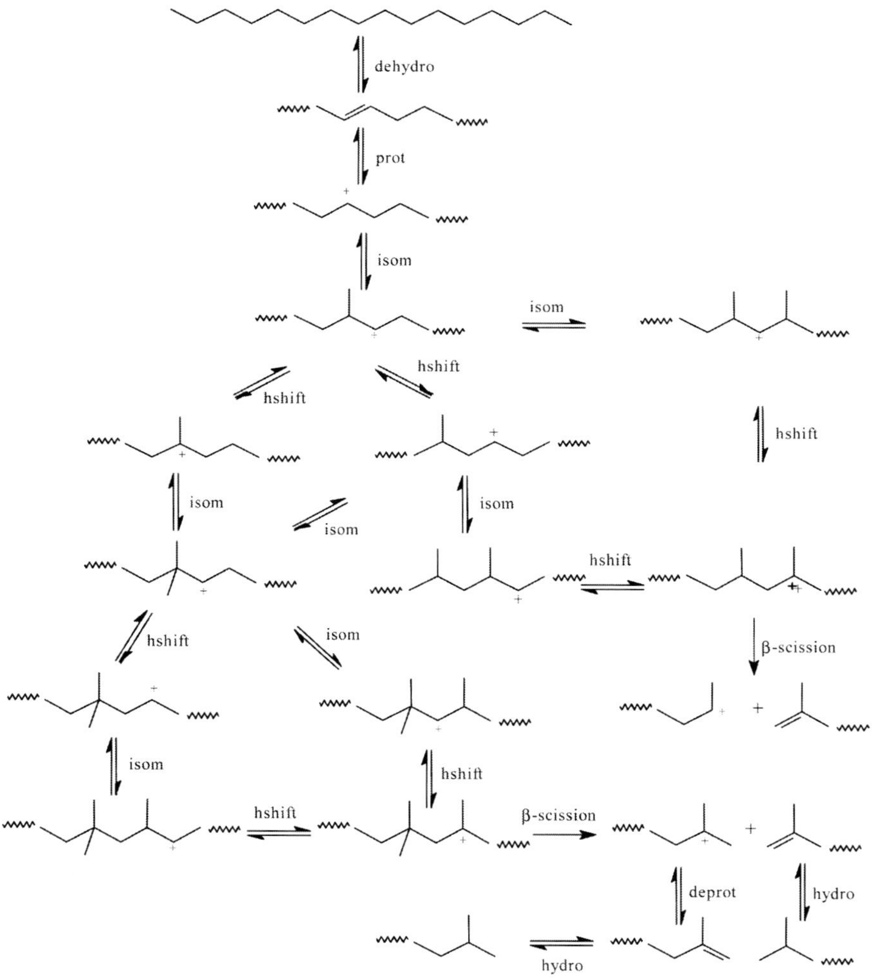

Figure 4. Representative Reaction Network of Paraffin Hydrocracking at the Mechanistic Level

The mechanistic paraffin hydrocracking model builder has a database of all reaction matrices and rules for all mechanistic paraffin hydrocracking reactions, involving both metal and acid chemistries, with the flexibility of changing them. A detailed library of paraffins, olefins, and ion intermediates consisting of their connectivity and thermodynamic information was built using the MOPAC computational chemistry package. The reactor balances equations were generated for a plug flow reactor (PFR) with molar expansion.

3.4   Kinetics: Quantitative Structure Reactivity Correlations

The rate constants within each reaction family were described in terms of a reaction family-specific Arrhenius A factor and the Polanyi relationship parameters that related the activation energy to the enthalpy change of reaction, as shown in Eq. 1.

The Polanyi relationship and the Arrhenius expression can be combined to represent the rate constant, k_{ij}, where i denotes the reaction family and j denotes the specific reaction in the family, and is shown in Eq. 4:

$$k_{ij} = A_i \exp(-(E_{o,i} + \alpha_i * \Delta H_{rxn,j}) / RT) \qquad (4)$$

The acidity of the catalyst was captured by a single parameter $\Delta H_{stabilization}$, signifying the relative stabilization of the H+ ion as compared to other carbenium ions. Since the reactions on the acid site are rate controlling, this was a useful way to capture the catalyst property (acidity) in the rate constant formalism, as shown in Eq. 5.

$$k_{ij} = A_i \exp\left(-\left(E_{o,i} + \alpha_i * \left(\Delta H_{rxn,j} - \Delta H_{stabilization}\right)\right) / RT\right) \qquad (5)$$

In short, each reaction family could be described with a maximum of three parameters (A, E_o, α). Procurement of a rate constant from these parameters required only an estimate of the enthalpy change of reaction for each elementary step. In principle, this enthalpy change of reaction amounted to the simple calculation of the difference between the heats of formation of the products and reactants. However, since many model species, particularly the ionic intermediates and olefins, were without experimental values, a computational chemistry package, MOPAC,[20] was used to estimate the heat of formations "on the fly". The organization of the rate constants into quantitative structure-reactivity correlations (QSRC) reduced the number of model parameters greatly from $O(10^3)$ to $O(10)$.

3.5   The C_{16} Paraffin Hydrocracking Model Diagnostics

A C_{16} hydrocracking model with 465 species and 1503 reactions was built automatically in only 14 CPU seconds on a Pentium Pro 200 PC. The corresponding plug flow reactor (PFR) model with molar expansion was then generated automatically and solved in 76 CPU seconds in once-through mode. Table 4 summarizes the characteristics of the C_{16} model.

The species distribution in Table 4 shows there were more intermediate olefins and ions than paraffins. The reaction distribution in Table 4 shows that each reactant molecule went through various rearrangements to form the right configuration before the β-scission. In this model, all type A, B1, and B2 β-scission reactions were allowed; type C reaction was ignored in the final model after the optimization with experimental results showed the C-type cracking to be insignificant. Hydride shift, methyl shift, and isomerization reactions were restricted for hydrocarbons having more than eight carbon atoms. All *n*-paraffins, and selected iso-paraffins were used to represent the portion of the feedstock larger than C_9. This not only helped to keep the model size reasonable, but also resulted in the inclusion of components with different reactivities in the feedstock where detailed characterization was not available.

Table 4. Characteristics of the C_{16} Paraffin Hydrocracking Model

Species	#	Reactions	#
Molecule		Dehydrogenation	233
Hydrogen	1	Hydrogenation	233
Paraffins	64	Deprotonation	328
Olefins	233	Protonation	328
Ion		Hydride and Methyl Shift	168
H-ion	1	PCP Isomerization	174
Carbenium Ions	165	β-Scission	37
Inhibator	1	Inhibition	2
Total Number of Species	465	Total Number of Reactions	1503

4. MODEL RESULTS AND VALIDATION

The C_{16} models were tuned with pilot plant data on a commercial Pt-Zeolite catalyst using the optimization program GREG [21] coupled with an equation solver. The objective function was the square of the difference between predicted and experimental yields weighted by the experimental standard deviation, as shown in Eq. 6,

$$F = \sum_{i=1}^{M} \sum_{j=1}^{N} (\frac{y_{ij}^{model} - y_{ij}^{exp}}{\omega_j})^2 \qquad (6)$$

where i is the experiment number and j is the species or lump number, and ω_j is the experimental measurement deviation.

The A factors for the acid chemistry were obtained from the literature[7,8] and held constant during optimization. The A factors for reactions on the metal sites were constrained to fall between $1 \le A_j \le 20$. The Polanyi α's in the QSRC formalism were held at 0.5, based on guidelines from literature.[7-9]

Further, the E_o's were constrained by the relation between activation energies and heat of reaction shown in Eq. 7.

$$E_{backward} - E_{forward} = \Delta H_{rxn} \qquad (7a)$$

$$E_{oj,forward} = E_{oj,backward} \qquad (7b)$$

$$\alpha_{j,forward} = 1 - \alpha_{j,backward} \qquad (7c)$$

In summary, only the E_o's, and one catalyst stabilization parameter were optimized by matching with the experimental data.

The model was optimized with the lumped data from experiments (the unconverted C_{16}, the mono-branched and multi-branched C_{16}, and all the cracked products). Then the model was used to predict the carbon number distribution and iso-to-normal ratio for the various carbon numbers. Figures 5a and 5b show the parity plots for the C_{16} paraffin hydrocracking for both (a) the lumped observations and (b) the carbon number distribution at various conditions (T, P, LHSV and NH_3). The agreement between the predicted and experimental data sets for all the high and low yield compounds is good, as all predictions are within the experimental error. The good prediction validates the fundamental nature of the model.

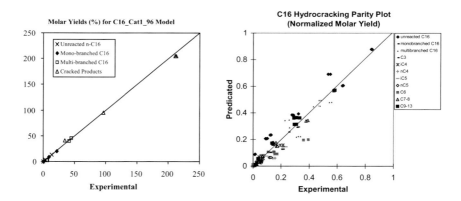

Figure 5. Parity Plots for C_{16} Paraffin Hydrocracking Model at Various Operation Conditions (T, P, LHSV, and Inhibition)

Various insights were obtained from the optimization results. For this set of paraffin hydrocracking experiments on a commercial bifunctional Pt-Zeolite catalyst, the results showed that the reaction rates ranked as: A-type Cracking > H/Methyl Shift > PCP Isomerization > B1-type Cracking > B2-type Cracking >> C-type Cracking. Thus the cracking products actually evolve from A-type cracking of tri-branched ions or B-type cracking of di-branched ions. Wherever several reaction pathways are possible, the one leading to the formation and subsequent cracking of a tertiary carbenium ion is preferred. Furthermore, the cracking of smaller paraffins via β-scission is less likely to occur, which explains their high yields even at high conversions. Also, from the product molecular structure point of view, PCP isomerization always leads to branching, A-type cracking always leads to branched isomers, and B-type cracking always leads to normal or branched isomers.

Practically no methane and ethane formation were observed from the experiments. This confirmed our modeling rule that no primary ions were allowed because of their instability compared to more-stable secondary and tertiary ions. This modeling rule basically eliminated the formation of methane and ethane via the carbenium ion mechanism in the reaction network. This also partially explains why the long chain paraffins tend to crack in or near the center.

The mechanistic notion that secondary carbenium ions are isomerized to more stable tertiary ions prior to cracking, as well as the high rate of H-shift to the tertiary carbenium ion, explains the high iso-to-normal ratio for paraffins in the product. The iso-to-normal ratio in the product paraffins increases with decreasing reaction temperature because at higher temperatures the cracking rate of isoparaffins increases faster than that of the n-paraffins.

The ammonia inhibition reduces not only the cracking activity but also the iso-to-normal ratio in the product paraffins because of its partial neutralization of the acid sites on the hydrocracking catalyst.

The tuning results indicated that the rate-determining reactions (isomerization and β-scission) occurred at the acid sites, whereas the metal sites served only the rapid hydrogenation and dehydrogenation function. This confirmed that the hydrocracking process studied was "ideal".

5. EXTENSION TO C$_{80}$ MODEL

Various model building and control strategies were exploited to construct a high carbon number model based on the leanings from the C$_{16}$ model. These strategies are described elsewhere.[1] In short, by restricting the isomorphic criteria to the carbon and branch number level in the generalized isomorphism algorithm, molecules with the same carbon number and branch number were lumped into a representative molecular structure. This allowed construction of a molecule-based paraffin hydrocracking model at the pathways level for a feedstock up to C$_{80}$.

Figure 6 summarizes the essence of this model at the carbon and branch number level. The complexity is reduced to several PCP isomerizations (normal, 1, 2, 3 branch) and cracking (type A and type B) reactions. This captures all the key observations from the paraffin hydrocracking reaction mechanism: paraffins go through isomerizations before cracking; PCP isomerization always leads to branching, A-type cracking always leads to branched isomers, and B-type cracking always leads to normal or branched isomers; all the cracking products actually come from A-type cracking of tri-branched isomers or B-type cracking of di-branched isomers, and the C-type cracking can be neglected.

The thus-deduced C_{80} paraffin hydrocracking model containing 306 species and 4671 reactions was automatically generated on the computer. This model can thus be further tuned to capture the carbon number distribution and iso-to-normal ratio in the paraffin hydrocracking process.

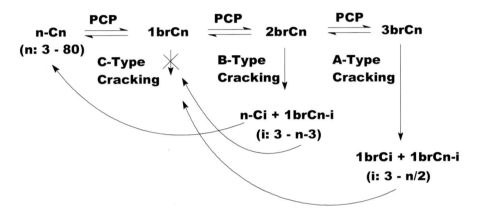

Figure 6. Mechanism Deduced Reaction Pathways at the Carbon Number and Branch Paraffin Hydrocracking Reaction Number Level

6.   SUMMARY AND CONCLUSIONS

Graph theoretic concepts were used to construct a paraffin hydrocracking mechanistic model builder incorporating bi-functional catalysis with metal and acid functions. The complete automated approach was successfully demonstrated for a C_{16} paraffin hydrocracking model with 465 species and 1503 reactions at the mechanistic level.

The reaction family concept was exploited by representing the reactions by various reaction families incorporating the metal function (dehydrogenation/hydrogenation) and the acid function (protonation/deprotonation, H/Me-shift, PCP isomerizations, and β-scission). The optimized C_{16} model provided excellent parity between the predicted and experimental yields for a wide range of operating conditions. This shows that the fundamental nature (feedstock and catalyst acidity independent) of the rate parameters in the model.

Various insights were obtained from the optimization results of the detailed C_{16} model: the skeletal isomerizations precede the cracking reactions; PCP isomerization led to branching, A-type cracking led to branched isomers, and B-type cracking led to normal or branched isomers; all the cracking products normally come from A-type cracking of tri-branched isomers or B-type cracking of di-branched isomers.

The generalized isomorphism algorithm was applied at the carbon and branch number level to reduce the complexity and explosion of modeling heavy paraffin hydrocracking. A thus-deduced C_{80} paraffin hydrocracking model with 306 species and 4671 reactions at the pathways level was developed from the teachings of the C_{16} model.

7. REFERENCES

1. Hou, G., Integrated Chemical Engineering Tools for the Building, Solution, and Delivery of Detailed Kinetic Models and Their Industrial Applications, Ph.D. Dissertation, University of Delaware, 2001.
2. Broadbelt, L. J.; Stark, S.M.; Klein, M. T., Computer Generated Pyrolysis Modeling: On-the-Fly Generation of Species, Reactions and Rates, *Ind. Eng. Chem. Res.*, **1994**, *33*, 790-799.
3. Broadbelt, L. J.; Stark, S. M.; Klein, M. T., Computer Generated Reaction Networks: On-the-fly Calculation of Species Properties using Computational Quantum Chemistry, *Chem. Eng. Sci.*, **1994**, *49*, 4991-5101.
4. Broadbelt, L. J.; Stark, S. M.; Klein, M. T., Computer Generated Reaction Modeling: Decomposition and Encoding Algorithms for Determining Species Uniqueness, *Comput. Chem. Eng.*, **1996**, *20*, 2, 113-129.
5. Korre, S. C., Quantitative Structure/Reactivity Correlations as a Reaction Engineering Tool: Applications to Hydrocracking of Polynuclear Aromatics, Ph.D. Dissertation, University of Delaware, 1994.
6. Russell, C.L., ,Molecular Modeling of the Catalytic Hydrocracking of Complex Mixtures: Reactions of Alkyl Aromatic and Alkyl Polynuclear Aromatic Hydrocarbons, Ph.D. Dissertation, University of Delaware, 1996.
7. Watson, B. A.; Klein, M. T.; Harding, R. H., Mechanistic Modeling of n-Heptane Cracking on HZSM-5", *Ind. Eng. Chem. Res.*, **1996**, *35(5)*, 1506-1516.
8. Watson, B. A.; Klein, M. T.; Harding, R. H., Catalytic Cracking of Alkylcyclohexanes: Modeling the Reaction Pathways and Mechanisms, *Int. J. Chem. Kinet.*, **1997**, *29(7)*, 545.
9. Dumesic, J. A.; Rudd, D. F.; Aparicio, L. M.; Rekoske, J. E.; Trevino, A. A. *The Microkinetics of Heterogeneous Catalysis*, American Chemical Society, Washington D.C. (1993).
10. Mochida, I.; Yoneda, Y., Linear Free Energy Relationships in Heterogeneous Catalysis I. Dealkylation of Alkylbenzenes on Cracking Catalysts, *J. Catal.*, **1967**, *7*, 386-392.
11. Mochida, I.; Yoneda, Y., Linear Free Energy Relationships in Heterogeneous Catalysis II. Dealkylation and Isomerization Reactions on various Solid Acid Catalysts, *J. Catal.*, **1967**, *7*, 393-396.
12. Mochida, I.; Yoneda, Y., Linear Free Energy Relationships in Heterogeneous Catalysis III. Temperature Effects in Dealkylation of Alkylbenzenes on the Cracking Catalysts, *J. Catal.*, **1967**, *7*, 223-230.
13. Mills, G. A.; Heinemann, H.; Milliken, T. H.; Oblad, A. G., *Ind. Eng. Chem.*, **1953**, *45*, 134.
14. Weisz, D.B., *Adv. Catal.*, **1962**, *13*, 137.
15. Langlois, G.E.; Sullivan R.F., *Adv. Chem. Ser. 97*, ACS, Washington, D.C., 1970; p.38.
16. Weitkamp, J., *Hydrocracking and Hydrotreating*, ACS Symp. Series 20, Ward, J.W.; Qader, S.A. (Eds), Washington, D.C., 1975; p. 1.
17. Steinberg, K.H.; Becker, K.; Nestler, K.H., *Acta Phys. Chem.*, **1985**, *31*, 441.
18. Brouwer, D. M.; Hogeveen, H., *Prog. Phys. Org. Chem.*, **1972**, *9*, 179.

19. Martens, J.A.; Jacobs, P.A.; Weitcamp, J., *Appl. Catal.*, **1986**, *20*, 239.
20. Stewart, J. J. P., Semiempirical Molecular Orbital Methods in *Rev. Comput. Chem.*, Lipkowittz, K. B.; Boyd, D D. B. (Eds.), VCH Publishers: New York, 1989.
21. Stewart, W. E.; Caracotsios, M.; Sorensen, J. P., Parameter Estimation From Multiresponse Data, *AIChE J.*, **1992**, *38(5)*, 641-650.

Chapter 21

MODELING OF REACTION KINETICS FOR PETROLEUM FRACTIONS

Teh C. Ho
Corporate Strategic Research Labs.
ExxonMobil Research and Engineering Co.
Annandale, NJ 08801

1.   INTRODUCTION

The petroleum industry will face a confluence of new challenges in the decades to come, owing to such factors as (1) rapid changes in market demand for high-performance products, (2) mounting public concerns over emissions and toxicity, and (3) dwindling supply of high-quality crudes. As a result, refiners will have to constantly stretch the limits of existing processes with minimum capital outlays. To this end, a low-cost approach is to develop predictive and robust process models that can take full advantage of modern advances in analytical chemistry, computing, control, and optimization.

Two distinguishing features of petroleum refining are that the number of reacting species is astronomically large and that the feedstock properties are continually changing. Refiners have long wanted to have at their disposal process models that can predict the effects of feedstock/catalyst properties and reactor configuration/conditions. But the daunting complexities of the composition and reactivity of petroleum fractions are such that until recently process models have largely been empirical, requiring constant and costly updating.

It is hardly surprising that model developers have been compelled to drastically simplify model development along two lines. One is what may be called *partition-based lumping*, while the other *total lumping*. In the former case, the reaction mixture is represented by a finite number of lumps and the reactions among them are tracked. The lumped system aims to capture essential features of the real system so that it has sufficient predictive power and robustness over ever-changing feeds and catalysts. In the latter case, the approach was motivated by the fact that in many situations what really matters is the aggregate, not the individual behavior of the reacting species. For instance, in hydrodesulfurization (HDS) one cares only about the reduction of total sulfur, not of the individual organosulfur species. Basically, all that is needed is to add up all sulfur species into a single total lump whose behavior is the focus of attention.

The purpose of this chapter was to provide a broad-brush survey of available theoretical tools for the two types of kinetic lumping problems. The emphasis will be on the general concept. As such, literature citations are merely illustrative rather than comprehensive. And there are very few examples of applying these tools; the reader should consult the original references for details. Throughout the chapter, scalars are denoted by italic letters, vectors by lower case boldface letters and matrices by capital boldface letters.

2.   OVERVIEW

2.1   Partition-Based Lumping

It is easier to think of petroleum fractions in such top-down terms as specific gravity, average molecular weight, boiling range, solubility, polarity, adsorptivity, and basicity, rather than, say, distinct individual gasoline molecules. The traditional PONA analysis lumps tens of thousands of molecules into just four lumps: <u>p</u>araffins, <u>o</u>lefins, <u>n</u>aphthenes, and <u>a</u>romatics. As a result, up until the '80s, modeling of petroleum reaction kinetics has largely focused on selecting a small number of easily measurable lumps based on gross properties and developing reaction networks in terms of the preselected lumps. There are many examples of such partition-based lumping approaches for various refining processes.[1-3] For a given process, there can be many lumped models, depending on the objectives and the desired level of detail. For instance, the number of lumps for fluid catalytic cracking (FCC) has been as small as three[1] and as large as 34.[4]

There is a limit to what can be gained by using the foregoing top-down approach in light of the growing need to develop high-resolution models for managing refinery streams at the molecular level. In response to this, many have pioneered multiscale approaches from the bottom up through modeling of the interactions of enormous numbers of, say, elementary reactions. The view is that it is the collective microscopic interactions at the bottom that give rise to the macroscopic kinetic behavior observed at the reactor level. One thus splits the reaction mixture at the molecular level, examines individual reactions, and works all the way up to the global level. Along the way one performs lumping at the local level to keep the number of rate constants and adjustable parameters at bay, because it is prohibitively costly or near impossible to track individual molecules with today's analytical techniques and computing power. Local lumping is achieved by using many time-honored reaction network reduction techniques and quantitative structure-reactivity relationships (QSRR). Final global lumping is done to predict product quality and properties from molecular composition. In contrast to the top-down route, here lumping is done after significant splitting.

2.2   Total Lumping

Total lumping may be viewed as a limiting case of the partitioned-based lumping. Here one is primarily interested in the overall behavior of a petroleum fraction. For instance, refiners would very much like to be able to predict how the overall behavior (e.g., HDS level) changes as feeds vary. Process developers want to know how different reactor types affect the overall behavior. And it is important for catalyst developers to rank exploratory catalysts based on their activities for the overall conversion of the feed. Consider first-order reactions. The concentration of the total lump at time t is $C(t) = \Sigma_i c_{if} \exp(-k_i t)$ where c_{if} is the feed concentration of the ith reactant and k_i the corresponding rate constant. The task is to find an overall kinetics $dC/dt = R(C)$, a one-lump model, that can be used for reactor modeling. It is highly desirable to be able to determine $R(C)$ and C *a priori* from information on how c_{if} and k_i are distributed in the feed. The inverse problem is to infer feed microscopic properties from experimentally determined $R(C)$ through fitting the C vs. t data. Given the vast number of species involved, a mathematically expedient approach has been to treat the mixture as a continuum. In practice, the chemical-analytical characterization data for petroleum fractions are often measured as a continuous function of, say, boiling point, molecular weight, or carbon number.

2.3   Reaction Network/Mechanism Reduction

The term kinetic lumping implies some kind of simplification and consolidation of reaction networks. This is essential to both the top-down and bottom-up approaches, To do so, kineticists have long used intuitive approaches based on such concepts as (1) rate-limiting step, (2) quasi-steady state approximation (QSA), (3) quasi-equilibrium approximation (QEA), (4) relative abundance of catalyst-containing species, (5) long-chain approximation, (6) on-plus constitutive equation, and (7) reaction shortsightedness. Helfferich[5] gives comprehensive recipes for these approaches. In the following we dwell on only two of these concepts.

Consider the first-order, two-dimensional reaction system A $\leftrightarrow$ B $\rightarrow$ C. Let k_1, k_2, and k_3 be the rate constants for the A-to-B, B-to-A, and B-to-C reactions, respectively. In the QSA, $k_2 \sim k_3 \gg k_1$, the ultra reactive B is maintained at a quasi steady state after the initial transient, so the system becomes one dimensional in that it is essentially governed by the disappearance of A. In the QEA, $k_1 \sim k_2 \gg k_3$, hence A and B quickly reach equilibrium after startup. The resulting one-dimensional reduced model simply states that the system is dictated by the slow depletion of the equilibrium pool of $(A + B)$ due to a "small leak" resulting from k_3.

Essentially, each of the above systems has two widely different time scales. If the initial transient is not of interest, the systems can be projected onto a one-dimensional subspace. The subspace is invariant in that no matter where one starts, after a fast transient, all trajectories get attracted to the subspace in which A and B are algebraically related to each other. In essence, what one achieves is *dimension reduction* of the reactant space through time scale separation. For large, complex systems such as oil refining, it is difficult to use the foregoing ad hoc approaches to reduce system dimensionality manually. Computer codes are available for mechanism reduction by means of the QSA/QEA[6-8] and sensitivity analysis.[9]

2.4   Mathematical Approaches to Dimension Reduction

The development of mathematical techniques for kinetic lumping is important in its own right. The reason is that even after significant simplification and reduction, the size of the problem may still be impractically large. Constructing lower dimensional models while retaining salient features of the original system can provide both fundamental insights and computationally feasible means for reactor design (especially hydrodynamically complex ones), control, and optimization. For instance, a widely used FCC 10-lump model [10] can be further reduced to a five-lump model through mathematical means such as projective transformation[11], sensitivity analysis[12], and cluster analysis.[13,14]

Research on mathematical lumping has focused on constructing kinetic lumps and determining the conditions under which the lumped system can at least approximate the underlying unlumped system. In so doing, one often needs to impose some constraints. Take catalytic reforming as an example. Kinetically and analytically, it makes sense to lump iso and normal paraffins together, but these hydrocarbons have so different octane numbers that they should be separated.

3. PARTITION BASED LUMPING

3.1 Top-down Approach

This traditional approach starts with a number of preselected, measurable kinetic lumps and determines the best reaction network and kinetics through experimental design and parameter estimation. The number of lumps depends on the level of detail desired. The lumps, satisfying the conservation law and stoichiometric constraints, are usually selected based on known chemistry, measurability and physicochemical properties (boiling range, solubility, etc.).

To give a flavor of the approach, we consider a simple lumped kinetic model for resid thermal cracking. The model needs to account for an induction time for coke formation, which is due to the phase separation of a second liquid phase formed from partially converted asphaltene cores. That is, coke forms when the solubility limit is exceeded. Wiehe[15] developed a simple first-order kinetic model based on solubility classes, as follows

$$M \xrightarrow{k_m} aA* + (1 - a)V \qquad (1)$$

$$A \xrightarrow{k_a} bA* + cM* + (1 - b - c)V \qquad (2)$$

$$A*_{max} = S_1(M + M*) \qquad (3)$$

$$A*_{ex} = A* - A*_{max} \qquad (4)$$

and at long reaction times,

$$A*_{ex} \longrightarrow yC + (1 - y)M* \qquad (5)$$

where a, b, c, and y are stoichiometric coefficients for maintaining carbon and hydrogen balance. M and A are maltenes (n-heptene soluble) and asphaltenes (toluene soluble/n-heptane insoluble) in the feed, M* and A* are the corresponding products, and V is volatiles. The solubility limit S_1, given as a fraction of total maltenes, determines the maximum cracked asphaltene concentration in the solution, $A*_{max}$. The excess asphaltene $A*_{ex}$ separates out as a second phase and produces coke C (toluene insoluble). With a single set of stoichiometric coefficients, the model correctly predicts the observed induction period and product slates for pyrolysis of both maltenes and whole resid.

Mosby et al.[16] reported a seven-lump residue hydroconversion model. Lumped models for steam cracking of naphtha and gas oils can be found in Dente and Ranzi's review.[17] The literature abounds with FCC kinetic models, with the number of lumps being three[1], four[18-20], five[21-25], six[26], eight[21], ten[10],

11^{27}, 17^{28}, and 34^4. The 17- and 34-lump models incorporate a fair amount of cracking chemistry. Figure 1 shows the 10-lump model developed by Weekman et al.[10], which has been incorporated in an FCC process simulator called CATCRACKER[29]. Dewachtere, et al.[30] combined the 10-lump model with a fluid dynamics model to simulate the feed injection zone in an FCC riser. They found that the feed nozzle with an included angle of 45° produces the most uniform temperature profile at the riser bottom.

Catalytic reforming models developed by Kmak[31] and Ramage et al.[32,33] give a fairly detailed description of the process. In Powell's hydrocracking model[34], the chemical composition parameters are continuous functions of carbon numbers. Quann and Krambeck[35] developed a kinetic model for olefin oligomerization over the ZSM-5 catalyst.

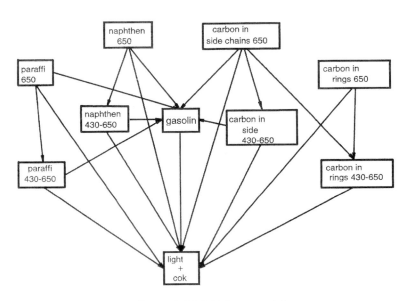

Figure 1. FCC 10-lump Model

The lumped kinetic models developed via the top-down route have limited extrapolative power. The rate "constants" generally depend on feedstock, catalyst, and reactor hardware configuration. As such, the models must be continually retuned for new situations. Moreover, the lack of sufficient molecular information makes it difficult to predict subtle changes in product qualities and properties. The bottom-up approach attempts to rectify this state of affairs, as discussed next.

3.2   Bottom-up Approach

Here one deals with individual molecules. The resultant reaction network describes the process chemistry in molecular detail. Broadly speaking, the approach has the following key elements.

(1). *Bookeeping of Molecules.* Hydrocarbon molecules in petroleum fractions, while huge in number, have only a small number of compound classes (e.g., aromatics, paraffins, etc.) or structural moieties. Their transformation, involving breaking and forming bonds, can be conveniently tracked by matrix operations with a computer.[36-39]

(2). *Splitting and Preprocessing.* The splitting of the mixture can be so fine that individual reactions are formulated in terms of elementary steps or of the single events of which an elementary step consist. These may include surface species such as carbocations in acid-catalyzed reactions. The rate constants are obtained from a variety of sources, such as analogies with liquid- and gas-phase reactions, transition state theory, semiempirical rules, quantum chemical calculations, empirical correlations, existing data bank, etc. Senken[40] reviewed conventional and quantum chemical methods for estimating thermochemical and reaction rate coefficients. Following the splitting is preprocessing of the huge number of individual reactions. This involves simplifying the networks of elementary reactions or single events with such tools as rate-controlling step, QEA, QSA, sensitivity analysis, thermodynamic constraints, etc. The dimension of the thus-reduced networks can still be hopelessly large, necessitating further reduction and parameterizations as discussed next.

(3). *Lumping based on Reaction Families.* That hydrocarbons in petroleum mixtures belong to a limited set of compound classes suggests that they give rise to a finite number of reaction families. For instance, reaction families in hydrocracking include hydrogenation, ring opening, dealkylation, HDS, hydrodenitrogenation (HDN), isomerization, and cracking. The reactions in any given family may be lumped in such a way that a small number of kinetic parameters can define the behavior of the entire family. To achieve this, one usually invokes QSRR, of which linear free energy relationships (LFER) are a prime example. QSRR can be developed from group contribution methods, reaction shortsightedness principle, empirical correlations, etc. The resulting reduced model, compatible with current analytical techniques, retains only the essential chemistry of the process and has a manageable number of parameters to be determined experimentally with judiciously chosen model compounds.

Below we briefly describe two specific bottom-up approaches.

3.2.1 Mechanistic Modeling

The models developed here account for unmeasurable intermediates such as adsorbed ions or free radicals. Microkinetic analysis, pioneered by Dumesic and cowokers[41], is an example of this approach. It quantifies catalytic reactions in terms of the kinetics of elementary surface reactions. This is done by estimating the gas-phase rate constants from transition state theory and adjusting these constants for surface reactions. For instance, isobutane cracking over zeolite Y-based FCC catalysts has 21 reversible steps defined by 60 kinetic parameters.[42,43] The rate constants are estimated from transition state theory

$$k = \frac{k_B T}{h} \exp(\Delta S^{\neq} / R) \exp(-\Delta H^{\neq} / RT) \qquad (6)$$

where k_B is the Boltzmann constant, h the Planck constant, $\Delta S^{\neq}$ the standard entropy change, and $\Delta H^{\neq}$ the enthalpy change from reactants to the transition state. The superscript double dagger denotes a property associated with the transition state. If k follows the Arrhenius law $k = A\exp(-E_\alpha/RT)$, then one can identify the preexponential factor A with $\Delta S^{\neq}$ and the activation energy E_α with $\Delta H^{\neq}$.

The entropies of all gas species are obtained from tables of thermodynamic properties. Data on gas phase basicity and proton affinities are used to estimate entropies of carbocations in the gas phase. The entropies of surface species and transition states are expressed in terms of standard entropy changes of adsorption. Some key assumptions are as follows. (1) The coverage of the Brønsted acid sites by carbocations is so low that all of the vibrational and rotational entropy is maintained during adsorption. (2) For C_n species with $n \leq 6$, translational entropy corresponding to 1/2 of one degree of freedom is maintained after adsorption, vs. 1/3 with $n \geq 7$. (3) For transition states of reactions involving gas phase and surface species, an additional degree of freedom of translational entropy is maintained upon adsorption compared to a similar surface species. (4) Hydride ion transfer reactions proceed through bulky transition states, and one degree of translational freedom of the corresponding gas phase species is maintained. These assumptions are consistent with adsorption entropy measurements made for NH_3 on H-mordenite and H-ZSM-5.

An Evans-Polanyi type of correlation is used to parameterize activation energies E_α based on known thermodynamic properties of gas phase hydrocarbons and carbocations. That is, $E_\alpha = E_o + \alpha\Delta H$ where ΔH is the reaction enthalpy and E_o and α are constants for a given reaction family. The reaction families considered are: (a) olefin adsorption/desorption, (b) carbocation isomerization involving tertiary and secondary ions, (c) oligomerization involving rearrangements between tertiary and primary carbocations, (d) β-scission with the occurrence of the isomerization of a secondary to tertiary carbocation and (e) β-scission. The protolysis reactions are not grouped into one family because each reaction involves the breaking of a different bond (C-H and C-C bonds) and has a different transition state. For hydride ion transfer, the Evans-Polanyi constants are allowed to change slightly. Ethylene production has a separate Evans-Polanyi parameter.

A significant aspect of the microkinetic analysis is the introduction of a catalyst factor ΔH_+, the heat of stabilization of a carbocation relative to that of a proton in the zeolite framework. The value of ΔH_+, assumed to be constant for all carbocations, is related to the Brønsted acid strength: the lower the value, the stronger the acid site. It may also be defined as the enthalpy of a carbocation transition state relative to the enthalpy of stabilization of a surface proton. The resulting kinetic model reproduces data obtained from a wide range of cracking conditions. The key parameter that changes when modeling calcined USY and steamed USY is ΔH_+. It explains why catalysts with lower Brønsted acid strength have a lower coverage of surface carbocations.

The mechanistic modeling approach for large reaction systems (e.g. oil refining) is advanced by the single-event theory developed by Froment and coworkers.[37,44-47] The theory allows the rate constants to be calculated from only a limited set of elementary step rates, which in turn are combinations of more basic molecular rearrangements. The central idea is to link rate constant to molecular structure. The preexponential factor A (hence the entropy change term; only the internal rotational entropy changes are relevant here) for a given reaction family can be systematically estimated from symmetry change[48,49] in going from the reactant to the transition state. This is done by factoring out a structural parameter n_e from A to account for possible chirality. So the rate constant of the elementary step (Eq.1) becomes $k = n_e \hat{k}$ where n_e is the number of single events for the elementary step and $\hat{k}$ the single-event rate constant. A global symmetry number σ_{gl} is defined as $\sigma_{gl} = \sigma/2^{n^*}$ where σ is the internal rotational symmetry number and n^* the number of chiral centers. From this, $n_e = \sigma_{gl,r}/\sigma_{gl,\neq}$ where the subscript r refers to the reactant.

As an example, Fig. 2 shows the methyl shift from 2-methyl-3-heptyl carbocation (2m3h) to 3-methyl-2-heptyl (3m2h) carbocation over a solid acid. Both the forward and reverse shift reactions involve secondary carbocations. However, two methyl groups can shift in the forward direction, while only one in the reverse direction. Therefore, different methyl shift rate constants are required to reflect this structural effect. Referring to Fig. 2, each of the 3 methyl groups in 2m3h can be superimposed on itself by a 120 degree rotation and there are no chiral carbon atoms, so $\sigma = 3 \cdot 3 \cdot 3 = 27$ and $\sigma_{gl,r} =$

$27/2^0 = 27$. The transition state is assumed to have a bridging structure, with two chiral carbons atoms. Presumably, the bridging methyl group loses the degree of freedom of internal rotation, so $\sigma_{gl,\neq} = 3 \cdot 3/2^2 = 9/4$. Thus $n_e = 27/(9/4) = 12$ for the forward shift. For the reverse shift, $n_e = [(3 \cdot 3 \cdot 3)/2]/(9/4) = 6$. The difference between the forward and reverse rate constants is due solely to the number of single events (or reaction path degeneracy[48]). Importantly, some major assumptions here are: (1) the structure of the transition state is known, (2) n_e for free carbocations can be used for adsorbed carbocations[50], and (3) there are no energy differences between the diastereomers. Baltanes et al.[44] gave examples of calculating n_e for hydroisomerization.

The foregoing discussion deals with the structure effect on A (or entropy change). The structure effect on the activation energy (or reaction enthalpy change) is described by the Evans-Polanyi relation, with just two parameters (E_o and α) for each single event type, which generally are obtained from model-compound experiments.

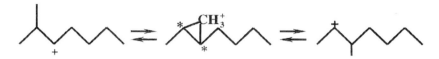

Figure 2. Methyl shift of 2-methyl-3-heptyl carbocation to 3-methyl-2-heptyl carbocation

The number of $\hat{k}$'s can be drastically reduced through a number of reaction "rules" for various reaction families. For instance, major reaction families in n-octane hydrocracking are protonation, deprotonation, hydride shift, methyl shift, protonated cyclopropane (PCP) branching, and β-scission. Examples of reaction rules are: (1) methyl and primary carbenium ions, being much less stable than secondary and tertiary ions, are neglected, (2) the protonation rate constant is independent of the olefin and depends on the carbocation type, not on its chain length, and (3) for homologous reference olefins, the deprotonation constants are equal. These rules reduce 383 single event rate constants to 22.

Armed with the reaction network, one can derive the rate equations for the feed components, intermediates and final products. The coupling between the adsorbed and gas-phase species are described by the Langmuir isotherm. Assuming that the rate-determining step is on the acid sites, one then develops the net formation rates for paraffins, olefins and diolefins, as well as those for hydride shift, methyl shift, PCP branching, ring contractions/expansions, and β-scission. These rate equations are combined to develop relations for the net formation of carbocations, which is set equal to zero via the QSA. An overall balance on the acid sites completes the problem statement. Here the lumps contain all molecules and carbocations, not just "key components." Froment's group has used the single event approach to develop kinetic models for FCC of vacuum gas oil.[51]

Klein's group[52] developed a mechanism-based lumping scheme for hydrocarbon pyrolysis involving free radicals. The model has two submodels. One is a five-component "training set" mixture (5CM) that calculates radical concentrations in terms of 42 representative free radical intermediates. The other is a module in which a feed mixture of many components reacts with the 5CM kernal to provide detailed rates and selectivities. The model retains the essence of pyrolysis chemistry with reasonable CPU demand.

Applying the mechanistic modeling approach to the petroleum refining scale is a big challenge because of the size and stiffness (wide range of time and length scales) of the problem. It is not obvious how this approach can be used to describe coke formation which is an important yet incompletely understood process. A computationally less intensive approach is to sidestep transient intermediates (carbocations or free radicals) and focus on dominant reaction pathways.

3.2.2 Pathways Modeling

Quann and Jaffe[39,53] developed what they called the structure-oriented-lumping (SOL) approach. A hydrocarbon molecule is described by a vector whose components, referred to as structural increments, represent the number of specific structural features. These increments are building blocks for constructing molecules. For example, A6 represents a six-carbon aromatic ring present in all aromatics, while A4, a four-carbon aromatic ring, is an incremental structure that must be attached to an A6 or another A4. Figure 3 shows the structure vector and its 22 structural increments that comprise three aromatic (A6, A4, A2) and six napththenic (N6-N1) ring structures, a -CH$_2$-group (R) to specify the carbon number of alkyl chains or paraffins, number of branches in a chain (br), number of methyl substituents (me) on rings, bridging between rings (AA), hydrogen deficiency (H), and heteroatom groups. Each group in the vector has carbon, hydrogen, sulfur, nitrogen, and oxygen stoichiometry and hence molecular weight.

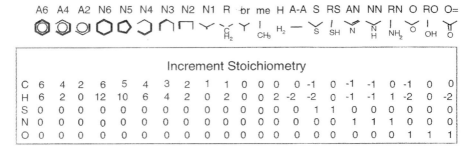

	A6	A4	A2	N6	N5	N4	N3	N2	N1	R	br	me	H	A-A	S	RS	AN	NN	RN	O	RO	O=
\multicolumn{23}{c}{Increment Stoichiometry}
C	6	4	2	6	5	4	3	2	1	1	0	0	0	0	-1	0	-1	-1	0	-1	0	0
H	6	2	0	12	10	6	4	2	0	2	0	0	2	-2	-2	0	-1	-1	1	-2	0	-2
S	0	0	0	0	0	0	0	0	0	0	0	0	0	0	1	1	0	0	0	0	0	0
N	0	0	0	0	0	0	0	0	0	0	0	0	0	0	0	0	1	1	1	0	0	0
O	0	0	0	0	0	0	0	0	0	0	0	0	0	0	0	0	0	0	0	1	1	1

Figure 3. Structure vectors and 22 structural increments defined in the structure-oriented-lumping (SOL) approach.

Figure 4 gives examples of the use of this vector. The vector components indicate the number of increments in each group. So benzene is represented by A6, with the first component of the vector having a value of one and all others zero. A homologous series is a set of vectors having identical structure increments except for the alkyl chain length and possibly the branching and methyl substituent indicators. This accounting system does not distinguish isomers. A key assumption then is that the isomers of a molecular class for a given carbon number have similar physicochemical and thermodynamic properties. In short, a petroleum mixture is represented by a large set of structure vectors of different weight percents; each vector is a molecule or a collection of isomers.

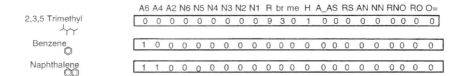

	A6	A4	A2	N6	N5	N4	N3	N2	N1	R	br	me	H	A_AS	RS	AN	NN	RNO	RO	O=
2,3,5 Trimethyl	0	0	0	0	0	0	0	0	0	9	3	0	1	0	0	0	0	0	0	0
Benzene	1	0	0	0	0	0	0	0	0	0	0	0	0	0	0	0	0	0	0	0
Naphthalene	1	1	0	0	0	0	0	0	0	0	0	0	0	0	0	0	0	0	0	0

Figure 4. Examples of SOL molecules

The next step is to define a finite number of reaction rules for the thousands of hydrocarbon components represented by the above 22 structural groups. For a given reaction family, a reaction rule determines if the reactant vector has the required increment(s) for that reaction to occur and then transforms the reactant vector to the product vector according to known chemistry and kinetics. Typically, petroleum process chemistry can be defined by 20-30 reaction families resulting in over 50,000 reactions.[53] Each refining process (e.g., FCC) has its own reaction rules, although all petroleum mixtures have the same vector representation. The lumping part of the approach is done through consolidation of reaction families. The reactions in the same family are assumed to have the same kinetic constants. In situations where this assumption becomes untenable, a perturbation analysis about equal reactivity may be done to obtain correction terms.[54,55] A more direct approach is to use available QSRR, as discussed in Section 3.2.3.

Many refining process models have been developed using the SOL approach. For FCC, Christensen et al.'s apply 60 reaction rules to more than 30,000 reactions involving over 3000 molecular species.[56] With a single set of kinetic parameters and reactivity relationships, the model can predict both bulk yields/properties and product composition for a range of feeds and catalysts. A smaller size (450 species and 5500 reactions) FCC model was developed by Klein's group.[57] In this model, a stochastic approach was used to describe the molecular structure of the gas oil feed (more on this later). The model reveals that significant protolytic cracking rate defines much of the product distribution for the light catalytic cycle oil, while significant hydride-abstraction defines the gasoline product distribution. Klein's group also developed a pathways-level model for catalytic reforming.[58]

3.2.3   Quantitative Correlations

A prominent example of the QSRR is the linear free energy relationship of the form $\ln k_i = a + bRI_i$ where k_i and RI_i are the rate constant and reactivity index for molecule i, respectively. The constants a and b are generally determined from experiments so that the rate constants for all members of the reaction family can be calculated from RI_i. The choice of RI_i reflects the controlling mechanism of the reaction, such as carbocation stability in catalytic cracking, or aromatics' electronic properties in hydrogenation[59,60], or reaction enthalpy in pyrolysis. Mochida and Yoneda[61-63] first demonstrated the use of LFERs for catalytic reactions. Klein and coworkers[64,65] have used LFERs extensively for modeling of reaction kinetics in oil refining. For instance, the number of independent rate constants for modeling individual gas oil reaction is of the order 10^4, whereas that of LFER parameters is of the order ten.[65]

Ho et al. developed a correlation for the poisoning effects of nitrogen compounds on FCC catalysts[66] and a scaling law for estimating the HDS reactivities of middle-distillates in terms of three key properties.[67] For solid acid catalyzed reactions, Sowerby and coworkers[68,69] developed a method for estimating adsorption equilibrium constants from an integrated form of van Hoff's equation.

3.2.4   Carbon Center Approach

Allen and coworkers[70-72] introduced a structural lumping approach based on group contribution concepts and pure-compound data. An oil fraction is assembled with a finite number of selected compound classes to capture key structural features of the oil. They calculated the number of CH, CH_2, CH_3 as well as terminal and nonterminal olefinic and aromatic carbons. They then followed the evolutions of carbon number distributions and carbon types in each compound class.

To illustrate, the FCC feed components are divided into *n*-paraffins, isoparaffins, naphthenes, olefins, and aromatics. Each of the five compound classes can have hundreds of surrogate compounds, each being constructed from the carbon centers (or functional groups) in that class. For instance, *n*-paraffins are built from CH_2 and CH_3. The cracking rates of these carbon centers are deduced from a small number of model-compound studies. Using the group contribution method reduces the problem of determining cracking kinetics for hundreds of compounds to that of estimating just two dozen parameters. For each compound class, group contribution parameters are determined for each of the following reaction families: cracking, isomerization, cyclization, dehydrogenation, ring opening, dealkylation, and coking. Reaction families neglected are hydrogen transfer and condensation.

The resulting model for cracking over amorphous SiO_2/Al_2O_3 has about 25 adjustable parameters that are estimated from pure-compound data. The collective behavior of these hundreds of surrogate compounds predicts the cracking behavior of the FCC feed. The product composition is used to estimate such bulk properties as vapor pressures and octane numbers. A sensitivity analysis shows that it is unnecessary to split each compound class into subsets of different carbon numbers and calculate a carbon center distribution for each carbon number. Using average carbon center distributions gives equally good product composition. In selecting surrogate compounds of a given carbon number within a compound class, one can ignore all molecules of concentrations lower than 10% of the maximum concentration regardless of chemical structure.

3.2.5   Lumping via Stochastic Assembly

This lumping approach, introduced by Klein and coworkers[73-75], deals with reaction systems involving large molecules whose structures are very difficult, if not impossible, to characterize. A case in point is asphaltenes, the portion of a resid that is aromatic-soluble and paraffinic-insoluble. Lumping is achieved by assuming that the reactivity of such large molecules is dominated by the reactivities of a small number of functional groups on the molecules. A similar lumping approach was also pursued by Gray[76] in his study of hydrotreating kinetics for heavy distillates. This line of approach, similar to that used in the carbon center approach, may be justified by reaction shortsightedness. This principle, as Helfferich[5] put it, says: "The reactivity of a group depends strongly on the local configuration, but little or not at all on the size and shape of the reactant molecules more than two atoms away." Where applicable, this principle can considerably simplify kinetic modeling by using the rate constants for a group of reactions.

For the pyrolysis of asphaltenes, Klein's approach basically consists of the following steps: (1) construction of an ensemble of reactant molecules based on characterization data, (2) development of lumped kinetics for a set of reactive functionalities, and (3) coupling of ensemble with lumped kinetics. This first step is stochastic while the rest are deterministic. Each of the steps is briefly described below.

The construction of a large ensembles of molecules, in which a finite number of functional groups are randomly incorporated in different molecules, is done stochastically with a computer. Since a real asphaltene feed comprises tens of thousands of molecules, its physicochemical properties are best described in terms of distributions rather than average values. The criterion for constructing such ensemble is that distributions of some key properties of the computer-generated ensembles must match those of the asphaltene feed in question. Examples of such distributions are the number of core systems/molecules, number of aromatic carbons/core, number of naphthenic carbons, number of pendants on a core, length of pendants, type of link attaching pendant to core or core to core, length of core-core link, to name a few.

A simple way of constructing each molecule in the ensemble is to connect ring systems (aromatic or naphthenoaromatic) with alkyl chains in a linear fashion subject to stoichiometric constraints. It is easier to construct an ensemble of liquid molecules than solid molecules. The latter requires modeling of three-dimensional structures.

The second step assumes that the reactivity of the ensemble is dominated by a few selected functionalities. The task then is to determine the reaction kinetics for each of the functional groups. Here the art of lumping applies in order to keep the number of kinetic lumps small. Information on reaction pathways and kinetics can be independently obtained from experiments using representative model compounds. For example, butyl benzene pyrolysis[77] may serve as a model system for the pyrolysis of alkyl aromatics moieties in resids.

The final step is to couple the lumped kinetics with the stochastically constructed ensemble. One then follows the reactions by updating the composition of the ensemble as the reactions proceed. The initial reaction network and pathways are allowed to run to a few conversions. The ensemble is then updated to reflect chemistry by examining concentration of key products. It is important to test the sensitivity of model prediction to various distribution functions.

4.   MATHEMATICAL REDUCTION OF SYSTEM DIMENSION

Suppose that one has already used intuitive ad hoc approaches to prune a large reaction model. A natural question then is, can the model be further reduced via some mathematical techniques? What follows is an overview of progress made in this area. Let us first consider the simplest conservation equation governing the fate of n reacting species

$$\frac{d\mathbf{c}}{dt} = \mathbf{f}(\mathbf{K};\mathbf{c}), \qquad \mathbf{c}(0) = \mathbf{c}_f \tag{7}$$

where t is a spatial or time variable, $\mathbf{c}$ the $n \times 1$ composition vector, $\mathbf{c}_f$ the feed composition vector and $\mathbf{K}$ the matrix of rate constants k_{ij}. The task is to find a lumped system with dimension $\hat{n} < n$, which can at least approximately portray the behavior of the underlying unlumped system. We first consider first-order kinetics with $\mathbf{f} = -\mathbf{Kc}$ where $\mathbf{K}$ is the $n \times n$ rate constant matrix.

4.1   Sensitivity Analysis

A kinetic model for oil refining invariably has a multitude of parameters and dependent variables, all of which can be strongly coupled among themselves. Sensitivity analysis can simplify the model by identifying which model parameters or dependent variables are unimportant so they can be lumped or removed from the model on the fly. As a simple example, consider a single first-order reaction with $c = c_o\exp(-kt)$. The sensitivity coefficient is $s = \partial c/\partial k = -c_o t\exp(-kt)$. So c is not sensitive to k at very short or very long times but is very sensitive to k in the neighborhood of $t = 1/k$. For large systems, one calculates the sensitivity matrix $s_{ij} = \partial c_i/\partial k_j$ at the nominal value (or point of interest) of $\mathbf{K} = \mathbf{K}_o$ where the sensitivity analysis is carried out.[78-80] Dumesic and coworkers[81] applied sensitivity analysis to catalytic cracking of isobutane. They identified 31 kinetically significant steps out of the 367 steps considered. Another example is that the 20 rate constants in the 10-lump FCC model[10] can be further lumped into 5 coarser lumps for an FCC unit operated in the gasoline mode.[12] Kramer et al.[82] developed a computer code "CHEMSEN" for sensitivity analysis of elementary reaction models.

4.2 Time Scale Separation

Suppose that the process time scale (or the time window of interest) is bounded between t_{min} and t_{max} ($t_{min} < t < t_{max}$) and is a small subset of the entire reaction time scale spectrum. Then the species reacting on time scales longer than t_{max} remain dormant; their concentrations are hardly different from their initial values. The state of these species may be treated as system parameters. On the other hand, species reacting with time scales shorter than t_{min} may be considered relaxed. The relaxed state of these fast-reacting species may be treated as system initial conditions. These considerations naturally help reduce the system dimensionality.

Briefly, a formal treatment is as follows. Let $\mathbf{X}$ be the matrix of the eigenvectors $\mathbf{x}_i$ of $\mathbf{K}$ and $\mathbf{\Lambda}$ be the diagonal matrix of the corresponding eigenvalues λ_i (i.e., $\mathbf{Kx}_i = \lambda_i\mathbf{x}_i$). The linear transformation[83] $\overline{\mathbf{c}} = \mathbf{X}^{-1}\mathbf{c}$ provides a set of pseudospecies called modes, which takes the form $\overline{\mathbf{c}}(t) = \overline{\mathbf{c}}_f \exp(\mathbf{\Lambda}t)$; that is, each of the pseudospecies $\overline{c}_i$ reacts independently with its own characteristic time scale $\tau_i = 1/|\text{Re}\lambda_i|$. If we are interested only in events occurring after $t >> 1/|\text{Re}\lambda_k|$, we can cross off those fast, relaxed modes and project the system to an $(n - k)$ dimensional subspace spanned by linearly independent eigenvectors $\mathbf{x}_{k+1}$, $\mathbf{x}_{k+2}$, $\cdots$, $\mathbf{x}_n$. The relaxed modes provide algebraic relations among species. This treatment is commonly called modal analysis. For nonlinear systems such as Eq.(7), one can still gain considerable insights through linearization at a reference state to obtain $d\mathbf{c}/dt = \mathbf{Jc}$ where $\mathbf{J}$ is the Jacobin $\partial\mathbf{f}/\partial\mathbf{c}$.

4.3 Projective Transformation

4.3.1 First Order Reactions

The treatment here, due to Wei and Kuo[84], lumps first-order reactions with $\mathbf{f} = -\mathbf{Kc}$. It projects the system onto a lower dimensional space via a linear transformation $\hat{\mathbf{c}} = \mathbf{Mc}$ where $\mathbf{M}$ is an $\hat{n} \times n$ lumping matrix. Thus $\mathbf{M}$ transforms the n-tuple vector $\mathbf{c}$ into an $\hat{n}$-tuple vector $\hat{\mathbf{c}}$ of a lower rank $\hat{n}$ ($\hat{n} < n$). If the system is exactly lumpable by $\mathbf{M}$, then one finds an $\hat{n} \times n$ matrix $\hat{\mathbf{K}}$ such that

$$\frac{d\hat{\mathbf{c}}}{dt} = -\hat{\mathbf{K}}\hat{\mathbf{c}} \qquad (8)$$

This is achievable if and only if [84]

$$\mathbf{MK} = \hat{\mathbf{K}}\mathbf{M} \tag{9}$$

The consequence of the lumpability on the eigenvalues and eigenvectors of the system is that the vector $\mathbf{Mx}_i$ either vanishes or is an eigenvector of $\hat{\mathbf{K}}$ with the same eigenvalue λ_i, that is,

$$\mathbf{Mx}_i = \mathbf{0} \text{ or } \hat{\mathbf{K}}(\mathbf{Mx}_i) = \lambda_i(\mathbf{Mx}_i) \tag{10}$$

Hence the matrix $\hat{\mathbf{K}} = \hat{\mathbf{X}}\Lambda\hat{\mathbf{X}}^{-1}$ has only $\hat{n}$ eigenvectors $\hat{\mathbf{x}}_i = \mathbf{Mx}_i$. Of the original n eigenvectors, $n - \hat{n}$ eigenvectors vanish after the projective transformation.

If $\mathbf{M}$ is known in advance, $\hat{\mathbf{K}}$ can be found from Eq.(9). To construct $\mathbf{M}$, one rewrites Eq.(9) (the superscript T denotes transpose)

$$\mathbf{K}^T\mathbf{M}^T = \mathbf{M}^T\hat{\mathbf{K}} \tag{11}$$

Viewing $\mathbf{K}^T$ as a mapping, Eq.(11) says that the action of $\mathbf{K}^T$ on $\mathbf{M}^T$ is just to create another matrix $\mathbf{M}^T\hat{\mathbf{K}}$ that still belongs to the same vector space ($\mathbf{M}^T$ and $\mathbf{M}^T\hat{\mathbf{K}}$ are both $n \times n$ matrices). With this property, the mapping $\mathbf{K}^T$ is said to have invariant subspaces. The spaces spanned by the eigenvectors of $\mathbf{K}^T$ are invariant subspaces because $\mathbf{K}^T\mathbf{x}_i = \lambda_i\mathbf{x}_i$; i.e., $\lambda_i\mathbf{x}_i$ coincides with $\mathbf{x}_i$. A straight line containing $\mathbf{x}_i$ is a one-dimensional invariant subspace of $\mathbf{K}^T$. The j-dimensional subspace spanned by $\mathbf{x}_1, \mathbf{x}_1 \cdots, \mathbf{x}_j$ ($j < n$) are also invariant subspaces. One thus can see that any eigensubspace of $\mathbf{K}^T$ can be used to construct $\mathbf{M}^T$, that is, by letting $\mathbf{M} = [\mathbf{x}_1, \mathbf{x}_1 \cdots, \mathbf{x}_{\hat{n}}]$, as demonstrated by Li and Rabitz.[85] First-order reactions are therefore always exactly lumpable based on the invariant subspaces of $\mathbf{K}^T$. As will be seen later, the existence of invariant subspaces is essential to lumping nonlinear kinetics.

Li and Rabitz[11] used the above approach to further contract the well-known 10-lump FCC model to a five-lump model with essentially the same predictive power for product slate changes. Moreover, if gasoline is the only unlumped species, then three lumps suffice. For nonane reforming, this approach reduces the number of lumps from 14 to 5 without significant errors if the total aromatics is kept unlumped.[86]

In practice, most reaction systems are not exactly lumpable. Hence, approximate lumping is of greatest practical importance. To do so, Kuo and Wei[87] defined an error matrix $\mathbf{E}$ as $\mathbf{E} = \mathbf{MK} - \hat{\mathbf{K}}\mathbf{M}$. The problem then becomes that of minimizing $\mathbf{E}$. They gave a detailed account for estimating $\mathbf{E}$. Liao and Lightfoot[88] gave a simpler framework for approximate lumping.

4.3.2 Non-Linear Systems

Li and Rabitz extended the Wei-Kuo approach to nonlinear kinetics.[11,85,89-92] With $\hat{\mathbf{c}} = \mathbf{Mc}$, Eq (7) becomes $d\mathbf{Mc}/dt = \mathbf{Mf}(\mathbf{c})$ and $\mathbf{c}(0) = \mathbf{Mc}_f$. A system is exactly lumpable if $\mathbf{Mf}(\mathbf{c})$ is a function of $\mathbf{Mc}$; that is, $\mathbf{Mf}(\mathbf{c}) = \hat{\mathbf{f}}(\mathbf{Mc})$ for every $\mathbf{c}$. The resulting lumped system of dimension of $\hat{n}$ is

$$\frac{d\hat{\mathbf{c}}}{dt} = \hat{\mathbf{f}}(\hat{\mathbf{c}}) \qquad (12)$$

Intuitively, one may expect that lumpability suggests that the system should have some degree of "partial linearity," which is related to the Jacobian matrix $\mathbf{J}[\mathbf{c}(t)] = \partial \mathbf{f}(\mathbf{c})/\partial \mathbf{c}$. Indeed, the system is exactly lumpable if and only if for any $\mathbf{c}$ in the composition space, the transpose of $\mathbf{J}[\mathbf{c}(t)]$ has nontrivial fixed (i.e., $\mathbf{c}$ independent) invariant subspaces. It can be shown that the exactly lumped system is of the form

$$\frac{d\hat{\mathbf{c}}}{dt} = \mathbf{M}\hat{\mathbf{f}}(\overline{\mathbf{M}}\hat{\mathbf{c}}) \qquad (13)$$

where $\overline{\mathbf{M}}$ is one of the generalized inverses of $\mathbf{M}$ satisfying $\mathbf{M}\overline{\mathbf{M}} = \mathbf{I}$. Whether or not the system is exactly lumpable depends on $\mathbf{f}$ and $\mathbf{M}$. When $\mathbf{f}$ is linear, $\mathbf{J}^T = \mathbf{K}^T$, a constant matrix. And the fixed invariant subspaces are nothing but the eigenspaces. While the number of the lumping matrices is infinite, the number of the invariant subspaces is finite. The nonuniqueness of the lumping matrix does not affect the form of the lumped equations. When $\mathbf{f}$ is nonlinear, not all of the $\mathbf{J}^T(\mathbf{c})$-invariant subspaces can be used to construct the lumping matrices. Some of the resulting lumped kinetic equations may have different forms from those for the underlying unlumped system.

In practice, some constraints are usually imposed on the lumped model. For example, certain chemical species (e.g., coke precursors) need to be kept unlumped. In this and similar situations, the lumping matrix needs to reflect the constraint *a priori*. Li and Rabitz also extended their analysis to include nonisothrmal effect[91] and intraparticle diffusion.[92]

Li and Rabitz[93,94] developed nonlinear mappings between the original and reduced dimension spaces. That is, $\hat{\mathbf{c}} = \mathbf{H}(\mathbf{c})$, which results in a self-contained lumped system of the form $d\hat{\mathbf{c}}/dt = \hat{\mathbf{f}}(\hat{\mathbf{c}})$. A necessary and sufficient condition for this to work is that $\hat{\mathbf{f}}[\mathbf{H}(\mathbf{c})] = \mathbf{f}(\mathbf{c})\partial \mathbf{H}/\partial \mathbf{c}$. The advantage of this approach is yet to be demonstrated.

4.3.3   Chemometrics

Chemometrics[95] is a convenient mathematical tool for developing QSRR or scaling laws that can be embedded in process models.[66,67] It is somewhat analogous to Wei and Kuo's treatment in that it contracts system dimensions through a projective transformation. Each lump is a linear combination of all original variables. The widely used principal component analysis (PCA)[95,96] and partial least squares (PLS)[97] are two examples of chemometric modeling.

The PLS analysis in many cases gives mechanistic insights into the underlying process chemistry, making it possible to construct QSRR from dominant descriptors. In the study of the poisoning effect of orgononitrogen compounds on FCC catalysts, Ho et al.[66] compressed 24 structural descriptors into two, each being patient of straightforward physical interpretation. The resulting correlation expresses the poisoning potency as a function of proton affinity and molecular weight. A similar approach gives a property-reactivity correlation for the HDS reactivity of middle distillates.[67] Finally, we remark that chemometrics is just one of many correlation-building tools. The latest developments in this area have been reviewed by Davis et al.[98]

4.4   Other Methods

There are other lumping techniques for linear kinetics, such as cluster analysis[13,14], observer theory[99-100], singular perturbation[101], and intrinsic low-dimensional manifolds.[102] Recently, Androulakis[103,104] treated kinetic mechanism reduction as an integer optimization problem with binary variables denoting the existence and nonexistence of reactions and species. The technique is amenable to uncertainty analysis.

5.   TOTAL LUMPING: OVERALL KINETICS

The problem addressed here can be stated as follows. Let $c_i(t)$ be the concentration of the individual reactant and $C(t) = \sum_{i=1}^{N} c_i(t)$ be the total concentration of all reactants. The aim is to predict the dependence of $C(t)$ on feed properties and reactor type. It is also of interest to know if an overall kinetics $R(C)$ can be found for the reaction mixture as a whole. If so, the thus-found $R(C)$ can be included in the conservation equations for modeling the coupling between kinetics and transport processes in a reactor. The selectivity toward a specific class of reaction intermediates (e.g., gasoline) is also of interest in many situations.

There have been four main lines of attacks to the problem posed here. The first addresses the question of whether one can find an overall kinetics for, say, a batch or plug flow reactor in the form $dC/dt = -R(C)$. There is no fundamental reason why $R(C)$ should always exist. In fact, $R(C)$ exists only for certain special reaction mixtures. The second approach is to obtain exact or approximate expressions for $C(t)$ when $R(C)$ cannot be found. Both approaches require complete information on the feed composition and reactivity spectra *a priori*; that is, to be working with a fully characterized feed. However, fully characterized petroleum fractions are hard to come by in practice. One has to settle for less, hoping to be able to say something about $C(t)$ and/or $R(C)$ with minimum information, that is, for a mixture that has only partially been characterized. This leads to a third approach aimed at finding $R(C)$ asymptotically for some specific regime of interest. The advantage is that the asymptotic kinetics are not sensitive to the details of the feed properties, thus yielding some rather general results. A fourth approach is to find upper and lower bounds on $C(t)$, the tightness of which depending on the quality of available information. These bounds can help choose empirical models and design reactor at least on a conservative basis.

Given the vast number of species in petroleum fractions, for mathematical tractability it pays to approximate the mixture as a continuum. The idea of continuous mixtures, first conceived by deDonder[105], was propounded in the context of reaction kinetics by Aris and Gavalas.[106] As will be seen later, the continuum approach gives considerable insights into the mixture's behavior. For instance, it provides a theoretical framework for explaining many peculiar behaviors observed in catalytic hydroprocessing (HDS, HDN, and hydrodearomatization). Some examples: (1) the overall HDS or HDN reaction order for the mixture as a whole is higher than that for individual sulfur or nitrogen species; (2) high-activity catalysts show lower overall order than low-activity catalysts; (3) refractory feeds show higher overall order than reactive feeds; (4) the overall HDS order decreases with increasing temperature; and (5) the overall order depends on reactor type (e.g., plug flow vs. stirred tank reactor).

5.1 Continuum Approximation

Aris[107] was the first to address the theoretical aspects of total lumping of first-order reactions. Luss and Hutchinson[108] later noticed that serious problems arise if one extends the continuum approach to nonlinear kinetics. Ho and Aris[109] discussed the origin of the difficulties in lumping nonlinear kinetics in continuous mixtures. They proposed a single-component-identity that must be satisfied by any continuum treatment in order to overcome the difficulties. Other aspects of the mathematical and conceptual difficulties have also been examined.[110-112] Krambeck[113] addressed thermodynamic issues. Ocone and Astarita[114] reviewed many aspects of continuous mixtures.

At least three approaches have emerged for total lumping of nonlinear kinetics. Astarita and Ocone's approach[110] is based on the notion that in lumping very many reactions, one should not ignore the interactions between the reactants. They proposed a specific class of interactive kinetics that includes Langmuir-Hinshelwood and bimolecular reactions. Aris[112] used a two-index approach consisting of two steps. First-order reactions (or, more generally, the Astarita-Ocone kinetics) are lumped in the first step to generate nonlinear kinetics which in turn are lumped in the second step. Chou and Ho[111] introduced a reactant-type distribution function that serves as a link between the continuous mixture and its underlying discrete mixture.

Without loss of generality, we use the rate constant k as the reactant label. The concentration of the total lump $C(t)$ can be expressed as[111]

$$C(t) = \int_0^\infty c(k,t)D(k)dk \qquad (14)$$

where $c(k,t)$ is the concentration of species k and $D(k)$ is the reactant-type distribution function. The slice $D(k)dk$ is the total number of reactant types with k between k and $k + dk$. For any given finite volume of the mixture, it is the number of reactant types, not the concentration of each reactant type that justifies the continuum hypothesis. Petroleum fractions of similar nature (e.g., gas oils) usually have the same, say, sulfur or nitrogen compound types (i.e., D is usually feed invariant), but the concentration of each compound type, $c(k,t)$, is feed dependent. For first-order reactions, $c(k,t) = c_f(k)\exp(-kt)$. It follows from Eq.(14) that $C(t)$ is uniquely defined by the product $D(k)c_f(k) \equiv h(k)$. This is not the case for nonlinear kinetics in general.

In the following we highlight some significant results obtained from the four approaches mentioned earlier. The majority of them were obtained from the continuum treatment.

5.1.1 Fully Characterized First Order Reaction Mixtures

For first-order reactions in a plug flow reactor (PFR), $C(t) = \int_0^\infty h(k)\exp(-kt)dk$. To portray a wide variety of feed properties with just two parameters, Aris[107] considered a class of feeds that can be characterized by the gamma distribution

$$h(k) = \gamma^\gamma(k/\bar{k})^{\gamma-1}\exp(-\gamma k/\bar{k})/[\bar{k}\Gamma(\gamma)] \qquad \gamma > 0 \qquad (15)$$

where Γ is the gamma function and γ a feed quality index; the smaller the γ value, the more refractory is the feed. When $\gamma = 1$ the distribution is exponential, indicating the feed contains a finite amount of unconvertible species. If $\gamma > 1$, the distribution is monomodal. Substituting Eq.(15) into

Eq.(14) gives $C(t) = (1 + \bar{k}t/\gamma)^{-\gamma}$, implying an overall $\bar{n}$ th-order kinetics with an overall rate constant $\bar{k}$, namely,

$$dC/dt = -R(C) = -\bar{k}C^{\bar{n}}, \quad \bar{n} = 1 + 1/\gamma > 1 \qquad (16)$$

Here the overbar indicates kinetic parameters associated with the overall kinetics $R(C)$. Thus, the mixture behaves like a single species with a higher-than-first order kinetics. As t increases, the reactive species disappear rapidly and the mixture becomes progressively more refractory, thus giving rise to a higher overall order. When the spread is broad (small γ), $\bar{n}$ is high. When the spread is narrow, $R(C)$ approaches first order. Equation (16) says that a tough feed with $\gamma = 1$ gives an overall order of two, while an easy feed with $\gamma = 2$ gives $\bar{n} = 1.5$.

The continuum treatment yields an explicit expression for the overall kinetics whose parameters relate macroscopic behavior to microscopic details of the feed. In the discrete mixture case, C is the sum of a bunch of exponential functions with a very large number of parameters. All one can say is that the $\ln C$ vs. t curve is concave upward[115], implying that $C(t)$ can generally be fitted with an overall order higher than one. Sau et al.[116] extended the continuum approach to include the effects of H_2S inhibition, H_2 partial pressure, and nonisothermality. In many petroleum refining processes, k can be related to the boiling point distribution.[116-118]

The foregoing results are obtained from a single gamma distribution. In practice, the distributions of sulfur compound types can be monomodal or bimodal. One may use a pair of gamma functions to describe a bimodal distribution. Distributions other than the gamma function have been considered.[119,120] In such cases, one can often obtain explicit expressions for $C(t)$ but not for $R(C)$.

5.1.2   Practical Implications

In catalyst exploratory studies, one normally screens a wide variety of experimental catalysts with a standard feed. A catalyst that does not activate the most refractory portion of the feed might well be evaluated by a high-order kinetics. But if the performance of the catalyst is improved (for instance, by the incorporation of a more effective promoter) so that there is no longer a refractory component, then a low overall order would have to be used in evaluating the improved catalyst.[121]   In process research work, one generally runs different feeds on a selected catalyst over a wide range of conditions. The overall reaction order is expected to increase when switching from an "easy" feed to a "hard" feed. The extent of this increase could be viewed as an index of feed refractoriness. The overall order is expected to increase with decreasing reaction severity. In short, the overall reaction order may be viewed as reflecting whether or not (1) the catalyst is active enough to attack the "unconvertibles" or (2) the conditions are severe enough to convert the "unconvertibles."

Let us cite some experimental results. The HDS rate with a single sulfur compound can often be described by pseudo first order kinetics.[122] By contrast, the HDS rate with a petroleum fraction typically has an overall order between 1.5 and 2.5.[123-125] Sonnemans[125] reported that the overall HDS order increased from 1.5 to 2 when switching from a blend of light coker gas oil and virgin gas oil to a light catalytic cracked oil (LCO). Beuther and Schmid[126] observed that the HDS of petroleum resids followed second-order kinetics. Sonnemans[125] reported an overall order of two for the HDN of LCO and an order of unity for the HDN of virgin gas oils. Heck and Stein[127] found that HDN and hydrodeoxygenation of coal liquids were best described by second-order kinetics. Some of the nitrogen and oxygen compounds in coal liquids are highly aromatic and therefore very refractory.[128] Ozaki et al.[129] observed that the overall HDS order increased with decreasing temperature. Stephan et al.[117] reported that removing H_2S in the reactor decreased the overall order of HDS of gas oil. Hensley et al.[130] changed $\bar{n}$ by modifying the support acidity of the hydroprocessing catalysts. The continuum approach used by Inoue et al.[131] allows the overall HDS order to be a variable. Breysse et al.[132] found that the overall HDS orders for three diesel feeds of increasing refractoriness are 2.2, 2.9, and 3.8 over a sulfided catalyst. Catalytic cracking of gas oils is commonly reported to be second order.[133]

The inverse problem is to obtain $h(k)$ empirically from measured $C(t)$. Gray[2] analyzed Trytten et al.'s HDS data[134] and found that $h(k)$ followed a gamma distribution with $\gamma = 2$. Others have also addressed the inverse problem.[117,135,136]

5.1.3 Partially Characterized First Order Reaction Mixtures

A partially characterized mixture (PCM) in the present context means that one has some information on the most refractory part of the feed. A problem of practical interest is the total lump's behavior at large times, or in the high-conversion regime. In oil refining one often wants to achieve the highest permissible conversion. In exploring or developing new catalysts, it is important that competing catalysts be evaluated at high conversions. As will be shown later, developing such long-time, overall asymptotic kinetics allows one to say something rather general about the mixture's behavior in the absence of detail information. In what follows the subscript a signifies long-time asymptotic kinetics.

5.1.3.1 Plug Flow Reactor

For irreversible first-order reactions in a PFR, Krambeck[137] showed that the asymptotic kinetics $R_a(C) \propto C^2$ whenever the feed contains a finite amount of unconvertible species. This is also true for reversible first-order reactions.[111] The case where the feed may or may not contain unconvertibles was treated by Ho and Aris.[109] The general treatment of the PCM starts with the expectation that the long-time behavior of the mixture should be governed by the most refractory part of the feed (as will be seen later, this is not always true). To find what goes on at large t, $D(k)$ and $c_f(x)$ near $k = 0$ can be expanded as follows[109,138]

$$D(k) \sim k^\mu(d_o + d_1 k + \cdots\cdots) \sim d_o k^\mu \qquad (17a)$$
$$c_f(k) \sim k^\nu(c_o + c_1 k + \cdots\cdots) \sim c_o k^\nu \qquad (17b)$$

in which $\nu \geq 0$ and $1 + \mu > 0$. When the feed has a finite amount of unconvertible species, $\mu = \nu = 0$. When $\mu \leq 0$, the mixture has a finite number of nearly unreactive reactants. The asymptotic behavior of $C(t)$ is completely dictated by the feed parameter γ defined as

$$0 < \gamma \equiv \mu + \nu + 1 \qquad (18)$$

This parameter characterizes the number of refractory reactant types and their concentrations. A small γ means a refractory feed. For first-order reactions, $h(k) \sim c_o d_o k^{\gamma-1} = h_o k^{\gamma-1}$ near $k = 0$. Then $C \sim h_o \Gamma(\gamma)/t^\gamma$ as $t \to \infty$, implying an asymptotic power law of the form

$$R(C) \sim R_a(C) = \gamma \left[h_o \Gamma(\gamma) \right]^{-1/\gamma} C^{\bar{n}_a}, \quad \bar{n}_a = 1 + 1/\gamma > 1 \qquad (19)$$

Thus, $\bar{n}_a > 1$, the tougher the feeds, the higher the order. Ho and White[139] determined the asymptotic power-law kinetics for deep HDS of LCO over three different catalysts, with $\gamma = 0.65$ and hence $\bar{n}_a = 2.5$. Sie[124] found that $\bar{n}_a = 2$ for both gas oil and prehydrotreated gas oil over a sulfided CoMo/Al$_2$O$_3$ catalyst. Ma et al.[140] fractionated a gas oil into five fractions; all give rise to second-order HDS kinetics, whether over a NiMo or a CoMo catalyst. An advantage of the asymptotic analysis is that only local information ($k \sim 0$) is needed for determining the overall rate in a PFR. This is not the case with a CSTR, as discussed below.

5.1.3.2 CSTR

Kinetics and catalyst exploratory studies are sometimes conducted in a CSTR. Slurry-phase reactors or ebullating beds, used for upgrading of heavy oils and resids, can be approximated as a CSTR. In a CSTR, $c(k,t) = c_f(k)/(1 + kt)$ for first-order reactions and

$$C(t) = \int_0^\infty \frac{h(k)dk}{1 + kt} \tag{20}$$

The system is more complex in that for each reactant type, there is an exponential distribution of residence times among all the molecules of that reactant. No $R(C)$ can be found and $R_a(C)$ admits three possibilities.[141] The case $\gamma < 1$ (refractory feeds) is similar to the PFR in that $C \propto 1/t^\gamma$ at large t, which is dominated by the most refractory reactants. Specifically, $R_a(C) \sim C^{1/\gamma}$. For $\gamma > 1$, $\bar{n}_a = 1$ and $C \sim I/t$ at large t (similar to that of its constituents) with $I \equiv \int k^{-1} h(k) dk$. Hence the long-time behavior of the mixture is governed by all species – not just by refractory species. Here the feed has a small amount of refractory species, the majority of which have short residence times. As a result, the refractory species cannot be dominating. To calculate C requires characterizing the whole feed. The situation is very different for the transition case $\gamma = 1$, which gives rise to $C \propto \ell n\, t/t$. This means that the decay of C, while governed by the refractory species, is accelerated by the reactive species. One can only define an instantaneous asymptotic order $\bar{n}_a = $ d $\ell n\, R_a$/d $\ell n\, C$, which decreases slowly with conversion, consistent with experiment.[2] At high conversions, $1 < \bar{n}_a = 1 + 1/|\ell n\, C| < 2$. For perspective, $\bar{n}_a$ for a PFR is always higher than that for a CSTR. The CSTR's size is larger than PFR's at constant conversion.[141] The PFR/CSTR reactor size ratio is less sensitive to conversion than in the single-component case.[141]

Experimentally, Rangwala et al.[142] found that the overall kinetics of HDS, HDN, and hydrocracking of a coker gas oil are all first order in a CSTR free of mass transfer effects. Trytten et al.'s CSTR HDS data showed an overall order greater than one.[134] And the PFR gave rise to a higher overall order than the CSTR. Dongen et al.[143] studied hydrodemetallization of heavy residual oils with $\gamma = 1$. $C(t)$ followed 1.5 order kinetics in a CSTR, while the same feed showed an overall order of two in a PFR. All these observations are consistent with the asymptotic kinetics. Although the asymptotic kinetics are developed for long times, they are very useful for modeling the mixture's behavior by power law kinetics.[138]

The overall kinetics represents an averaging over the reactivity and composition spectra. Such averaging should be different in reactors with different mixing characteristics. Golikeri and Luss[144] were the first to warn that the overall kinetics obtained from the PFR cannot be directly carried over to the CSTR. To the extent possible, it is important to develop if-then rules that can help translate data for different reactors.

Summarizing, a reaction mixture's overall behavior in a CSTR is the result of the interplay of the spreads of reactor residence times, reactant reactivity, and concentration. While all reactants are slowed down compared to those in the PFR, the fast-reacting ones are hampered more than the slow-reacting ones. As a result, the disparities among the species become smaller, making the mixture more homogeneous. A relatively homogeneous feed (high γ) can be further homogenized in the CSTR to such an extent that its long-time behavior approaches that of a single reactant.

5.1.3.3 Diffusional Falsification of Overall Kinetics

A tacit assumption used in the foregoing development is that the system is kinetically controlled. When a single species undergoes an nth-order reaction, the effect of a severe diffusion limitation is to shift the order from n to $(n + 1)/2$. The order remains intact when $n = 1$. Then there remains the question, "Would the overall order of many first-order reactions remain intact if all the constituent reactions become severely diffusion limited?"

For a PFR, $c(k,t) = c_f(k)\exp[-k\eta(k)t]$ where $\eta(k)$ is the catalyst effectiveness factor. Denoting $\bar{n}_d$ as the overall asymptotic order for the mixture controlled by diffusion, Ho et al.[121] showed that $\bar{n}_d = (\bar{n}_a + 1)/2$, a relationship similar to that in the single-reactant case. Hence diffusional falsification occurs for mixtures with $\bar{n}_a \neq 1$. Since $\bar{n}_a > 1$ in general, so $\bar{n}_d < \bar{n}_a$. Gosselink and Stork[145] found for the HDS of a gas oil that the overall order changed from two to 1.4 in going from ground-up to 3 mm catalyst particles. Stephan et al.[117] observed that the overall order for powder catalyst is three, vs. two for 5 mm pellets.

For a CSTR, $\bar{n}_d = \bar{n}_a = 1$ when $\gamma > 1$, $\bar{n}_d = \gamma\bar{n}_a = 1$ if $1/2 < \gamma < 1$, and $\bar{n}_d = \bar{n}_a/2 = 1/(2\gamma)$ for $0 < \gamma < 1/2$.[141] Thus, diffusion intrusion lowers $\bar{n}_d$ only for tough feeds ($\gamma < 1$). For easy feeds ($\gamma > 1$) the asymptotic order is one, with or without severe diffusion limitation. Here diffusion intrusion reduces species reactivity disparity, resulting in a wider region of "single-reactant" behavior (i.e., no shift in reaction order). Before addressing more complex kinetics, let us address the validity and limitations of the continuum theory.

5.1.3.4 Validity and Limitations of Continuum Approach

The concept of a continuum applied to reaction mixtures, arguably, is on a less firm foundation relative to its use in fluid mechanics. A simple thought experiment can show that the continuum approximation must break down for very long times. Consider first-order reactions in a PFR. At sufficiently large t, the mixture behaves as if there were only one species decaying exponentially, since the concentrations of all other species are exponentially smaller. This counters the power-law decay predicted by the continuum theory. Thus, in order for the continuum theory to be valid, time cannot be unconditionally large, even though the asymptotic kinetics are developed for large times. The issue then becomes one of determining the condition under which the continuum theory and its long-time limit are *both* valid.

For first-order reaction mixtures, the condition is $1/k^* \ll t \ll 1/\Delta k$.[139] Here k^* is a characteristic rate constant for a moderately refractory species and Δk is the difference between the rate constants for two species whose reactivities are adjacent to each other ($\Delta k \ll k^*$). From Houalla et al.'s HDS data[146], k^* can be taken as the rate constant for the HDS of dibenzothiophene (DBT). And Δk can be estimated by the rate constants for the HDS of 4-methyl DBT and of 4,6-dimethyl DBT at 300°C and 10.5 MPa over a sulfided Co-Mo/Al$_2$O$_3$ catalyst. The resulting region of validity is $3.87 \cdot 10^{-3} \ll t \ll 161 \cdot 10^{-3}$ h·g·cat./cm^{3}feed. This is not stringent at all for the HDS of refractory middle distillates. However, if the regime of very long times ($t > 1/\Delta k$) is of interest, one should use a discrete approach.

5.1.3.5 First Order Reversible Reactions

Hydrogenation of aromatic petroleum fractions is an example of reversible reactions. Wilson et al.'s experiments[147] show how various aromatic ring-number fractions approach equilibrium. Also, the distribution of equilibrium constants within each ring-number fraction is not wide. The hydrogenation of coal extract in a CSTR shows an overall order of unity.[148]

There are a few theoretical studies of the reaction mixture represented by $c(k, t) \, \Phi \, c'(k, t)$.[111,141,149] To examine the mixture's near-equilibrium behavior[141], let $c'(k, 0) \sim k^{\gamma-1}(c_0' + c_1'k + \cdots)$ for small k and define $h'(k) = c'_f(k)D(k)$. The equilibrium constant K is related to k by the Polyanyi equation $K(k) = k^\beta/\alpha$ (α, β are constants). Note that k and K change in the opposite directions ($\beta > 0$) for hydrogenation of mononuclear aromatics on metal sulfides. The reverse is true for noble metal catalysts.[150]

Both the PFR and CSTR exhibit a much wider variety of asymptotic behaviors than in the irreversible case.[141] Specifically, the PFR admits 13 possibilities, vs. CSTR's 21 possibilities. Still, R_a for both the PFR and CSTR follows power law in most cases. A striking difference between the two reactors is that when $\beta < -1$, the oil's near-equilibrium behavior in a PFR is governed by species of intermediate reactivities, whereas that in a CSTR is governed by either the most refractory species or all species.

5.1.3.6 Independent *n*th Order Kinetics

In heterogeneous catalysis, *n*th-order kinetics may be the result of adsorption on a nonideal catalyst surface. In homogeneous systems, *n*th-order kinetics may represent the overall rate of the underlying elementary reactions, e.g., the classical Rice-Herzfeld mechanism for thermal cracking of hydrocarbons.[49] For simplicity, n is assumed to be constant for all species. This is not a strong assumption for many petroleum processes.[128,133,151]

Aris[112] showed that $\bar{n} = 2$ for a special fully characterized mixture containing a finite amount of unconvertibles. Ho et al.[138] examined the overall kinetics asymptotically and numerically. There exist two critical values of n, denoted by n^* and n_*, which depend on the properties of the most refractory part of the feed; that is: $n_* = 1 - 1/v$ and $n^* = 1 + 1/(1 + \mu)$. At large times, $R_a \sim \bar{k}_a C^{\bar{n}_a}$ in most cases. Specifically, (1) $\bar{n}_a = n$ when $n > n^*$, (2) $\bar{n}_a = (nv + \mu + 2)/(1 + \mu + v)$ for $n_* < n < n^*$, and $\bar{n}_a > n$ and 1, (3) $R_a \propto C^{\bar{n}_a}/|\ell n C|$ when $n = n^*$, implying an instantaneous overall order of higher than two, and (4) the total lump depletes in a finite time when $n \leq n_*$.

The long-time behavior of C in cases (2) and (3) is governed by the most refractory species, whereas in case (1) by all species. Case 1 deals with a mixture whose constituents have relatively uniform reactivities. As a result, the overall, asymptotic rate exhibits the same order as that of the individual reactions. But $\bar{k}_a$, an average over the entire reactivity spectrum, weighs more heavily toward the refractory end. The asymptotic kinetics can provide guidance for empirical fitting of $C(t)$ by an $\bar{m}$ th-order model for all t. This is because $\bar{m}$ and $\bar{n}_a$ behave similarly in many respects. Numerical experiments[138] have shown that a continuum of zero-order reactions collectively can give rise to an apparent order of higher than unity if the feed contains high levels of refractory species, consistent with HDN data.[128]

5.1.3.7 Uniformly Coupled Kinetics

We have thus far considered systems in which the disappearance of a reactant depends only on the concentration of that reactant. To consider the interactions between reactants, Astarita and Ocone[110] proposed a class of kinetics of the form $r_i = -k_i c_i F(\Sigma_j w_j c_j)$, which is a special form of the pseudo-monomolecular system discussed by Wei and Prater[152] in a different context. Since the dimensionless function F depends only on $y \equiv \Sigma_j w_j c_j$; as such, its influence on all reactants is uniform. This assumption, while restrictive, reduces the problem to that of first-order reaction mixtures on a warped time scale. When $F(y) = 1/(1 + y)^m$ (m is assumed to be constant), w is the adsorption constant in a Langmuir-Hinshelwood mechanism. The $m = 1$ case is sometimes called the Eley-Rideal mechanism. If $F(y) = y^{n-1}$, one speaks of bimolecular power law kinetics. A special case is $n = 2$ and $w = k$, corresponding to mass-action bimolecular kinetics for which each of the bimolecular rate constants is separable. Many useful results are available for this type of continuous mixtures[112, 153-156]; the reader is referred to Ocone and Astarita's review.[114] In the following, we focus on asymptotic kinetics.

Again, w is also expanded as $w(k) \sim w_0 k^{\xi}$ as $k \to 0$ ($\xi + \gamma > 0$). For Langmuir-Hinshelwood reactions, competitive adsorption should become increasingly unimportant as time increases, thus degenerating $R_a(C)$ to that for first-order reactions regardless of reactor type. Indeed, $R_a(C) \propto C^{\overline{n}_a}$ for a PFR at large t with $\overline{n}_a = 1 + 1/\gamma$.[141] For a CSTR, $R_a(C)$ degenerates to the three cases for first-order kinetics discussed earlier. For bimolecular power-law kinetics in the PFR, $\overline{n}_a = [n(\gamma + \xi) - \xi + 1]/\gamma$.[157] In particular, if the feed has some unconvertibles ($\mu = \nu = 0$) with a nonzero w, then $\overline{n}_a = n + 1$. The corresponding CSTR case is more complex: it admits nine possibilities.[141] In all cases, the PFR gives rise to a higher $\overline{n}_a$ than the CSTR.

5.1.4   Upper and Lower Bounds

It is useful to develop bounds on $C(t)$ without detailed information on the feed. From the initial behavior of $C(t)$, Hutchinson and Luss obtained the following bounds for first-order reaction mixture in a PFR[158]

$$\exp(-M_1 t) \le C(t) \le \frac{\chi^2}{M_2} + \frac{M_1^2}{M_2} \exp(-t M_2 / M_1) \qquad (21)$$

where χ is the standard deviation of $h(k)$ and M_n are moments of $h(k,0)$, $M_n = \int_0^\infty k^n h(k) dk$. If $h(k)$ is unknown, one can measure $M_0 = C(0)$, $M_1 = -(dC/dt)_{t=0}$, and $M_2 = (d^2C/dt^2)_{t=0}$. Since M_n are based on early-time ($t \to 0$) information, the gap between the bounds increases with t and approaches χ^2/M_2 asymptotically. The upper bound is rather conservative at large t since it approaches χ^2/M_2. For instance, for the gamma distribution, $\chi^2/M_2 = 1/(\gamma + 1)$, the bound is especially conservative for small γ (refractory feeds). Tighter bounds can be obtained if one also knows that $c_f(k)$ is monomodal with a known mode.[158] Gray[2] used Eq.(21) to analyze Trytten et al.'s HDS data[134] and concluded that the bounds are indeed very conservative. The lower bound is the effluent concentration of a single pseudospecies undergoing first-order reaction with the rate constant M_1. Thus, using the average rate constant would overestimate the overall decay rate of the mixture. Similar bounds have been obtained for a series of N CSTRs[141] and for independent irreversible nth-order reactions in a PFR.[108]

For a variety of kinetics, Ho[157] developed a much improved upper bound that requires information on the most refractory part of the feed. The general form of the upper bound is

$$C \le C_u \equiv 1/(1 + t/\lambda^{1/\tilde{\gamma}})^{\tilde{\gamma}} \tag{22}$$

Table 1 lists $\tilde{\gamma}$, a generalized feed refractoriness index, and λ for various kinetics.

Table 1. Parameters for upper bound shown in Eq.(22)

Kinetics, $-r_i$	$\tilde{\gamma}$	λ
$k_i c_i$ and $k_i c_i/(1 + \Sigma_j w_j c_j)$	γ	$h_o \Gamma(\gamma)$
$k_i c_i (1 + \Sigma_j w_j c_j)^{n-1}$	$\gamma/\varpi*$	$h_o \Gamma(\gamma)/(\varpi^{n-1}\varpi*)^{\gamma/\varpi*}$
$\Sigma_j k_{ij} c_i c_j **$	1	$1/\,\overline{k}_a = \mathbf{e}^T \mathbf{K}_s^{-1} \mathbf{e}$
$k_i c_i^n$, $n_* < n < n^*$	$\gamma/(\nu + 1 - n\nu)$	$B(\tilde{\gamma}, \overline{\beta} - \overline{\gamma})d_o c_o^n \overline{\beta}^{\,\overline{\gamma}+1}/\,(\nu+\overline{\beta})$

$\varpi* \equiv 1 + (n-1)(\gamma + \xi)$, $\varpi \equiv w_o h_o \Gamma(\gamma + \xi)$, $\overline{\beta} = 1/(1 - n)$, B is the beta function
** This case is discussed in Section 5.2.2.

Note that C_u itself implies a power law; that is, $dC_u/dt = -\tilde{\gamma}\lambda^{-1/\tilde{\gamma}}C_u^{1+1/\tilde{\gamma}}$. As such, if $R(C)$ is of a power-law form. With gamma distributions for first-order and uniform kinetics in a PFR, the upper bound C_u becomes exact.[157]

5.1.5   One Parameter Model

In many practical situations, one is mainly interested in $C(t)$, which for most kinetics (power law, Langmuir-Hinshelwood, bimolecular reactions, etc.) can be accurately estimated or exactly calculated from the following simple expression[141,157]

$$C_q(t) = \frac{1}{(1+t^{zq}/\sigma^q)^{1/q}} \quad q > 0 \tag{23}$$

Here $C_q = C(t)/C_f$ and the subscript q signifies that C_q depends on the adjustable parameter q. Both z and σ can be determined from information on the most refractory portion of the feed. The parameter q should be determined experimentally at an intermediate conversion (say, between 45 to 60%) at or in the neighborhood of $t = t^* = \sigma^{1/z}$. Once $C(t^*)$ is known, then $q = -\ell n\,2/[\,\ell n\,C(t^*)/C_f]$. This expression, valid for both PFR and CSTR, is obtained from combining the large and small t asymptotes.

The point here is that Eq.(22) is all that is needed for an estimate of $C(t)$, which requires characterization of only the most refractory portion of the feed. A much improved estimate needs a data point on $C(t)$ at an intermediate conversion, Eq.(23).

5.1.6   Intraparticle Diffusion

Here the attention is focused on the behavior of the overall effectiveness factor $\bar{\eta}$ as a function of an overall Thiele modulus. For a given particle size, the extent of diffusion intrusion depends on the interplay of the reactivities and diffusivities of the constituent species. As discussed before, lumping very many species with widely different reactivities results in a steeper concentration profile than that in the single-reactant case. If the Thiele modulus ϕ follows a gamma distribution with $<\phi>$ being its average, Golikeri and Luss[159] showed for parallel first-order isothermal reactions that the $\bar{\eta}$ vs. $<\phi>$ curve is similar to that for a single reactant. The asymptotic similarity of catalyst geometry also holds in terms of a generalized Thiele modulus. When the diffusion limitation is severe, $\bar{\eta}$ is insensitive to how ϕ is distributed. Calculations based on the gamma distribution show that $\bar{\eta}$ is smaller than that calculated for the single-reactant case. Thus one would overestimate $\bar{\eta}$ if the conventional η - ϕ plot for a single reactant is used. This practice apparently is not uncommon.[115,134] Golikeri and Luss also obtained bounds for $\bar{\eta}$.

Ocone and Astarita[160] considered uniformly coupled reactions in an isothermal slab catalyst. All species are assumed to have the same diffusivity, which may correspond to a petroleum fraction of narrow boiling range. A significant result is that $\bar{\eta}$ can be greater than unity at small Thiele moduli, since the overall kinetics at the catalyst external surface may have a negative order. For nonuniformly coupled bimolecular reactions of the form $r_i = \sum_i k_{ij} c_i c_j$, Upper and lower bounds on $\bar{\eta}$ can be obtained by reducing the problem to that of finding the effectiveness factor for a single second-order reaction[161]. The resulting bounds can be used to estimate $\bar{\eta}$ based on information about the average and the spread of the individual Thiele moduli. For a given average Thiele modulus, $\bar{\eta}$ is smaller than that calculated for the single-component case.

5.1.7 Temperature Effects

Suppose that the overall kinetics can be written as $R(C) = \overline{k}f(C)$. A natural question is: if the individual rate constants k_i are all Arrheniusic, what is the temperature dependence of the overall rate constant $\overline{k}$? If $\overline{k}$ happens to be a function of the products, ratios, or ratios of products of k_i, then it obeys Arrhenius' equation. But there is no a priori reason why this should often be the case in practice. Golikeri and Luss[162] considered $r_i = k_i c_i^n$ with activation energy E_i in a PFR. If $\overline{k}$ is forced to obey the Arrhenius equation with an overall activation energy $\overline{E}$, then $\overline{E} = \Sigma_i(r_iE_i)/\Sigma_ir_i$ at a constant overall conversion, implying that $\overline{E}$ depends on temperature and concentration. So $\overline{k}$ in general does not follow the Arrhenius form unless by chance all E_i are equal. Calculations show that $\overline{E}$ is sensitive to temperature and/or concentration when E_i are widely spread. This says that in general the activation energy determined at one conversion level should not be applicable to another conversion level.

Experimentally, however, the overall temperature responses of many refining processes (FCC[163,164], hydrotreating[148], hydrocracking[165,166]) can often be adequately described by the Arrhenius equation. It seems, then, that the activation energies do not vary widely, or the temperature range is not very broad, or one of the activation energies may dominate, or a combination of the above. In systems where there is a wide spectrum of activation energies (e.g., pyrolysis of coal, resids, asphaltenes, or tar sands), some researchers[167-170] used the activation energy as the continuous variable for labeling the reactants. The assumption is that all k_i have the same pre-exponential factor. They found that the overall temperature response could be reasonably described by the mean of a continuous activation energy distribution function (e.g., Gaussian distribution). The mean activation energy determined by the continuum approach is consistent with pyrolysis mechanisms.

5.1.8 Selectivity of Cracking Reactions

Laxminarasimhan et al.[171] treated the problem of maximizing the liquid yield in hydrocracking by introducing a distribution function $p(k, \kappa)$ characterizing cracking stoichiometry

$$\frac{dc(k,t)}{dt} = -kc(k,t) + \int_k^{k_{max}} p(k,\kappa)\kappa c(\kappa,t)D(\kappa)d\kappa \quad (24)$$

The hydrocracking rate constant k is assumed to be a monotonic function of boiling point. And $p(k, \kappa)$ describes the amount of species with reactivity k that is formed from the cracking of the species with reactivity κ. Its properties are: (1) $p(k, \kappa) = 0$ for $k \geq \kappa$; (2) $p(k, \kappa) > 0$ and has a finite, small value when $k \to 0$, and (3) $\int_0^\kappa p(k, \kappa) D(k) dk = 1$. The reactant-type distribution function $D(\kappa)$ accounts for the cracking of all species with reactivity κ; since the mass balance is written in the k space.

A skewed Gaussian-type distribution function can depict the yield distributions. The resulting model reproduces published pilot-plant and commercial data on vacuum gas oil hydrocracking.[118,172,173] This has led to the development of a hydrocracking process model.[118] Selective cracking in FCC has also been addressed.[174] Browarzik and Kehlen[175] used a fragmentation-based model for n-alkane hydrocracking. Similar approaches have been used for polymer reaction systems.[176-179]

5.1.9   Reaction Networks

The overall selectivity of a bunch of reaction networks can be quite different from that of a single network, as discussed by Golikero and Luss[180,181] for discrete mixtures and later by Astarita[182] for continuous mixtures. A simple example is the first-order parallel reaction system $A_i \to B_i$ and $A_i \to C_i$ ($i = 1, 2, \cdots, N$). Let $a_i(t)$, $b_i(t)$, and $c_i(t)$ be the individual concentrations and A, B, and C be the lumped concentrations, respectively (i.e. $A = \Sigma a_i$). In the single-reactant case, the selectivity $s(t) = b(t)/c(t)$ is independent of conversion. In the mixture case, however, the selectivity $S(t) = B(t)/C(t)$ in general depends on conversion and feed composition. Moreover, the equations governing the kinetic behavior of the lumped species are very different from those for individual species and are often of unconventional forms. Numerical experiments indicated that $S(t)$ may have at most $N - 2$ extremum points. For a two-species system, $S(t)$ is a monotonic function of conversion. Thus, while optimum residence time is an issue for the mixture, it is not for a single reactant. For the consecutive system $A_i \to B_i \to C_i$, $B(t)$ can exhibit multiple maxima, and the overall selectivity $S(t) = B(t)/[A(0) - A(t)]$ depends on the feed composition. Attempts have also been made to treat multi-step reactions in continuous mixtures, which have been reviewed by Ocone and Astarita.[114]

5.2   Discrete Approach: Nonuniformly Coupled Kinetics

An exemplary system that captures structural features of the interacting kinetics is described by the following irreversible bimolecular reactions:

$$\frac{dc_i}{dt} = -\sum_{j=1}^{N} k_{ij} c_i c_j, \quad \frac{dC}{dt} = -\sum_{i=1}^{N} \sum_{j=1}^{N} k_{ij} c_i c_j, \quad c_i(0) = c_{if} \geq 0 \qquad (25)$$

with $k_{ij} = k_{ji}$. A major source of coke in catalytic cracking is the bimolecular reactions between olefins and aromatics. Bimolecular reactions can also be found in pyrolysis, condensation, oligomerization, disproportionation, alkylation, etc.

Here the continuum approach does not offer much advantage because an explicit expression for c_i is not available. Finding C or $R(C)$ is necessarily more difficult than the uniformly coupled case where $k_{ij} = k_i k_j$. The problem has been attacked asymptotically along two paths, as highlighted below.

5.2.1   Homologous Systems

The system in question is weakly nonuniform in that k_{ij} do not span a wide range; that is, $k_{ij} = k_m + \varepsilon_{ij}$ with $\varepsilon \equiv \max |\varepsilon_{ij}| = \max |k_{ij}/k_m| << 1$ where k_m is a mean rate constant. Explicit expressions for $C(t)$ in a PFR and CSTRs have been obtained via perturbation about the equal-reactivity limit.[54] As expected, an overall second-order kinetics appear at lowest order. With no more than three terms the perturbation series is accurate for ε at least as high as 0.5. Upper and lower bounds for $C(t)$ and $c_i(t)$ have also been derived.[183] The same approach can be used for systems in which the coupling is weak, that is, $k_{ij} << k_{ii}$.[54]

5.2.2   Long-Time Behavior

A crucial question here is, what are the consequences of strong interactions between reacting species? To answer this question, we first consider a binary system that can remember its history; that is, its long-time behavior depends on where it starts.[183,184]   There are two governing parameters: $\delta_1 \equiv k_{11} - k_{12}$ and $\delta_2 \equiv k_{22} - k_{12}$. Thus the relative magnitude of k_{12} plays the key role in determining the interaction intensity. To reveal what goes on at large t, it pays to work with reactant mole fraction $x_i(t) = c_i(t)/C(t)$ ≥ 0 and $x_i(0) = x_{if}$ so $\Sigma_i x_i(t) = 1$ at all t. At large t, $R \sim \bar{k}_a C^2$ and $x_i(t)$ approaches a steady state $\bar{x}_i$. In order of increasing interaction intensity, the system's behavior can be classified as follows.[184]

Case I. Single steady state ($\delta_1 > 0$ and $\delta_2 > 0$): There is very little "cross talk" between the two species. Hence, both species coexist at all t and $\bar{k}_a = (k_{11}k_{22} - k_{12}^2)/(k_{11} + k_{22} - 2k_{12})$. The system is robust in that it has only one realizable steady-state composition ($\bar{x}_1 = \delta^*$ and $\bar{x}_2 = 1 - \delta^*$ where $\delta^* = \delta_2/[\delta_1 + \delta_2]$), whatever the feed composition.

Cases II. Single steady state (Case IIa. $\delta_1 \geq 0$ and $\delta_2 \leq 0$, but not both zero; Case IIb. $\delta_1 \leq 0$ and $\delta_2 \geq 0$, but not both zero): There is some "cross talk" between a very reactive species and a refractory species. The latter dominates the system's long-time behavior, which is robust in that $\bar{k}_a = k_{22}$ and $\bar{x}_2 = 1$ in Case IIa, and $\bar{k}_a = k_{11}$ and $\bar{x}_1 = 1$ in Case IIb.

Case III. Multiple steady states ($\delta_1 < 0$ and $\delta_2 < 0$): Here the cross-reaction dominates the system behavior, which is sensitive to its initial conditions. Specifically, $\bar{x}_1 = 1$ and $\bar{k}_a = k_{11}$ for any $x_{1f} > \delta^*$ and $\bar{x}_2 = 1$ and $\bar{k}_a = k_{22}$ for any $x_{1f} < \delta^*$. The interior steady state ($\bar{x}_1 = \delta^*$) is unstable and hence unrealizable.

Thus, as the interaction intensifies, steady-state multiplicity and stability come into play. In the case of uniformly coupled bimolecular reactions ($k_{ij} = k_i k_j$), the interaction is not strong enough to induce steady-state multiplicity; the system behaves similarly to Case II above.

The above treatment can be generalized to an N-component mixture.[184] At steady state, either $x_i = 0$ or $(\mathbf{Kx})_i = \mathbf{x}^{\mathrm{T}}\mathbf{Kx}$ for each i where $\mathbf{K}$ is the matrix formed by k_{ij}. Hence the system has two submixtures. One contains N_s refractory species that survive the reaction after a long time. Another contains $N - N_s$ reactive species that become exhausted at large t. Some of the steady states may not be stable and hence unobservable. Principal findings are: (1) $R(C) = k(t)C^2$ where the instantaneous rate constant $k(t)$ decreases with t and eventually approaches a constant $\bar{k}_a > 0$ if $k_{ii} > 0$ for every i, regardless of the feed composition. That is, $R_a = \bar{k}_a C^2$. That $d\bar{k}_a/dt \leq 0$ implies that the instantaneous order of the overall kinetics is greater or equal to two. While R_a is usually second order, exceptions do exist. (2) The uniqueness of $\bar{k}_a$ depends on the feed composition, although $\bar{k}_a$ can be chosen from only a finite number of possibilities. Thus, the system can exhibit a long-term memory. (3) In most cases, each possibility for $\bar{k}_a$ corresponds to a unique asymptotic mixture composition. However, under special circumstances, a continuum of steady-state compositions can possibly exist; even in this case the number of overall rate constants is always finite. (4) $\bar{k}_a = 1/(\mathbf{e}^{\mathrm{T}}\mathbf{K}_s^{-1}\mathbf{e})$ where $\mathbf{e}$ is an N_s-dimensional vector of all ones and $\mathbf{K}_s$ the $N_s \times N_s$ rate constant matrix formed from the surviving species. (5) The long-time behavior of the mixture can be affected by the fast reactions of the system. (6) When the refractory species decay at an asymptotic second-order rate, the reactive species get exhausted asymptotically at a power-law rate with an order less than two. (7) The behavior of the mixture is not oscillatory because $\mathbf{x}$ cannot return to the same point as time proceeds. (8) Stability plays a central role in determining the system behavior. One of the stability criteria can be defined in terms of the transition probability matrix of a Markov chain.

Numerical calculations demonstrate that a mixture with as few as three reactants can exhibit a rich variety of behaviors. The results also point to the importance of varying the feed composition in kinetics studies and that long-time rate data can be used for estimating kinetic constants.

6.   CONCLUDING REMARKS

Over the past two decades, much progress has been made on the theory and practice of kinetic lumping, largely in response to increasingly stringent requirements for fuels/lubes quality, crude slate flexibility, and reactor control and optimization. Several new approaches have been developed for constructing robust process models used in petroleum refining. Along the way, many QSSR and if-then rules have been developed. In certain areas, high-resolution models have proved their worth by making molecular management a reality.

Much remains to be done, however. One of the most challenging areas is to incorporate catalyst properties in kinetic models. For crystalline catalysts with well-characterized structures (e.g., zeolites), some limited progress has been made, but this is far from being the case for structurally complex catalysts such as highly amorphous metal sulfides.

Control and mitigation of coke formation is an integral part of improving existing and developing new refining technologies. Coke on catalyst not only decreases activity, but also greatly affects selectivity. To complicate matters, there are many types of coke for a given process chemistry (e.g., FCC, coking). Reactor fouling caused by carbonaceous deposit (higher H-to-C ratio than coke) remains an important operation issue in many cases. We really do not know much about the fundamentals of these processes. These areas are fertile grounds for further research.

The tradeoff between kinetics and hydrodynamics for hydrodynamically complex systems continue to be a big challenge. For instance, to address problems of fluid maldistribution in trickle bed reactors or the feed injection zone at the base of a FCC riser, one is compelled to sacrifice chemistry. The same is true when one needs to solve control and optimization problems.

Heavy feed processing is certain to become more important in years to come. This is so because refining processes in today's economic climate must be capable of responding to ever-changing feedstocks, cost and availability, as well as changes in market demand. Moreover, the supply of high quality cat feeds is surely dwindling. Many heavy feed catalytic processes involve multiphase flows with multiscale structure. Modeling of such systems requires consideration of linking different time/length scales (e.g., catalyst surface,

intraparticle transport, bubbles/droplets, sprays, etc.) and development of a
framework for predictions at the global (reactor, interprocess) scale.

7.   REFERENCES

1. Weekman, V. W., *Chem. Eng. Prog. Symp. Ser.* **1979**, *75*, 3.
2. Gray, M. R., *Upgrading Petroleum Resids and Heavy Oils*, Marcel Dekker, 1994.
3. Sapre, A. V., Kinetic Modeling at Mobil: A Histroical Perspective, In *Chemical Reactions in Complex Mixtures*, Sapre, A.V.; Krambeck, F. J. (Eds.), Van Nostrand Reinhold, 1991.
4. Meier, P. E.; Ghonasgi, D. B.; Wardinsky, M. D., US 6,212,488, **2001**.
5. Helfferich, F. G., *Kinetics of Homogeneous Multistep Reactions*, Elsevier, 2001.
6. Chen, J.-Y., *Combust. Sci. Technol.*, **1988**; *57*, 89-94.
7. Chen, J.-Y., *Workshop on Numerical Aspects of Reduction in Chemical Kinetics*, CERMICS-ENPC Cite Descartes-Champus Sur Marve, France, September 2, 1997.
8. Sung, C. J.; Law, C. K.; Chen, J.-Y., *Combust. Sci. Technol.*, **2000**, *156*, 201-220.
9. Lutz, A. E.; Kee, R. J.; Miller, J. A., SENKIN: A Fortran Program for Predicting Homogeneous Gas Phase Kinetics With Sensitivity Analysis, SAND87-8248, Sandia Nat. Labs, Livermore, 1988.
10. Jacob, S. M.; Gross, B.; Voltz, S. E.; Weekman, V. W., *AIChE J*, **1976**, *22*, 701-713.
11. Li, G.; Rabitz, H. *Chem. Eng. Sci.*, **1991**, *46*, 583.
12. Pareek, V. K.; Adesina, A. A.; Srivastava, A.; Sharma, R., *J. Mole. Catal.*, **2002**, *181*, 263-274.
13. Coxson, P. G.; Bischoff, K. B., *Ind. Eng. Chem. Res.*, **1987**, *26*, 1239-1248.
14. Coxson, P. G.; Bischoff, K. B., *Ind. Eng. Chem. Res.*, **1987**, *26*, 2151-2157.
15. Wiehe, I. A., *Ind. Eng. Chem. Res.*, **1993**, *32*, 2447.
16. Mosby, J. F.; Buttke, R. D.; Cox, J. A., *Chem. Eng. Sci.*, **1986**, *41*, 989-995.
17. Dente, M. E.; Ranzi, E. M., Mathematical Modeling of Pyrolysis Reactions. In *Pyrolysis: Theory and Industrial Practice*, Albright, L. F.; Crynes, B. L.; Corcoran, W. H. (Eds.), Academic Press: NY, 1983.
18. Lee, L. S.; Chen, Y. W.; Huang, T. N.; Pan W. Y., *Can. J. Chem. Eng.*, **1989**, *67*, 615-619.
19. Gianetto, A.; Faraq, H.; Blasetti, A.; deLasa, H. I., *Ind. Eng. Chem. Res.*, **1994**, *33*, 3053.
20. Dave, N. C.; Duffy, G. J.; Udaja, P., *Fuel*, **1993**, *72*, 1333-1334.
21. Hagelberg, P.; Eilos, I.; Hiltuneu, J.; Lipiäinen, K.; Niemi, V. M.; Attamaa, J.; Krause, A. O. I., *Appl. Catal.*, **2002**, *223*, 73-84.
22. Cerqueira, H.S.; Biscaia, E. C.; Sousa-Augias, E. F., *Appl. Catal.*, **1997**, *164*, 35-45.
23. Ancheyta-Juárez, J.; Lopez-Isunza, F.; Aguilar-Rodriquez, E.; Moreno-Mayorga, J., *I&EC Res.*, **1997**, *36*, 5170-5174.
24. Rosca, P.; Ionescu, C.; Apostal, D.; Ciuparu, D., *Prog. Catal.*, **1994**, *3*, 89-96.
25. Zhang, Y. Y., *Comput. Chem. Eng.*, **1994**, *18*, 39-44.
26. Takatsuka, T., Sato, S., Morimoto, Y., Hashimoto, H., *Int. Chem. Eng.*, **1987**, *27*, 107-115.
27. Sha, Y.; Deng, X.; Chen, X.; Weng, H.; Mao, X.; Wang, S., *Proc. Int. Conf. Pet. Refin. Petrochem. Proc.*, Hou, X. (Ed.), 1991; 3: 1517.
28. Pitault, I.; Nevicato, D.; Forissier, M.; Bernard, J., *Chem. Eng. Sci.*, **1994**, *49*, 4249-4262.
29. Kumer, S.; Chadha, A.; Gupta, R.; Sharma, R., *Ind. Eng. Chem. Res.*, **1995**, *34*, 3737.
30. Dewachtere, N. V.; Fromert, G. F.; Vasalos, I.; Markatos, N.; Skandalis, N., *Appl. Thermal Eng.* **1997**, *17*, 837-844.
31. Kmak, W. S. AIChE National meeting, Houston, TX, 1972.
32. Ramage, M. P.; Graziani, K. R.; Krambeck, F. J., *Chem. Eng. Sci.*, **1980**, *35*, 41
33. Ramage, M.P.; Graziani, K.R.; Schipper, P. H.; Choi, B.C., *Adv. Chem. Eng.*, **1987**, *13*, 193-266.

34. Powell, R. T., *Oil Gas J.*, **1989**, 64-65.
35. Quann, R. J.; Krambeck, F. J., Olefine Oligomerization Kinetics over ZSM-5, In *Chemical Reactions in Complex Mixtures*, Sapre, A. V.; Krambeck, F. J. (Eds.), Van Nostrane Reinhold, 1991.
36. Clymans, P. J.; Frement, G. F., *Comput. Chem. Eng.*, **1984**, *8*, 137.
37. Baltanas, M. A.; Froment, G. F., *Comput. Chem. Eng.*, **1985**, *9,* 71-81.
38. Broadbelt, L. J.; Stark, S. M.; Klein, M. T., *Ind. Eng. Chem. Res.*, **1994**, *33*, 790-9.
39. Quann, R. J.; Jaffe, S. B., *Ind. Eng. Chem. Res.*, **1992**, *31*, 2483-2497.
40. Senkan, S. M., Detailed Chemical Kinetic Modeling: Chemical Reaction Engineering of the Future, *Adv. Chem. Eng.*, **1992**, *18*, 95.
41. Dumesic, J. A.; Rudd, D. F.; Aparicio, L. M.; Rekoske, J. E.; Trevino, A. A., *The Microkinetics of Heterogeneous Catalysis*, Am. Chem. Soc., Washington, DC, 1993.
42. Yaluris, G.; Rekoske, J. E.; Aparicio, L. M.; Madon, R. J.; Dumesic, J. A., *J. Catal.*, **1995**, *153*, 54.
43. Yaluris, G.; Madon, R. J.; Dumesic, J. A., *J. Catal.*, **1999**, *186*, 134.
44. Baltanas, M. A.; Raemdonck, K. K. V.; Froment, G. F.; Mohedas, S. R., *I&EC Res.*, **1989**, *28*, 899.
45. Feng W.; Vynckier, E.; Froment G. F., *I&EC Res.*, **1993**, *32*, 2997.
46. Svoboda, G. D.; Vynckier, E.; Debrabandere, B.; Froment, G. F., *I&EC Res.*, **1995**, *34*, 3793-3800.
47. Vynckier, E.; Froment G. F., Modeling of Kinetics of Complex Processes Based Upon Elementary Steps. In *Kinetic and Thermodynamic Lumping of Multicomponent Mixtures*, Astarita, G. A.; Sandler, S. I. (Eds.), Elsevier, 1991.
48. Benson, S. W., *Thermochemical Kinetics*, 2nd. ed., John Wiley, 1976.
49. Laidler, K. J., *Chemical Kinetics,* Harper & Row: New York, 1987.
50. Kazansky, V. B.; Senchemya, I. N., *J. Catal.*, **1989**, *119*, 108-120.
51. Dewachtere, N. V.; Santaella, F.; Froment, G. F., *Chem. Eng. Sci.*, **1999**, *54*, 3653-3660.
52. Fake, D. M.; Nigam, A.; Klein, M. T., *Appl. Catal.*, **1997**, *160*, 191-221.
53. Quann, R. J.; Jaffe, S. B., *Chem., Eng. Sci.*, **1996**, *51*, 1615
54. Li, B. Z.; Ho, T. C., *Chem. Eng. Sci.*, **1991**, *46*, 273-280.
55. Ho, T. C., *J. Catal.*, **1991**, *129*, 524.
56. Christensen, G.; Apelian, M. R.; Hickey, K. J.; Jaffe, S. B., *Chem. Eng. Sci.*, **1999**, *54*, 2753- 2764.
57. Joshi, P. V.; Lyer, S. D.; Klein, M. T., *Rev. Process. Chem. Eng.*, **1998**, *1*, 111-140.
58. Joshi, P. V.; Klein, M. T.; Huebner, A. L.; Leyerle, R. W., *Rev. Process. Chem. Eng.*, **1999**, *2*, 169-193.
59. Korra, S.; Neurock, M.; Klein, M. T.; Quann, R. J., *Chem. Eng. Sci.*, **1994**, *49*, 4191-4210.
60. Neurock, M.; Klein, M. T., *Polycyclic Aromat. Compd.*, **1993**, *3*, 231-46.
61. Mochida, I.; Yoneda, Y., *J. Catal.,* **1967**, *7*, 223-230.
62. Mochida, I.; Yoneda, Y., *J. Catal.*, **1967**, *7*, 386-392.
63. Mochida, I.; Yoneda, Y., *J. Catal.*, **1967**, *7*, 393-396.
64. Nigam, A.; Klein, M.T., *Ind. Eng. Chem. Res.*,**1993**, *32*, 1297-303.
65. Watson, B. A.; Klein, M. T.; Harding, R. H., *Appl. Catal.*, **1997**, *160*, 13-39.
66. Ho, T. C.; Katritzky, A. R.; Cato, S. J., *I&EC Res.*, **1992**, *31*, 1589-1597.
67. Ho, T. C., *Appl. Catal, - A: General*, **2003**, *244*, 115-128.
68. Sowerby, B.; Becker, S. J.; Belcher, L. J., *J Catal.*, **1996**, *161*, 377.
69. Sowerby, B.; Becker, S. J., Modeling Catalytic Cracking Kinetics Using Estimated Adsorption Equilibrium Constants, In *Dynamics of Surfaces and Reaction Kinetics in Heterogeneous Catalysts*, Froment, G. F.; Waugh, K. C. (Eds.), Elsevier, 1997.
70. Liguras, D. K.; Allen, D. T., *Ind. Eng. Chem. Res.*, **1989**, *28*, 665.
71. Liguras, D. K.; Allen, D. T., *Ind. Eng. Chem. Res.*, **1989**, *28*, 674.

72. Allen, D. T., Structural Models of Catalytic Cracking Chemistry, In *Kinetic and Thermodynamic Lumping of Multicomponent Mixtures,* Astarita, G.; Sandler, S. I. (Eds.), Elsevier: Amsterdam, 1991.

73. Nigam, A.; Neurock, M.; Klein, M. T., Reconciliation of Molecular Detail and Lumping: An Asphaltene Thermolysis Example, In *Kinetic and Thermodynamic Lumping of Multicomponent Mixtures,* Astarita, G.; Sandler, S. I. (Eds.), Elsevier: Amsterdam, 1991.

74. Klein, M. T.; Neurock, M.; Nigam, A.; Libanati, C., Monte Carlo Modeling of Complex Reaction Systems: An Asphaltene Example, In *Chemical Reactions in Complex Mixtures,* Sapre, A. V.; Krambeck, F. J. (Eds.), Van Nostrand Reinhold: N.Y., 1991.

75. Trauth, D. M.; Stark, S. M.; Petti's, T. F.; Neurock, M.; Klein, M. T., *Energy Fuels,* **1994**, *8*, 576-88.

76. Gray, M. R., *I &EC Res.*, **1990**, *29*, 505-512.

77. Freund, H.; Olmstead, W. N., *Int. J. Chem. Kinet.*, **1989,** *21*, 561-574.

78. Rabitz, H., Systems Analysis at the Molecular Scale, *Science*, **1989**, *246*, 221.

79. Yetter, R.; Dryer, F.; Rabitz, H., *Combust. Flames*, **1985**, *59*, 107.

80. Hwang, J. T., *Int. J. Chem. Kinet.*, **1983**, *15*, 959.

81. Sanchez-Castillo, M. A.; Agarwal, N.; Miller, C.; Cortright, R. D.; Madon, R. J., Dumesic, J. A., *J. Catal.*, **2002**, *205*, 67-85.

82. Kramer, M. A.; Kee, R. J.; Rabitz, H., CHEMSEN: A Computer Code for Sensitivity Analysis Elementary Chemical Reaction Models. SAND82-8230, Sandia National Labs, Livermore, 1982.

83. Stewart, G. W., *Introduction to Matrix Computation,* Academic Press: NY, 1973

84. Wei, J.; Kuo, J. C. A., *Ind. Eng. Chem.*, **1969**, *8*, 114.

85. Li, G.; Rabitz, H., *Chem. Eng. Sci.*, **1989**, *44*, 1413.

86. Li, G.; Rabitz, H., *Chem. Eng. Sci.*, **1991**, *48*, 1903-1909.

87. Kuo, J.C.; Wei, J., *I &EC Fundam.*, **1969**, *8*, 124-133.

88. Liao, J. C.; Lightfoot, E. N., *Biotechnol. Bioeng.*, **1988**, *31*, 869-879

89. Li, G., *Chem. Eng. Sci.*, **1984**, *39*, 1261

90. Li, G.; Rabitz, H., *Chem. Eng. Sci.*, **1990**, *45*, 977.

91. Li, G.; Rabitz, H., *Chem. Eng. Sci.*, **1991**, *46*, 95.

92. Li, G.; Rabitz, H., *Chem. Eng. Sci.*, **1991**, *46*, 2041.

93. Li, G.; Rabitz, H.;Toth, J., *Chem. Eng. Sci.*, **1994**, *49*, 343-361

94. Li, G.; Tomlin, A. S.; Rabitz, H.; Toth, J., *J. Chem. Phys.*, 1994, *101*, 1172-1187.

95. Sharaf, M. A.; Illman, D. L.; Kowalski, B. R., *Chemometrics,* John Wiley, 1986.

96. Vajda, S.; Valko, P.; Turanti, T., *Int. J. Chem. Kinet.*, **1985**, *17*, 55.

97. Wold, S.; Geladi, P.; Esbensen, K.; Ohman J., *J. Chemom.*, **1987**, *1*, 41

98. Davis, J. F.; Plovuso, M. J.; Hoo, K. A.; Bakshi, B. R., Process Data Analysis and Interpretation, In *Adv. Chem. Eng. 25*, Academic Press, 2000; pp 2-102.

99. Liu, Y. A.; Lapidus, L., *AIChE J.*, **1973**, *19*, 467.

100. Coxson, P. G., *J. Math. Anal. Appl.*, **1984**, *99*, 435-446.

101. Lam, S. H.; Goussis, D. A., *Int. J. Chem. Kinet.*, **1994**, *26*, 441-486

102. Maas, U.; Pope, S. B., *Combust. Flame*, **1992**, *88*, 239-264.

103. Androulakis, I. P., *AIChE. J.*, **2000**, *46*, 361.

104. Androulakis, I. P., *AIChE. J.*, **2000**, *46*, 1769.

105. DeDonder, Th., Gauthiers-Villars: Paris, 1931.

106. Aris, R.; Gavalas, G. R., *Phil. Trans. Roy. Soc.,***1966**, *A260*, 351.

107. Aris, R., *Arch. Ratl. Mech. Anal.*, **1968**, *27*, 356.

108. Luss, D.; Hutchinson, P., *Chem. Eng. J.*, **1971**, *2*, 172.

109. Ho, T. C.; Aris, R., *AIChE J.*, **1987**, *33*, 1050.

110. Astarita, G.; Ocone, R., *AIChE J.*, **1988**, *34*, 1299.

111. Chou, M. Y.; Ho, T. C., *AIChE J.*, **1988**, *34*, 1519

112. Aris, R., *AIChEJ.*, **1989**, *35*, 539.
113. Krambeck, F. J., *Chem. Eng. Sci.*, **1994**, *49*, 4179-4189.
114. Ocone, R.; Astarita, G., Kinetics and Thermodynamics in Multicomponent Mixtures, In *Adv. Chem. Eng. 24*, 1-77, Academic Press: N.Y., 1998.
115. Yitzhaki, D.; Aharoni, C., *AIChE J.*, **1977**, *23*, 342.
116. Sau, M., C.; Narasimhan, S. L.; Verma, R. P., A Kinetic Model for Hydrodesulfurization, In *Studies in Surface Science and Catalysis*, Froment, G. F.; Delmon, B.; Grange, P. (Eds.), Elsevier, 1997.
117. Stephan, R.; Emic, G.; Hoffman, H., *Chem. Eng. Process.*, **1985**, *19*, 303-315.
118. Narasimhan, C. S. L.; Sau, M.; Verma, R. P., An Integrated Approach for Hydrocracker Modeling, In *Studies in Surface Science and Catalysis*, Froment, G. F.; Delmon, B.; Grange, P. (Eds.), Elsevier, 1997.
119. Harris, C. C.; Chakravarti, A., *Trans. of SME*, **1970**, *247*, 162.
120. Boudreau, B. P.; B. R. *Am. J. Sci.*, **1991**, *291*, 507-538.
121. Ho, T. C.; Chianelli, R. R.; Jacobson, A., *J. Appl. Catal.*, **1994**, *114*, 131.
122. Girgis, M. J.; Gates, B. C., *I&EC Res.*, **1991**, *30*, 2021.
123. Ammus, J. M., Androutsopoulos, G. P., *I&EC Res.*, **1987**, *25*, 494-501.
124. Sie, S. T., *Fuel Process. Technol.*, **1999**, *61*, 149-171.
125. Sonnemans, J. W. M., Hydrotreating of Cracked Feedstocks. Ketjen Catalyst Symposium, 1982.
126. Beuther, H.; Schmid, B. K., 6th World Petroleum Congress Section III, paper 20, PD7, 1963.
127. Heck, R. H.; Stein T. R., *ACS Symp., Ser.* **1977**, *22*, 948-961.
128. Ho, T. C., *Catal. Rev.-Sci. Eng.*, **1988**, *30*, 117.
129. Ozaki, H.; Satomi, Y.; Hisamitsu, T., *Proc. 9th World Pet. Cong.*, **1976**, 6 PD, *18 (4)*, 97.
130. Hensley, A. L., Jr.; Tait, A. M.; Miller, J. T.; Nevitt, T. D., US Patent 4,406,779, 1983.
131. Inoue, S.; Takatsuka, T.; Wada, Y.; Hirohama, S.; Ushida, T., *Fuel*, **2000**, *79*, 843.
132. Breysse, M.; Djega-Mariadassou, G.; Pessayre, S.; Geantet, C.; Vrinat, M.; Pérot, G.; Lemaire, M.; *Catal. Today*, **2003**, *84*, 129-138.
133. Venuto, P. B.; Habob, E. T., Fluid Catalytic Cracking with Zeolite Catalysts, Marcel Dekker: N.Y., 1978.
134. Trytten, L. C.; Gray, M. R.; Sanford, E. C., *I&EC Res.*, **1990**, *29*, 725-730.
135. Scott, K. F., *J. C. S. Faraday I*, **1980**, *76*, 2065-2079.
136. De Pontes, M.; Yokomizo, G. H.; Bell, A. T., *J. Catal.*, **1987**, *104*, 147-155.
137. Krambeck, F. J., ISCRE 8; I. *Chem. Eng. Symp. Ser.*, **1984**, *A260*, 351.
138. Ho, T. C.; White, B. S.; Hu, R., *AIChE J.*, **1990**, *36*, 685.
139. Ho, T. C.; White, B. S., *AIChEJ.*, **1995**, *41*, 1513.
140. Ma, X.; Sakanishi, K.; Isoda, T.; Mochida, I., *I&EC Res.*, **1995**, *34*, 748-754.
141. Ho, T. C., *AIChE J.*, **1996**, *42*, 214.
142. Rangwala, H. A.; Wanke, S. E.; Otto, F. D.; Dalla Lana, I. G., *Prepr. 10th Symp. Catal*, Kinston, Ont., June 15-18, 1986.
143. Van Dongen, R. H.; Bode, D.; van Der Eijk, H.; van Klinken, J., *I&EC Proc. Res. Dev.*, **1980**, *19*, 630.
144. Luss, D.; Golikeri, S. V., *AIChE J.*, **1975**, *21*, 865.
145. Grosselink, J. W.; Stork, W. H., *J. Chem. Eng. Process.*, **1987**, *22*, 157-62.
146. Houalla, M.; Broderick, D. H.; Sapre, A. V.; N. K.; deBeer, V. H. J.; Gates, B. C.; Kwart, H., *J. Catal.*, **1980,** *61*, 523.
147. Wilson, M. F.; Fisher, I. P. Kriz, J. F., *J. Cat.*, **1985**, *95*, 155.
148. Chen, J. M.; Schindler, H. D., *I&EC Res.*, **1987**, *26*, 921.

149. Aris, R., Multiple Indices, Simple Lumps, and Duplicitous Kinetics. In *Chemical Reactions in Complex Mixtures, The Mobil Workshop*, Sapre, A. V.; Krambeck, F. J. (Eds.), Van Nostrand Reinhold, 1991.
150. Stanislaus, A.; Cooper, B. H., *Catal. Rev. - Sci. Eng.*, **1994**, *36*, 75-123.
151. Shabtai, J.; Yeh, G. J. C., *Fuel*, **1988**, *67*, 314-20.
152. Wei. J.; Prater, C. D., *Advances in Catalysis*, Vol. 3, 1962.
153. Astarita, G., *AIChE J.*, **1989**, *35*, 529.
154. Chou, M. Y.; Ho, T. C., *AIChE J.*, **1989**, *35*, 533.
155. Aris, R. The Mathematics of Continuous Mixtures, In *Kinetics and Thermodynamic Lumping of Multicomponent Mixtures*, Astarita, G. A.; Sandler, S. I. (Eds.), Elsevier: Amsterdam, 1991.
156. Astarita, G.; Nigam, A., *AIChE J.*, **1989**, *35*, 1927.
157. Ho, T. C., *Chem. Eng. Sci.*, **1991**, *46*, 281.
158. Hutchinson, P.; Luss, D., *Chem. Eng. J.*, **1970**, 1, 129.
159. Golikeri, S. V.; Luss, D., *Chem. Eng. Sci.*, **1971**, *26*, 237.
160. Ocone, R., Astarita, A., *AIChE J.*, **1993**, *39*, 288.
161. Ho. T. C.; Li, B. Z.; Wu, J. H., *Chem. Eng. Sci.*, **1995**, *50*, 2459.
162. Golikeri S. V.; Luss, D., *AIChE J*, **1972**, *18*, 277.
163. Pachovsky, R. A.; Wojciechowski, B. W., *J. Catal.*, **1975**, *37*, 120.
164. Pachovsky, R. A.; Wojciechowski, B. W., *J. Catal.*, **1975**, *37*, 358.
165. Köseoglu, R. O.; Phillips, C. R., *Fuel*, 1988, *66*, 741.
166. Köseoglu, R. O.; Phillips, C. R., *Fuel*, 1988, *67*, 906, 1411.
167. Anthony, D. B. Howard, J. B., *AIChE J.*, **1976**, *22*, 625.
168. Schucker R. C., *ACS Symp. Div., Fuel. Chem.*, **1982**, *27*, 214.
169. Braun, R. L.; Burnham, A. K., *Energy Fuels*, **1987**, *1*, 153.
170. Lin, L. C.; Deo, M. D.; Hanson, F. V.; Oblad, A. G., *AIChEJ.*, **1990**, *36*, 1585.
171. Laximinarasimhan, C. S.; Verma, R. P.; Ramachandran, P. A., *AIChEJ.*, **1993**, *42*, 2645.
172. Bennett, R. N.; Bourne, K. H., ACS Symp. Advances Distillate and Residual Oil Technology, New York, 1972.
173. El-Kardy; F. Y., *Indian J. Technol.*, **1979**, *17*, 176.
174. Cicarelli, P.; Astarita, G.; Gallifuocco, A., *AIChE J.*, **1992**, *38*, 7.
175. Browarzik, D.; Kehlen, H., *Chem. Eng. Sci.*, **1994**, *49*, 923.
176. Teymour, F.; Campbell, J. D., *Macromolecules*, **1994**, *27*, 2460-2469.
177. Wang, M.; Smith J. M.; McCoy, B. J., *AIChEJ.*, **1995**, *41*, 1521-1533.
178. Kodera, Y.; McCoy, B. J., *AIChE. J.*, **1997**, *43*, 3205-3214.
179. Kruse, T. M.; Woo, O. S.; Broadbelt, L. J., *Chem. Eng. Sci.*, **2001**, *56*, 971-979.
180. Golikeri, S. V.; Luss, D., *Chem. Eng. Sci.*, **1974**, *29*, 845.
181. Luss, D.; Golikeri, S. V., *AIChE J.*, **1975**, *21*, 865.
182. Astarita, G., Chemical Reaction Engineering of Multicomponent Mixtures: Open Problems, In *Kinetics and Thermodynamic Lumping of Multicomponent Mixtures*, Astarita, G. A.; Sandler, S. I. (Eds.), Elsevier: Amsterdam, 1991.
183. Li, B. Z.; Ho, T. C., An Analysis of Lumpting Bimolecular Reactions, In *Lumping Kinetics and Thermodynamics*, Astarita, G. A.; Sandler, S. I. (Eds.), Elsevier: Amsterdam, 1991.
184. White, B. S.; Ho, T. C.; Li, H. Y., *Chem Eng. Sci.*, **1994**, *49*, 781.

Chapter 22

ADVANCED PROCESS CONTROL

Paul R. Robinson[1] and Dennis Cima[2]
1. *PQ Optimization Services, 3418 Clear Water Park Drive, Katy, TX 77450*
2. *Aspen Technology, Inc., 1293 Eldridge Parkway, Houston, TX 77077*

1.   INTRODUCTION

At relatively low cost, model-predictive control improves the capability of process units by increasing throughput, improving fractionator performance, decreasing product quality giveaway, reducing operating costs, and stabilizing operations. Figure 1 presents an overview of the scope of refinery software applications and indicates the frequency at which they typically run.

2.   USEFUL DEFINITIONS

A *proportional-integral-derivative* (PID) controller is a feedback device. A *process value* (PV) is measured by a field transmitter. A controller compares the PV to its *setpoint* (SP) and calculates the required change – for example, a new *valve opening position* (OP) – to bring the PV closer to the SP. The required change is calculated with a PID algorithm. In practice, proportional, integral, and derivative constants are parameters used for tuning.

Advanced process control (APC) applications involve the use of control algorithms to provide improved process control when compared to regulatory PID controllers in loops or cascades.

Traditional Advanced Control (TAC) employs the use of advanced control algorithms combined with regulatory control functions (i.e., lead/lag, ratio, high/low selectors, etc) to implement a control strategy.

A *programmable logic controller* (PLC) is a small computer used to automate real-world processes. A PLC receives input from various sensors and responds to changes by manipulating actuator valves according to pre-programmed logic stored in the memory of the PLC.

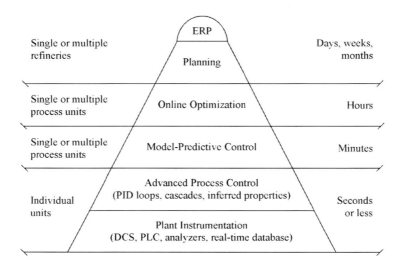

Figure 1. Overview of computer control applications. ERP stands for enterprise resource planning, a software solution that integrates planning, manufacturing, sales, and marketing.

In a ***distributed control system*** (DCS), process measurements and control functions such as multiple PID loops are connected to application processors, which are networked throughout the plant. A ***graphical user interface*** (GUI) makes it easier for operators to view data, create plots, change setpoints, and respond to alarms. In addition to process control, modern DCS software includes sophisticated trending and data storage.

Multivariable process control (MVPC) applications use one or more independent variables to control two or more dependant variables. If a variable is truly independent, its value is not affected by other process variables. There are two types of independent variables. ***Manipulated variables*** (MV) can be changed by an operator to control the process. These include setpoints for regulatory PID controllers and valve positions. ***Feed-forward*** (FF) and ***disturbance variables*** (DV) affect the process but can't be manipulated. These include ambient temperature, the quality of an external feed, etc.

The value of a dependent variable can be calculated using the values of independent variables and an appropriate dynamic model. ***Controlled variables*** (CV) are maintained at a desired steady-state target. Constraint variables are maintained between high and low limits. Many variables are dependent, but not all dependent variables are important enough to be controlled by the APC application.

Model-predictive control (MPC) uses process models derived from past process behavior to predict future process behaviour.[1] The predictions are used to control process units dynamically at optimum steady-state targets. MPC applications may also include the use of predicted product properties (***inferential analyzers***) and certain process calculations. Model-predictive controllers almost always include multiple independent variables.

In this chapter, when we say "model-predictive control" we mean constrained, multivariable, model-predictive control, which is arguably the predominant type of "advanced control" application in the petroleum refining industry.

3. OVERVIEW OF ECONOMICS

Table 1 presents typical benefits for applying model-predictive control to refinery units.[2] After installation of the requisite infrastructure – distributed control system, analyzers, operator terminals, process computer(s) – model-predictive control projects on major refinery units can be completed within 2 to 4 months. The return on investment is quite high; typical payback times usually are 4 to 12 months.

Table 1. Typical Benefits of Multivariable Model-Predictive Control

Process Unit	Typical Benefits $U.S./bbl	Source(s)
Crude distillation	0.015 to 0.03	Higher feed rate
		Reduced product quality giveaway*
		Increased energy efficiency
Fluid Catalytic Cracking	0.15 to 0.30	Higher feed rate
		Reduced ΔP across slide valves
		Reduced product quality giveaway*
		Increased energy efficiency
Catalytic reformer	0.10 to 0.20	Higher feed rate
		Maximum coke on catalyst (CCR units)
		Reduced product quality giveaway*
		Increased energy efficiency
Hydrocracker	0.10 to 0.20	Higher feed rate
		Improved control of gas/oil ratio
		Increased T and P stability
		Reduced product quality giveaway*
		Increased energy efficiency
Gas Oil Hydrotreater	0.02 to 0.10	Higher feed rate
		Improved control of gas/oil ratio
		Increased T and P stability
		Reduced product quality giveaway*
		Increased energy efficiency
Gas plant	0.05 to 0.10	Higher feed rate
		Reduced product quality giveaway*
		Increased energy efficiency
Product blending	0.10 to 0.20	Reduced product quality giveaway*

*Equivalent to increased production of desired products

The applications that benefit most from model-predictive control have one or more of the following characteristics:

- High production capacity
- Competing control objectives
- Highly interactive processes
- Unusual dynamic behavior
- Day/night or seasonal variation
- A need to operate close to constraints
- A need to closely track optimization system targets
- A need to transition smoothly from one set of targets to another

In refineries, the largest benefits of model-predictive control come from crude distillation units and gasoline blenders, for which the throughput is high, and from fluid catalytic cracking (FCC) and other conversion units, for which the difference in value between feeds and products usually is high.

4.   SOURCES OF BENEFITS

The benefits of APC and model-predictive control come from one or more of the following:

- Reduced process variability
- Maximizing throughput against multiple process constraints
- Increased yield of high-value products
- Reduced product quality giveaway
- Reduced production losses[3]

These benefits accrue from two sources – increased process capability due to reduced variability, and constraint pushing.

Variability is a characteristic of all continuous processes. As shown in Figure 2, simply reducing variability provides little (if any) benefit. Benefits start to accumulate when operators run the plant closer to true process constraints.

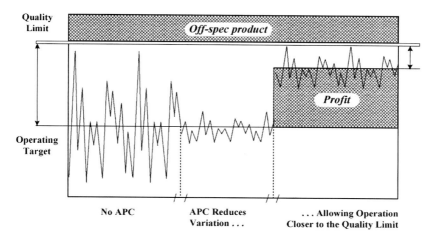

Figure 2. Reduced variation: one source of APC benefits

For example, if a low-pressure hydrotreater must produce off-road diesel fuel with ≤500 wppm sulfur to meet product specifications, and if there is a severe economic penalty for exceeding the specification, the refiner may set a process target of 350 wppm to ensure that the plant never exceeds the limit. In this case, the difference between the target and the limit – i.e., the cushion – is 150 wppm. A cushion of this size is not atypical for coastal refineries where feed quality and/or hydrogen composition can change significantly from one day to the next, and where there is limited ability to correct a specification violation with blending. But a 30% cushion is expensive, resulting in higher operating cost and accelerated catalyst deactivation. If increased stability allows the process target to be raised from 350 wppm to 450 wppm, hydrogen consumption will decrease, heater firing will decrease and catalyst cycle life will increase significantly.

Figure 3 illustrates how model-predictive control can achieve additional benefits by pushing against process constraints. Plant operators tend to run complex units within a certain comfort range. This comfort range gives an operator extra time or cushion to respond to process changes.

An APC application responds quickly to process changes. A well-tuned model-predictive control application can run outside the comfort range of a human operator while pushing simultaneously against multiple process constraints. More significantly, a model-predictive control application can calculate moves for each MV every minute, which a plant operator cannot. Special techniques, such as move suppression, are used to prevent the plant from moving too far too fast.

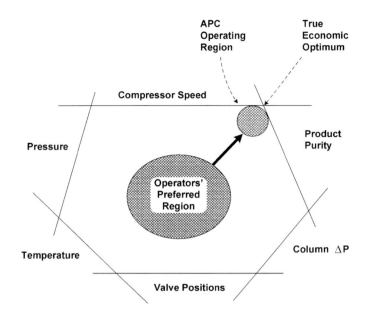

Figure 3. Operating near the right constraints: major source of APC benefits

In certain special circumstances – emergencies, startups, and shutdowns – model-predictive control cannot be used. It is more realistic to say that well-tuned, well-maintained model-predictive control application can emulate the plant's best operator – every minute of every day. Figure 4 compares the temperature response to a soot-blowing disturbance under three types of control. The solid line shows the open-loop (manual) response. The heavily dotted line shows better response with a PID (feedback) controller. The lightly dotted line shows superb response with model-predictive control.

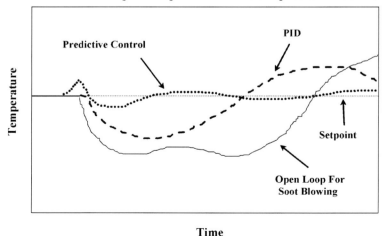

Figure 4. Comparison between open-loop, PID and model-predictive control

5. IMPLEMENTATION

The implementation of model-predictive control requires four main steps – plant response testing, model analysis, commissioning, and training.

During the plant response testing phase of a project, independent variables are moved and dependant variable responses are captured electronically. Obtaining good plant test data from which accurate models can be regressed is the key to a successful model-predictive control project. For this reason, special care must be taken to ensure that the underlying instrumentation – meters, analyzers, and PID controllers – are properly tuned and calibrated. Traditionally, independent variables were moved manually during the plant response test, but with recently developed software, an engineer can obtain equivalent plant test data using closed loop testing methods. Proprietary identification software converts response-test data into dynamic models for the plant. Response-test models can be used to predict future plant behavior with the following control equation:

$$\delta CV = [A] \times \Delta I$$

where δCV is the predicted change in a given CV, $[A]$ is the gain matrix obtained during the model analysis phase, and ΔI is a matrix of independent variable changes. Figure 5 shows an example of the control equation in matrix form.

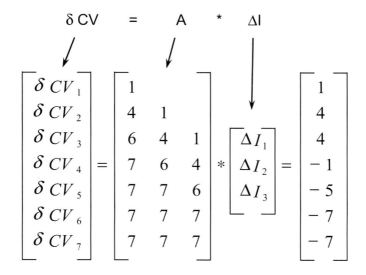

Figure 5. Example of the control equation in matrix form

Model predictions are used to control the plant against constraints, as shown in Figure 3. This is not a trivial matter, because the application must cope successfully with the following:

- Plant/model mismatch
- Instrument failure
- Unmeasured Disturbances
- Input data error
- Diverse process dynamics
- Changing process objectives

Despite these challenges, when implemented by qualified personnel, model-predictive control applications provide considerable value for refiners throughout the world.

6.   COSTS

Figure 6 illustrates the infrastructure required for computer control. APC and model-predictive control software usually runs on a separate process computer, which uses a "data highway" to communicate with the DCS, the laboratory information management system (LIMS), and a real-time database. Advanced applications receive process values from the DCS, calculate the sizes of MV moves, and send setpoints back to the DCS.

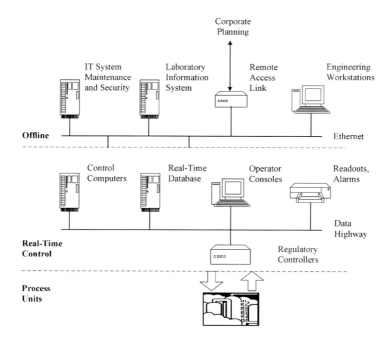

Figure 6. Automation infrastructure

Figure 7 compares the relative costs and benefits of computer applications. As is the case for a personal computer, hardware accounts for most of the cost, but software provides most of the benefit.

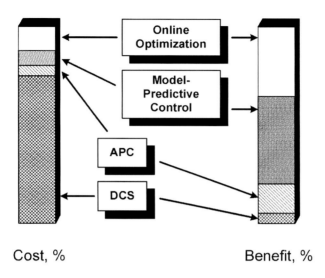

Cost, %   Benefit, %

Figure 7. Relative costs and benefits of computer applications in oil refineries

7. REFERENCES

1. Cutler, C.R.; Ramaker, B.L. "Dynamic Matrix Control – A Computer Algorithm," AIChE National Meeting, Houston, TX, April 1979.
2. Richard, L.A.; Spencer, M.; Schuster, R.; Tuppinger, D.M.; Wilmsen, W.F., "Austrian Refinery Benefits from Advanced Control," Oil Gas J., March 20, 1995.
3. Vermeer, P.J.; Pedersen, C.C.; Canney, W.M.; Ayala, J.S., Blend Control System All But Eliminates Reblends for Canadian Refiner, Oil Gas J., 95(30), July 28, 1997.

Chapter 23

REFINERY-WIDE OPTIMIZATION WITH RIGOROUS MODELS

Dale R. Mudt,[1] Clifford C. Pedersen,[1] Maurice D. Jett,[2] Sriganesh Karur,[2] Blaine McIntyre,[3] and Paul R. Robinson[4]
1. Suncor, Inc, Sarnia, Ontario, Canada
2. Aspen Technology Inc., Houston, Texas
3. AspenTech Canada Ltd., Calgary, Alberta, Canada
4. PQ Optimization Services, Katy, Texas

1.   INTRODUCTION

One reason for writing this chapter is to report the success of real-time, online refinery-wide optimization (RWO) at Suncor-Sarnia using rigorous process models. Another reason is to show how the same rigorous models can be used offline to quantify key non-linear relationships during the evaluation of project ideas, especially those related to the production of clean fuels.

2.   OVERVIEW OF SUNCOR

Suncor operates a 70,000 barrels-per-day refinery at Sarnia, Ontario, Canada. The refinery processes of feeds from the following sources:
–   Synthetic crude oil from Suncor's oils sands processing plant at Fort McMurray, Alberta, Canada
–   Conventional crude oil
–   Condensate
–   VGO from a nearby refinery

Figure 1 shows an outline of the plant, which started up in 1953. Synthetic crude and conventional crude oil come to the refinery through the Interprovincial Pipeline, which runs from Edmonton, Alberta to Sarnia,

Ontario. The transit time is about one month. Other crudes are also available from nearby facilities.

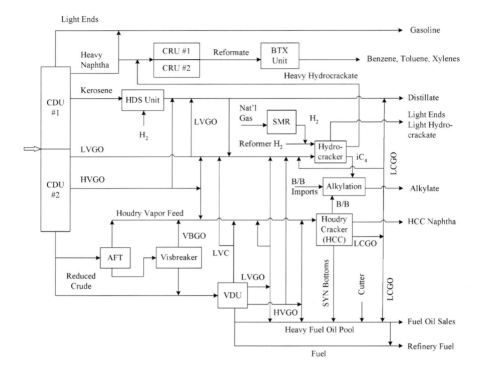

Figure 1. Overview of the Suncor Refinery in Sarnia, Ontario

The refinery includes the following major process units:
− Crude distillation (CDU) — 2 units
− Vacuum distillation (VDU)
− Houdriflow catalytic cracking unit (HCC)
− Catalytic reforming (CRU) — 2 units
− Alkylation unit
− Aromatics recovery unit (BTX)
− Unsaturated gas plant
− Saturated gas plant
− Naphtha hydrotreater
− Diesel / gas oil hydrotreater
− Hydrocracker complex

3.   REFINERY-WIDE OPTIMIZATION (RWO)

The refinery-wide optimization (RWO) initiative began in July 1998. Three phases were defined for the project:

Phase 1.   Build a solid and sustainable advanced control and optimization foundation

Phase 2.   Ensure model consistency between planning, scheduling, and real-time optimization

Phase 3.   Integrate planning, scheduling, and real-time optimization by linking the optimization models to the refinery LP

Expected benefits from the RWO initiative included:
−   Increased refinery throughput
−   Improved refinery yields
−   Improved gas oil management
−   Improved hydrogen management
−   Optimum product recoveries
−   Reduced energy costs
−   Increased ability to respond to market changes

Rigorous models for stand-alone units also can provide significant benefits. Previously,[1] we reported benefits of US$3,000 per day (US$0.15 per barrel) for the initial optimizer on the hydrocracking complex; these benefits were in addition to those provided by model-predictive DMC control. For RWO, a revised model based on Aspen Hydrocracker (AHYC) was developed. It includes a catalyst deactivation block, which enhances maintenance turnaround planning by predicting future catalyst activity, product yields and product properties for a variety of assumed feeds and specified operating conditions. This information also is used to impose constraints on present-day operation.

The RWO initiative was driven in part by the need to upgrade or replace software applications that were not Y2K compliant. In addition, the project reduced application maintenance costs because the same software versions are used throughout the plant, and because modern control and optimization software is more user-friendly than it was in the 1980s, when Suncor first began implementing model-predictive control and online optimization.

Table 1 summarizes the scope of the major RWO applications. A large DMCplus application—64 manipulated variables (MV) x 16 feed forward variables (FF) x 168 controlled variables (CV)—controls Plant 1. For Plant 1, the process scope of the controller is equivalent to that of the optimizer, which sends 34 targets to the controller. In Plant 3, separate controllers are used for the hydrogen plant, the HYC, the HYC gas plant, and low-sulfur diesel (LSD) tower, but a single optimizer sends targets to these controllers.

Table 1. Scope of Refinery-Wide Optimization Applications

	Scope
Plant 1	
Crude and HCC	Preheat Train, Crude Heaters, Crude Tower, Atmospheric Flash Tower, Syn Tower, HCC Heaters, Reactor, Kiln, Steam Coils, Air Blower, Cyclone, Main Fractionator, Unsaturated Gas Plant
Plant 2	
Crude and Vacuum Reformer 2 and BTX	Preheat Train, Heaters, Atmospheric Tower, Vacuum Tower Heaters, Compressors, Pretreater, Reactors, Flash Drums, Depropanizer, Naphtha Splitter, Absorber/Deethanizer, BTX Complex
Plant 3	
SMR, HYC, Gas Plant	H_2 Plant, Hyodrocracker Reactor Section, Preflash, Main Tower, Absorber/Stripper, Debutanizer, Jet Tower, Diesel Tower, Saturated Gas Plant

The RWO initiative is using the hardware and software in Table 2.

Table 2. Hardware and Software for RWO

Item	Vendor
DCS Systems	Honeywell TDC 2000/3000, Foxboro I/A
Online Computers	DEC Alpha (VMS)
Offline Computers	Compaq (Windows NT)
Real-Time Database	Aspen Infoplus, Honeywell PHD
Control Software	DMCplus, Aspen IQ
Optimization Software	Aspen RT-Opt
Refinery LP	Haverly GRTMPS

Each optimizer includes an offline version, which is periodically updated with live plant data. Offline models are used for tuning and monitoring the performance of the optimizers. They can also be used for maintenance planning and design studies.

The success of RWO depended heavily upon proper integration of the several separate applications. Integration points fell into three categories (Table 3). Intermediate tanks presented special problems in part because they tend to stratify, but mostly because they complicate time-scale issues. For several critical tanks, open-equation models calculate tank compositions on a regular basis—about every 5 minutes—to provide feed-forward inputs to other optimizers or possibly DMCplus controllers. Special logic is used to compensate for stratification. To date, these tank composition models are performing well.

Table 3. Integration Point categories

Type of Integration Point	Time Scale
Feed forward information required by the model predictive controllers	Minutes
Steady state information (inputs to and outputs from) the online optimizers	Hours
Higher level information (inputs to and outputs from) the refinery LP	Daily

Other key integration points are shown in Table 4:

Table 4. Key Integration Points

Plants 2 (H$_2$ from reformer)	=>	Plant 3 (HYC)
Plants 1, 2, and 3 (reformates and HYC light naphtha)	=>	Gasoline blender
Plant 3 (HYC heavy naphtha)	=>	Plant 2 (Reformer 2)
Plant 3 (isobutane)	=>	Plant 1 (alky feed)

Integrating the real-time optimizers into Suncor's planning and scheduling methodology will require consistency between the optimizers and the refinery LP in the following areas:

- Yields and product qualities
- Degrees of freedom
- Constraints

At present, data are transferred manually between the LP and the optimizers. This is beneficial because the models allow us to increase the accuracy of LP shift vectors. A recent publication[2] describes the benefits of running steady-state models in recursion with Aspen PIMS, a widely used LP program. This method may offer the best of both worlds—the practicality of LP technology augmented by the rigor of non-linear, unit-specific models.

During RWO Phase 3, the transfer of data will be automated, section by section, as we gain confidence in the robustness of the individual optimizers, and in the fidelity of the LP solutions.

4. RIGOROUS MODELS FOR CLEAN FUELS

As mentioned above, one purpose of this paper is to show how rigorous models can be used offline to quantify key non-linear process relationships, including those related to manufacturing clean fuels.

At most North American refineries, including Suncor-Sarnia, most of the sulfur in the gasoline pool comes from a catalytic cracking unit. To make low-sulfur gasoline, an Axens Prime-G unit was installed to desulfurize gasoline from the HCC and visbreaker units. Another option would have been to pretreat all of the HCC feed.[3] (Already, hydrocracker bottoms comprise some of the HCC feed, but only a minor fraction.)

To summarize: our RWO experience has demonstrated the following:

- Rigorous optimizers provide significant economic benefits
- Open-equation optimizers can be linked, and linked optimizers can successfully represent non-linear interactions between process units.
- AFCC (modified) and AHYC provide high-fidelity representations of Suncor's HCC and hydrocracker, respectively.

–   Therefore, we can conclude that linking AHYC with AFCC will enhance the industry's understanding of FCC feed pretreating, and eventually to operate FCC-plus-pretreater installations as a single, optimized complex.

4.1   Feedstock and Product Characterization

For any rigorous model, success requires detailed, accurate feedstock and product characterizations. For the HCC and hydrocacker, GC/MS, ^{1}H NMR, ^{13}C NMR, HPLC, and standard ASTM methods were used to analyze a battery of feedstock and product samples. The analytical results were used to generate base-case distributions for the components used in the models and to tune reaction kinetics.

In the real world, feedstock properties and product slates are changing continually. In AFCC and AHYC, proprietary feed-adjust models skew the base-case component distribution to match the measured properties of feeds and products.

4.2   Aspen FCC Overview

Aspen FCC (AFCC) is an open-equation, flowsheet-based model designed for both online and offline applications. The model is constructed from individual blocks that represent the separate pieces of equipment—riser, standpipe, slide valve, cyclone, transfer line, etc.—found in commercial FCC units. Each building block is generic and can be configured with dimensions corresponding to a given commercial unit (Table 5).

Table 5. Aspen FCC Building Blocks

Riser model: hydrodynamics and kinetics
Reactor-vessel cyclones and plenum model
Reactor dilute-phase model
Reactor dense-bed model
Reactor stripping zone model
Catalyst stand-pipe model
Catalyst slide-valve two-phase flow model
Catalyst transfer-line model
Regenerator dense-bed model: coke burn kinetics
Regenerator freeboard model: coke burn kinetics
Regenerator cyclone model
Feed characterization, component-mapping model

A typical FCC unit configuration is presented in Figure 2. The diagram shows a simple version of the actual blocks used to model the feed, feed adjust and preheat systems.

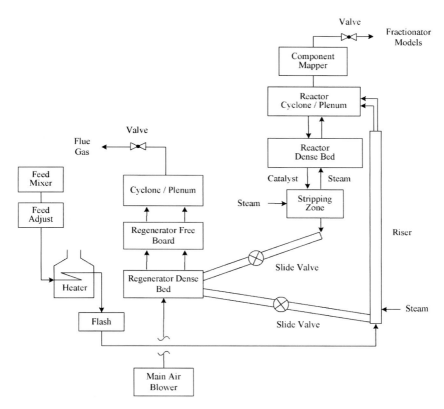

Figure 2. Model Components for Aspen FCC (Riser/Reactor/Regenerator)

In the riser model, detailed hydraulics and heat effects are captured by linking flow equations to kinetics. The riser model can be configured vertically, horizontally, or at any angle of inclination. Multiple risers can be used. The individual risers can process different feeds or the same feed. In essence, any commercial riser/reactor configuration can be simulated with the modular components of AFCC.

Hydrocarbon feed and regenerated catalyst models are connected to the inlet of the Riser model. The catalyst-stream model includes the mass flow, temperature, pressure, enthalpy of the catalyst/coke mixture, particle density of the catalyst/coke mixture, coke-on-catalyst weight fraction and coke composition.

Pressure drop calculations are based on head, acceleration, and frictional effects. Pressure drop through the riser is modeled as a combination of pressure drop due to vapor and pressure drop due to solids. For the vapor, the frictional contribution is based on friction-factor correlations with the Reynolds number. For catalyst flow, frictional contributions are based on correlations for solid-vapor flow in a conduit. Proper prediction of pressure drop along the riser has a significant impact on predicted yields.

The riser model uses vertical or horizontal slip factor correlations (depending on the riser orientation) to determine the differential velocity between the vapor and the solid phases. The slip calculation is critical for determining bulk density profiles, which govern the path dependence of cracking profiles.

As coke is differentially produced along the length of the riser, additional solids are differentially transferred from the vapor to the catalyst surface. The physical effects of this transfer are described by continuity equations. The molar heat of adsorption for coke laydown is included in the heat balance. The heat of coke desorption is calculated in the regenerator model. Slip-factor correlations are based on fully developed flow. The model applies a correction for the turbulent, high-slip zone at the inlet of the riser.

The riser model effluent is connected to the reactor vessel models. These include the reactor dilute phase, the dense phase, and the cyclone system, which represents the segregation of effluent vapors and catalyst. The cyclone model performs a two-phase, loading-based ΔP calculation, for which cyclone inlet and body diameters are used. The cyclone model can be configured with one or two riser inlet ports.

The reactor dilute phase model is identical to the riser model without entry-zone and frictional effects. The reactor dense bed model can be configured to the exact geometry of an operating unit. It is used to represent the inventory of spent catalyst above the stripping zone. For large vessels with significant holdup, this model can be configured as a fluidized bubbling bed system. For low holdup or riser-cracking systems, it is treated as a fluidized bed with a parameterized void fraction.

Spent catalyst and regenerator air are connected to the regenerator model. The regenerator includes three blocks, each of which can be configured for a specific unit. The first block (from the bottom up) is the regenerator dense bed. This model represents the bubbling, fluidized bed where the majority of coke burn occurs. The second block is the regenerator free board (disperse phase), which is the region between the surface of the bubbling bed (dense bed) and the inlet to the cyclones. The third block is the regenerator cyclone model.

The regenerator dense bed is modeled as a bubbling bed with heterogeneous coke burn and CO conversion to CO_2. Bulk density is modeled as a function of bed height and pressure. Promoters are modeled by updating selected coke-burn parameters. Combustion air composition is determined independently by the regenerator air model. Therefore, O_2 enrichment can be simulated and optimized.

The regenerator freeboard model represents the vessel region between the surface of the dense bed and the cyclone inlets. Inlets to the model are entrained, regenerated catalyst from the dense-bed model and the combustion-gas stream from the dense bed. Outlets from the model are the freeboard combustion-gas stream and the catalyst stream from the cyclones. The freeboard is modeled as a simple plug-flow reactor with homogeneous CO-to-CO_2 after-burn. Temperatures may rise rapidly in the presence of excess O_2.

The regenerator cyclone model performs a two-phase, loading-based ΔP calculation. Flue-gas compositions are calculated and reported on a standard dry mole percent basis for parameterization purposes. The flue gas stream can be connected to the valve model between the regenerator and CO boiler, or to a downstream power recovery system.

Inputs and outputs for the blocks described below are compatible with the vapor and catalyst streams used in the reactor/regenerator models.

The stripping zone model performs heat, mass and pressure balance calculations around the stripping zone. A tunable stripping efficiency curve is included. The stripping efficiency is related to the ratio of catalyst flow to stripping-steam flow.

The catalyst standpipes create a standing head that drives catalyst circulation. Different FCC designs make more or less use of standpipe technology. Standpipes typically end at a slide valve, which is used to control catalyst circulation. Standpipe diameter and length are variables in the standpipe model.

The catalyst transfer line model represents the pneumatic transport regime in the catalyst transfer line. Transfer-line diameter and length are configurable. Like the riser model, the catalyst transfer line model can be configured as vertical, horizontal or inclined.

The slide valve model calculates ΔP for a two-phase, loading-based orifice. The valve coefficient is parameterized from measured slide valve position (percent open) and pressure differential across the slide valve.

Components with similar reaction kinetics are grouped into 21 lumps. These kinetic lumps are listed in Table 6. In practice, the 21 components provide sufficient granularity to model all of the important steady-state cause-and-effect relationships in the FCC reactor-regenerator complex. Kinetic parameters for the riser and reactor models are segregated from the hydraulics and heat balance relationships. This permits different FCC kinetic schemes to be implemented within the same rigorous riser/reactor models. The off-line version of the model includes a simplified fractionator model and a product-property model.

Table 6. Aspen FCC 21 Reactive Lumps

Gas lump
Gasoline lump
221-343°C (P, N, As, Ar1, Ar2)
343-510°C (P, N, As, Ar1, Ar2, Ar3)
510°C-plus (P, N, As, Ar1, Ar2, Ar3)
Coke (2 lumps)

The simplified fractionator includes a delumper model to convert the 21 kinetic lumps into >80 pure- and pseudo-components, which are then divided into user-specified boiling fractions. A non-linear distribution function generates ideal distillation curves with realistic fraction-to-fraction overlap. The fractionator can inter-convert distillation methods, so a user can calculate D-86, D-1160, D-2887, and/or TBP curves for gasoline and LCO.

The product property model generates the product properties listed in Table 7. During projects, other product roperty calculations are added as needed. The online application at Suncor-Sarnia does not use the simplified fractionator model. Instead, it uses fully rigorous tray-by-tray models for the HCC fractionation section.

Table 7. Aspen FCC Standard Product Properties

As-Fractionated Product Properties							
	API Gravity	Specific Gravity	Sulfur	Cloud Point	RON	MON	PONA
Gasoline	x	x	x		x	x	x
LCO	x	x	x	x			
HCO	x	x	x				
Bottoms		x	x				
Standard-Cut Product Properties							
Light Naphtha	x	x	x		x	x	x
Heavy Naphtha	x	x	x		x	x	x
LCO	x	x	x	x			
Bottoms	x	x	x				

4.3   Aspen Hydrocracker

In many respects, AHYC is simpler than AFCC, mainly because it doesn't have to model the flow of a solid phase. Complexity arises for the following reasons:

− A separate reactor is used for each catalyst bed.
− The same set of reactions is used for both fixed-bed hydrotreating and fixed-bed hydrocracking units.
− All recycle loops are closed.
− More components and reactions are required. The model employs 116 components and 195 reactions.

- Catalyst deactivation has a major impact on the economics of a fixed-bed hydrocracker.

AHYC is an update of SARCRACK, a hydrocracker model developed by Sun Oil Company,[4] and modified by Suncor Sarnia with assistance from DMC corporation for online optimization. The component slate and reaction network for AHYC are consistent with publications by Klein, et al., Stangeland,[5] Quann, et al.,[6,7] Filiminov, et al.,[8] and Jacobs.[9]

Figure 3 presents a diagram of the high-pressure reaction section of the hydrocracker at Suncor-Sarnia. The unit is a classical single-stage Unicracker with two reactors. In the first (R1), all three beds are loaded with high-activity hydrodenitrogenation (HDN) catalyst. In the second (R2), all four beds are loaded with a zeolite-based, distillate-selective hydrocracking catalyst. For both reactors, the incoming gas is heated separately. The unit has two separator drums. The hydrogen-rich gas from the overhead of the high-pressure separator (HPS) is recycled. The HPS bottoms flow through a power-recovery turbine (PRT) to the low-pressure separator (LPS). LPS bottoms go through a pre-flash tower to the main fractionator.

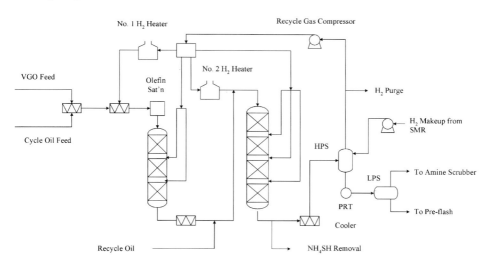

Figure 3. AHYC Model for the Suncor Sarnia Hydrocracker: Reaction Section

Heavy naphtha from the main fractionator goes to the large catalytic reformer, which is used to produce both gasoline and feedstock for the aromatics plant. A portion of the bottoms from the main fractionator is recycled to R2. The remainder goes to the first of two distillation towers for recovery of middle distillate products.

AHYC is an open-equation, flowsheet-based model, which is used both online and offline at Suncor Sarnia, where it was configured and tuned to rigorously represent the catalyst beds, quench valves, compressors, flash drums, heaters, exchangers, etc., in the commercial unit. A rigorous model for the hydrogen plant, a steam/methane reformer, was linked to the hydrocracker model. The entire fractionation section—comprising a pre-flash tower, main fractionator, all five gas plant towers, and two middle distillate recovery towers—was modeled rigorously.

After R1, a component-splitter model was used to simulate the removal of NH_4HS upstream from the HPS. This is a simplification, but it eliminates the need for a three-phase electrolytic flash model in the HPS, and it has no effect on reaction kinetics, the composition of recycle hydrogen, or the composition of the LPS bottoms.

A relatively simple extent-of-reaction model was used to simulate the saturation of olefins, including the associated consumption of hydrogen and generation of heat. It was assumed that 100% of the olefins in the feed are removed in the first bed of R1.

As mentioned above, each catalyst bed is modeled separately. Kinetic constituents are segregated from hydraulic and heat balance relationships, which permits different kinetic schemes to be implemented within the same mechanical framework.

Trickle-bed hydrodynamics are modeled with equations described by Satterfield.[10] Reaction rates and flash calculations are performed at multiple collocation points in each reactor bed. This enhances the ability of the model to perform an accurate heat-release calculation (see Table 9 below).

A customized version of the Langmuir-Hinshelwood-Hougen-Watson (LHHW) mechanism is used both for reversible (Figure 4) and irreversible reactions (Figure 5). The main steps in this mechanism are:
– Adsorption of reactants to the catalyst surface
– Inhibition of adsorption
– Reaction of adsorbed molecules
– Desorption of products

Several inhibitors are included in AHYC reaction kinetics. These include H_2S, ammonia, and organic nitrogen compounds. The inhibition of hydrodesulfurization (HDS) reactions by H_2S is modeled, and so is the inhibition of acid-catalyzed cracking reactions—dealkylation of aromatics, the opening of naphthenic rings, and paraffin hydrocracking—by ammonia and to a much greater extent by organic nitrogen.

$$\text{Rate} = A * k * \frac{((K_iC_i * K_{H2} * (P_{H2})^x / K_{eq}) - K_jC_j) * PF^y}{I_a * I_b * I_c}$$

where A = catalyst activity
k = rate constant
K_i = adsorption constants for hydrocarbons
C_i = concentrations of hydrocarbons
K_{H2} = adsorption constants for hydrogen
P_{H2} = concentrations of hydrogen
K_{eq} = equilibrium constant
PF = pressure factor
I_a, I_b, I_c, etc. = inhibition factors

Figure 4. LHHW rate equation for reversible reactions.

$$\text{Rate} = A * k * \frac{K_iC_i * K_{H2} * (P_{H2})^x * PF^y}{I_a * I_b * I_c}$$

where A = catalyst activity
k = rate constant
K_i = adsorption constants for hydrocarbons
C_i = concentrations of hydrocarbons
K_{H2} = adsorption constants for hydrogen
P_{H2} = concentrations of hydrogen
PF = pressure factor
I_a, I_b, I_c, etc. = inhibition factors

Figure 5. LHHW rate equation for irreversible reactions (C-C scission reactions such as paraffin hydrocracking, opening of naphthene rings, and ring dealkylation).

4.3.1 Reaction Pathways

The following reaction types were modeled in SARCRACK:
–   Hydrodesulfurization (HDS)
–   Hydrodenitrogenation (HDN)
–   Saturation of aromatics
–   Ring opening

- Ring dealkylation
- Paraffin hydrocracking

AHYC also models olefin saturation and (in an empirical way) paraffin isomerization.

The AHYC reaction scheme has the following characteristics:

- 45 reversible aromatics saturation reactions involving components in each major distillation range. The components include naphthene-aromatic compounds.
- 19 irreversible olefins saturation reactions. The model assumes that olefin saturation is complete within the first catalyst bed.
- Saturation and dealkylation for sterically hindered sulfur and nitrogen lumps. This allows AHYC to use different HDS and HDN rates for easy-to-treat and hard-to-treat to sulfur and nitrogen compounds.

Recent publications[11,12,13,14] confirm that a hydrotreater removes sulfur from organic sulfides, disulfides, and mercaptans with relative ease. Thiophenes, benzothiophenes, unhindered dibenzothiophenes and hindered dibenzo-thiophenes are successively harder to desulfurize.

Figure 6 illustrates the so-called "direct" mechanism for the HDS of dibenzothiophene, and Figure 7 shows a widely accepted mechanism for the HDS of hindered dibenzothiophenes. Note that the direct HDS of dibenzothiophene requires 2 moles of hydrogen per sulfur atom, while the HDS of hindered dibenzothiophenes requires 5 moles of H_2 per sulfur atom.

Structural differences between sulfur-containing compounds translate into significantly different HDS reaction rates. Table 8 shows pilot plant data for the HDS of small amounts of pure thiophenes in a clean-diesel solvent. The catalyst and temperature were the same for each compound. Note that HDS rates for the first three compounds, which are unhindered, are 30 to 100 times faster than rates for the last three compounds, which are hindered.

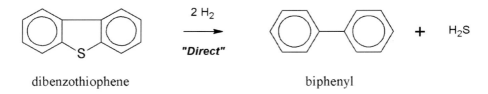

dibenzothiophene   biphenyl

Figure 6. Direct mechanism for the hydrodesulfurization of dibenzothiophene

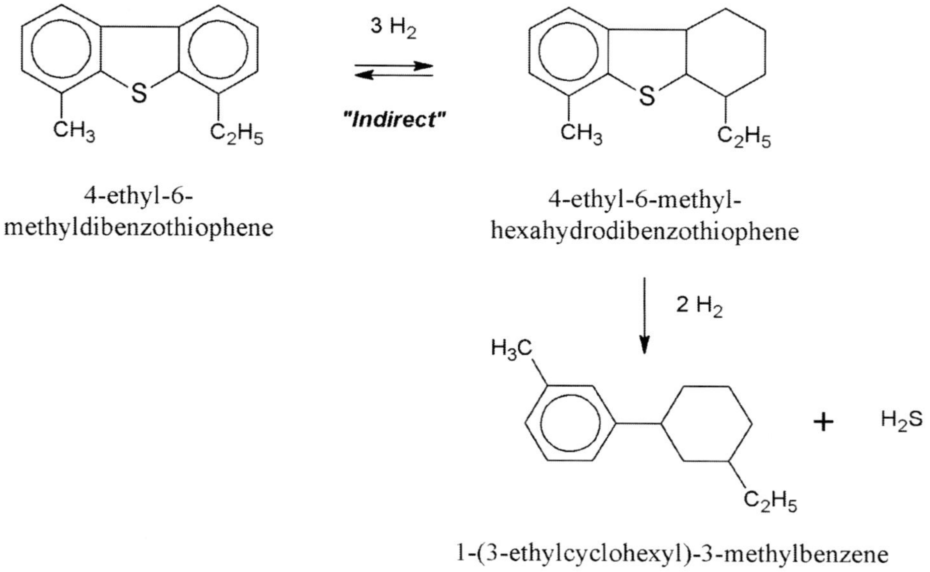

4-ethyl-6-
methyldibenzothiophene

4-ethyl-6-methyl-
hexahydrodibenzothiophene

1-(3-ethylcyclohexyl)-3-methylbenzene

Figure 7. Mechanism for the hydrodesulfurization of hindered dibenzothiophenes

Table 8. Realtive Reaction Rates of Hindered and Unhindered Sulfur Compounds

Sulfur Compound	Rel. HDS Rate	Boiling Point (°F)	Boiling Point (°C)
Thiophene	100	185	85
Benzothiophene	30	430	221
Dibenzothiophene	30	590	310
Methyldibenzothiophene	5	600 - 620	316 - 327
Dimethyldibenzothiophene	1	630 - 650	332 - 343
Trimethyldibenzothiophene	1	660 - 680	349 - 360

4.3.2 Catalyst Deactivation Model

In AHYC, the catalyst deactivation model calculates the deactivation rate as function of the following:
– Coke precursors in the feed
– Time on stream
– Average bed temperature for each catalyst bed
– H_2 partial pressure

The model predicts future catalyst activity, required temperature, yields, hydrogen consumption, and product properties. Alternatively, if the catalyst cycle life is fixed, the model can compute the optimal changes in feed rate, feed properties and/or conversion needed to reach the target end-of-run date.

Figure 10 shows how the deactivation model can be used to predict future temperatures required to hit a user-selected process objective. The process objective can be sulfur removal, nitrogen removal, or hydrocarbon conversion—for example, the conversion of 650°F-plus to 650°F-minus (343°C-plus to 343°C-minus) material.

At the hydrocracker at Suncor Sarnia, the deactivation model is applied to each catalyst bed. This enables the development of strategies to bring all catalyst beds to end-of-run at (roughly) the same time, and prediction of future yields, selectivity and product properties.

4.3.3 AHYC Model Fidelity

Table 9 presents data from an AHYC project at a U.S. East Coast refinery. The data show the excellent fit that can be achieved with the model. During that project, AHYC predictions were compared to step-test gains from a model-predictive controller. The optimizer matched most gains within +/- 20% (relative). This shows that the model does a good job of predicting process behavior under conditions well away from those at which it was tuned.

Table 9. Aspen Hydrocracker Model Fidelity

Variable	Measured Value		Predicted Value		Offset
Conversion	63.6 wt% feed		63.54 wt% feed		0.057
R1 WART*	724°F	384.4°C	724°F	384.4°C	0.55°F
R2 WART	723°F	383.9°C	723°F	383.9°C	-0.71°F
N at R1 exit	61 wppm		59 wppm		2 wppm
N at R2 exit	3 wppm		3 wppm		0 wppm
S at R1 exit	300 wppm		299 wppm		1 wppm
S at R2 exit	33 wppm		33 wppm		0 wppm
H2 makeup	40.5 million scf/d		47,800 Nm^3/hr		-1 Mscf/d
R1B1 outlet temp	707°F	375°C	707°F	375°C	0.001°F
R1B2 outlet temp	762°F	405.6°C	762°F	405.6°C	0.009°F
R2B1 outlet temp	761°F	405°C	761°F	405°C	0.008°F
R2B2 outlet temp	760°F	404.4°C	760°F	404.4°C	0.002°F
R3B1 outlet temp	725°F	385°C	725°F	385°C	0.001°F
R3B2 outlet temp	726°F	385.6°C	726°F	385.6°C	0.001°F
R3B3 outlet temp	743°F	395°C	743°F	395°C	0.003°F

* WART = weighted average reactor temperature

4.4   Clean Fuels Planning

4.4.1 Hydrogen Requirements for Deep Desulfurization

We ran a series of case studies to calculate the hydrogen requirements for deep desulfurization in a high-pressure gas-oil hydrotreater.

For the saturation of aromatics in a hydrotreating or hydrocracking unit, equilibrium effects, which favor formation of aromatics, start to overcome kinetic effects above a certain temperature. This causes a temperature-dependent "aromatics cross-over" effect, which explains the degradation of important middle distillate product properties—including kerosene smoke point and diesel cetane number—at high process temperatures near the end of catalyst cycles. The cross-over temperature is affected by feed quality and hydrogen partial pressure, so it can differ from unit to unit.

Figure 8 shows results from a case study in which the weighted average reactor temperature (WART) was changed over a wide range in a model based on the treating section of the Suncor Sarnia hydrocracker. Even though the WART was changed, the reactor temperature profile—equal outlet temperatures for the catalyst beds—was the same for every case.

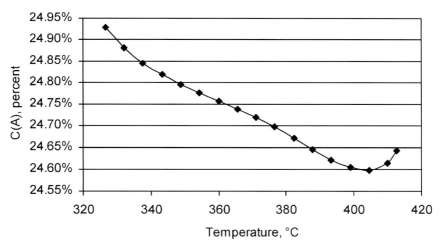

Figure 8. Product Aromatics versus Temperature

The y-axis shows the mole fraction of total aromatics remaining in product. For this case study, the feed rate, feed type, and catalyst activity were kept constant, which means that deactivation effects were not included. With less-active end-of-run catalysts, the rates of forward (saturation) reactions are inhibited, so the aromatics cross-over effect is amplified.

Figure 9 shows results of a case-study in which the product sulfur was kept constant (fixed) at different levels between 500 and 15 wppm. Required WART and hydrogen makeup flow are plotted on the y-axis. Because the product sulfur was fixed, another normally fixed variable can be allowed to vary. For this study, the chosen variable was WART.

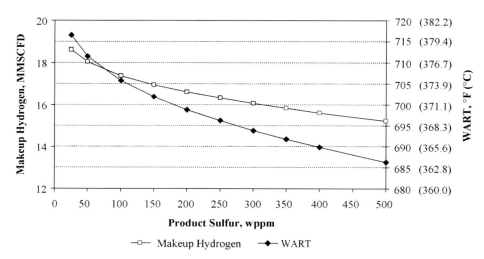

Figure 9. Aspen Hydocracker case study results: hydrogen consumption and weighted average reactor temperature (WART) versus product sulfur content.

Again, we ran the case study using the hydrotreating portion of the Suncor-Sarnia hydrocracker model with an equal-outlet temperature profile. For 500 wppm sulfur in the product, the makeup H_2 flow is 15.2 MMSCFD. For 15 wppm, the makeup flow is 18.6 MMSCFD, more than 40% higher. Higher WART leads to increased conversion, increased HDN, and (up to point) increased saturation of aromatics. This explains why the hydrogen consumption increased so much—and so non-linearly—when the target product sulfur content went from 500 wppm to 15 wppm.

We can also keep WART constant and achieve the desired level of HDS by allowing changes in LHSV. When this is done, the effect on hydrogen consumption is less severe. But in the real world, decreasing the LHSV is equivalent to lowering feed rate or adding a reactor, both of which are expensive. The WART-based trend shown here is more relevant to an existing unit.

Figure 9 is consistent with the existence of both hindered and unhindered sulfur compounds in the feed. As mentioned above, the last traces of sulfur in diesel oil are locked inside highly hindered hydrocarbon molecules, which have to be cracked open before the sulfur can be removed.

4.4.2 Effects of Hydrotreating on FCC Performance

Many refiners are including the hydrotreating of FCC feed in their plans to make clean fuels. To demonstrate the feasibility of running a rigorous model for an FCC-plus-pretreater complex, we connected AHYC to AFCC. To link the two, we created a component mapper model to compress the 97 reactive components from AHYC into the 21 lumps used in AFCC. We then connected the unconverted oil stream from the AHYC fractionator model to the feed-adjust block of AFCC.

Initially, the combined model was huge, containing more than 1.2 million non-zero terms in its matrix of variables. To allow the model to run in a reasonable amount of time on a Pentium III computer, we made some simplifications. In the reduced model, the four catalyst bed models are still fully rigorous. However, the hydrogen furnaces are represented with a heat-exchanger model, quench valves are modeled with mixers, a component splitter model is used for the wash-water system, and a group of component splitters is used for the fractionation section. These changes reduce the number of equations and non-zeros to 130,000 and 680,000 respectively. Despite these simplifications, the slimmed-down model remains, in our collective opinion, a useful tool for offline what-if studies and for economic comparisons of different process options.

We first looked at the effect of varying the amount of hydrotreated material in an FCC feed blend from 0.0 to 96.7 vol%. Table 10 presents selected process conditions for the feed hydrotreater, which achieved 90% desulfurization and roughly 23 wt% conversion of vacuum gas oil to middle distillate and naphtha.

Table 10. Feed Pretreater Operating Conditions

Variable	Units	Value
Feed rate	b/d	39,000
Number of catalyst beds		4
Weighted average reactor temperature	°F	720
	°C	382
LHSV	hr^{-1}	1.2
Hydrogen partial pressure	psig	2000
	barg	136
Hydrotreater feed sulfur	wt%	1.95
Sulfur in hydrotreated FCC feed	wt%	0.195
Conversion of 343°C-plus to 343°C-minus	wt%	23.0

We modeled the effect of feed hydrotreating in two ways. The first and most logical approach was to model the pretreating of the 100% of the FCC feed. We simply operated the pretreater model at several different severities and watched the response of the FCC model. Complications arose because upstream conversion in the pretreater affects the amount of feed going to the FCC model. When we compensated by increasing or decreasing the overall fresh feed rate to maintain a constant feed rate to the FCC, conversion in the pretreater changed due to the corresponding differences in LHSV. This is a classical—and very realistic—economic optimization problem, for which the combined AHYC/AFCC model is well suited. However, it doesn't easily lend itself to creating straight-forward graphs for publications.

In another approach to modeling the effects of FCC feed pretreating, the FCC model receives a blend of hydrotreated feed from AHYC with straight-run feed from a tank model. The latter case is discussed here.

Table 11 shows the properties of the four FCC feed blends that were used in the combined hydrotreater/FCC simulations. Table 12 presents FCC conversion and yield data. In each case, we held the FCC riser outlet temperature at 1030°F (544°C).

Table 11. FCC Feed Properties for Blends of Straight-run and Hydrotreated Feeds

Variable	Units	Case 1	Case 2	Case 3	Case 4
Straight-run VGO	b/d	30,000	20,000	10,000	1,000
Hydrotreated VGO (90% HDS)	b/d	0	10,000	20,000	29,000
Total Feed Rate	b/d	30,000	30,000	30,000	30,000
API gravity	°API	22.60	24.57	26.58	28.44
Specific gravity		0.9182	0.9067	0.8951	0.8847
Sulfur	wt%	1.95	1.38	0.80	0.26
Basic Nitrogen	wppm	417	380	344	310
CCR	wt%	0.730	0.525	0.315	0.122
Vanadium	wppm	3.1	2.1	1.1	0.2
Nickel	wppm	5.2	3.5	1.8	0.3
Sodium	wppm	1.3	0.9	0.5	0.1
Iron	wppm	2.6	2.1	1.5	1.1
Copper	wppm	0.2	0.2	0.1	0.1
UOP K Factor		11.67	11.79	11.92	12.03

Table 12. FCC Conversion and Yields

Variable	Units	Case 1	Case 2	Case 3	Case 4
Net conversion	Vol%	74.39	78.45	82.01	84.88
	Wt%	73.11	77.05	80.61	83.56
				Yields	
Fuel gas (H_2-C_2)	FOE*	5.68	5.83	5.95	6.03
C_3	Vol%	16.48	17.07	17.55	17.90
C_4	Vol%	25.38	26.71	27.90	28.88
C_5	Vol%	0.85	0.93	1.00	1.06
Naphtha	Vol%	44.46	47.13	49.55	51.53
LCO	Vol%	19.13	16.56	14.15	12.20
Bottoms	Vol%	4.56	2.92	1.62	0.63
Coke	Wt%	5.30	5.30	5.30	5.30
Total liquid		116.54	117.16	117.74	118.23
Flue gas SOx	wppm	390	296	180	60

Figure 10 shows that the sulfur content of FCC gasoline is essentially linear with respect to the percent of hydrotreated oil in the FCC feed blend. Figure 11 shows that relatively more sulfur goes into heavy FCC products (LCO, HCO and slurry oil) at higher percentages of hydrotreated oil in the feed.

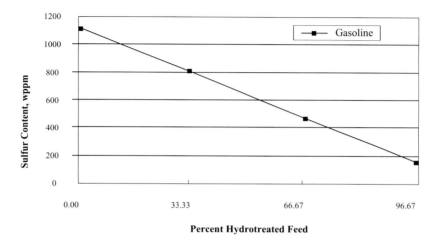

Figure 10. Sulfur in FCC gasoline versus percent hydrotreated oil in the FCC feed.

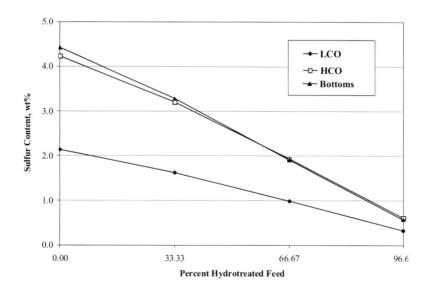

Figure 11. Sulfur in heavy FCC products versus percent hydrotreated oil in the FCC feed.

This is consistent with the easy-sulfur/hindered-sulfur hypothesis, which predicts that nearly all of the sulfur left behind after high-severity hydrotreating is encased in hindered compounds. These compounds contain multiple aromatic rings, so they more likely (relative to other sulfur compounds) to be incorporated into the coke that forms on FCC catalysts. When the coke is burned away, the sulfur is converted into sulfur oxides (SOx), which ends up in the regenerator flue gas.

5.   CONCLUSIONS

Rigorous models are being used for closed-loop real-time optimization at the Suncor refinery in Sarnia, Ontario. Single-unit rigorous models—and combinations of models similar to those being used in Sarnia—can be used offline to quantify non-linear relationships between important refinery units, such as the FCC and its associated feed pretreater. These relationships are important when planning for clean fuels, especially when selecting and designing processes for hydrogen production and deep desulfurization.

6.   ACKNOWLEDGEMENTS

We are pleased to thank the following people for their contributions to RWO and to the development of Aspen FCC or Aspen Hydrocracker: John Adams, John Ayala, Darin Campbell, Steve Dziuk, Fernando Garcia-Duarte, Steve Hendon, Nigel Johnson, Ajay Lakshmanan, Jiunn-shyan Liou, Skip Paules, Rosalyn Preston, and Charles Sandmann.

7.   REFERENCES

1.   Pedersen, C.C.; Mudt, D.R.; Bailey, J.K.; Ayala, J.S. "Closed Loop Real Time Optimization of a Hydrocracker Complex," NPRA Computer Conference, CC-95-121, November 6-8, 1995.
2.   English, K.; Dunbar, M. "Using Steady State Models to Generate LP Model Inputs," NPRA Computing Conference, CC-97-131, November 16-19, 1997.
3.   Shorey, S.W.; Lomas, D.A.; Keesom, W.H. "Improve Refinery Margins and Produce Low-Sulfur Fuels," World Refining Special Edition: Sulfur 2000, Summer 1999, p. 41.
4.   Lapinas, A.T.; Klein, M.T.; Gates, B.D.; Macris, A.; Lyons, J.E. "Catalytic Hydrogenation and Hydrocracking of Fluorene: Reaction Pathways, Kinetics and Mechanisms," *Industrial and Engineering Chemistry Research*, 30 (1), **1991**, 42.
5.   Stangeland, B.E. "Kinetic Model for the Prediction of Hydrocracker Yields," *Industrial & Engineering Chem., Process Des. Development*, 13 (1), **1974**, 71.

6. Quann, R.J. "Hydrocracking of Polynuclear Aromatic Hydrocarbons. Development of Rate Laws through Inhibition Studies," *Industrial & Engineering Chemistry Research*, 36 (6), **1997**, 2041.
7. Jacob, S.M.; Quann, R.J.; Sanchez, E.; Wells, M.E. "Compositional Modeling Reduces Crude-Analysis Time, Predicts Yields," *Oil & Gas Journal*, July 6, 1998, p. 51.
8. Filimonov, V.A.; Popov, A.A.; Khavkin, V.A.; Perezhigina, I.Ya.; Osipov, L.N.; Rogov, S.P.; Agafonov, A.V. "The Rates of Reaction of Individual Groups of Hydrocarbons in Hydrocracking," *International Chemical Engineering*, 12 (1), **1972**, 21.
9. Jacobs, P.A. "Hydrocracking of n-Alkane Mixtures on Pt/H-Y Zeolite: Chain Length Dependence of the Adsorption and the Kinetic Constants," *Industrial & Engineering Chemistry Research*, 36 (8), **1997**, 3242.
10. Satterfield, C.N. "Trickle-Bed Reactors,", *AIChE Journal*, 21 (2), **1975**, 209.
11. Callejas, M.A.; Martinez, M.T. "Hydrocracking of a Maya Residue. Kinetics and Product Yield Distributions," *Industrial & Engineering Chemistry Research*, 38 (9), **1999**, 3285.
12. Cooper, B.H.; Knudsen, K.G. "Ultra Low Sulfur Diesel: Catalyst and Processing Options," NPRA Annual Meeting, Paper No. AM-99-06.
13. Cooper, B.H.; Knudsen, K.G. "What Does it Take to Produce Ultra-Low-Sulfur Diesel?" World Refining, November 2000, p. 14.
14. Bjorklund, B.L.; Howard, N.; Heckel, T.; Lindsay, D.; Piasecki, D. "The Lower It Goes, The Tougher It Gets," NPRA Annual Meeting, Paper No. AM-00-16, March 26-28, 2000.

Chapter 24

MODELING HYDROGEN SYNTHESIS WITH RIGOROUS KINETICS AS PART OF PLANT-WIDE OPTIMIZATION

Milo D. Meixell, Jr.
Aspen Technology, Incorporated
Advanced Control and Optimization Division
Houston, Texas

1.   INTRODUCTION

Hydrogen for industrial use is manufactured primarily by steam reforming hydrocarbons, or is available as a byproduct of refinery or chemical plant operations.   Hydrogen demand usually is significantly greater than that available as a byproduct, and many times the plant's ability to supply hydrogen limits overall plant throughput.   An operating plant can exploit existing in-place steam reformer and related equipment designs with rigorous models that employ kinetics.   During plant design kinetics models are not necessarily used because experience and more simple models using "approach to equilibrium", space velocity, maximum heat flux, and other parameters allow designers to build hydrogen plants to supply whatever hydrogen amount, purity, and pressure that is desired.   The objective of suppliers of hydrogen plants is to establish an economic feasible design, given these predetermined criteria.   Additional hydrogen, or less expensive hydrogen can be manufactured from these facilities, after construction, when rigorous models are used in on-line, closed loop optimization systems.   These systems manipulate independent operating conditions (degrees of freedom), maximize or minimize an objective function (usually operating profit), while honoring numerous constraints, such as tube metal temperatures, pressure drops, pressures, damper and valve positions.   A good kinetic model is the cornerstone of the predictive capabilities of such on-line optimization systems. An on-line optimization system exploits equipment under operating demands and constraints that were not necessarily anticipated during design.

The steam reformer and associated models represent the most significant models in an overall hydrogen or synthesis gas plant real-time, closed-loop optimization project. The reformer does the great majority of converting the feedstock (i.e. natural gas, butane or naphtha, plus steam) to the primary reactant for downstream synthesis reactors, hydrogen. Consequently, detailed modeling is required to predict the correct product yields from varying feed stock composition at various operating conditions. Additionally, the primary reformer is subject to numerous constraints, such as tube metal temperatures, that also demand detailed modeling to predict. The primary reformer consumes significant energy, and the methane slip or hydrogen purity from it affects operating conditions in downstream equipment. The optimization system must be able to exploit these reactors if optimal plant operation is to be achieved.

The reformer reactions (reforming and water gas shift) are modeled using the best available heterogeneous kinetic relationships from the literature,[1-4] and have been validated with industrial data and literature data [28] over a wide range of conditions. Equilibrium relationships are appropriately incorporated into the rate relationships, but the model does not use the empirical "approach to equilibrium temperature" as the basis for predicting outlet compositions. Nor does the model use "pseudo-homogeneous" rate relationships.

The steam reformer model can handle feeds from methane to naphtha, with all the typical components that are present in natural gas, as well as recycled synthesis purge gas, or hydrogen recovery unit tail gases. Naphtha feed is characterized as about 30 chemical species, some of which are pure components, and some are hydrocarbon fractions (pseudo components). Each hydrocarbon species participates in a reaction that includes adsorption onto the catalyst, reforming, and desorption. The model includes diffusion effects within the catalyst, as well as heat transfer resistance from the bulk gas to the catalyst surface.

Since the model is based on mechanistic rate relationships, the composition profiles (as well as temperature and pressure) are calculated at positions from the inlet of the catalyst tube to the outlet. The differential reaction rate relationships, as well as the mass balances, heat balances, and pressure drop relationships are solved simultaneously. A global spline collocation algorithm is used to pose the problem, and an SQP algorithm (sequential quadratic programming) solves the relationships.

The heat transfer rate from the primary reformer fire box to the catalyst tubes is calculated along the tube using radiant and convective heat transfer relationships. Heat transfer is also calculated from the tube outer surface, across the tube metal, through the inside tube surface film to the bulk fluid, and finally from the bulk fluid to the catalyst particles. Measurements of the catalyst temperatures at several positions along the tube, as well as measured tube skin temperatures provide feedback that is used to update heat transfer parameters, as well as to validate the model.

Pressure drop calculations are included, which use the Ergun relationship.[5] This relationship predicts the pressure drop through packed beds, such as the primary reformer catalyst tubes and the secondary reformer (auto-thermal) bed. Measurements provide feedback that is used to update appropriate parameters in this relationship.

The kinetics developed for natural gas and naphtha steam reforming in fired and auto-thermal reactors were put to more stringent tests by applying them to two other reactors. One is a "pre-reformer", an adiabatic fixed bed typically used upstream of a fired tube reformer for debottlenecking. The other reactor is a fired furnace with high carbon dioxide content in the feed, making a 1:1 ratio hydrogen to carbon monoxide synthesis gas, at 3.5 bars (as opposed to the typical 40 bars pressure in many steam reformers), which is processed in downstream Fischer-Tropsch reactors, ultimately making oxo-alcohols. Calculated and measured yields and temperature profiles in these reactors, using the same kinetics, are in excellent agreement.

These rigorous models, developed from credible literature sources, and validated against industrial data over a wide range of conditions, are the foundation of several plant-wide optimization systems, including those for hydrogen, ammonia, methanol, and oxo-alcohol plants.

2. STEAM REFORMING KINETICS

The reaction scheme and associated kinetics are based on methane reforming and "heavier hydrocarbon" reforming studies reported in the literature. The methane reforming relationships must be included in any model processing heavier hydrocarbons since methane is generated from these feeds, and subsequently reacts. The primary source of information used for methane reforming is Xu and Froment's most recent literature articles on intrinsic kinetics and on using these along with diffusional resistance relationships to model industrial reformers.[1, 2, 6] The source of information for higher hydrocarbon steam reforming is based upon the work of Rostrup-Nielsen of Haldor-Topsøe A/S, as well as that of others, whose work Rostrup-Nielsen has referenced extensively.[3, 4, 7]

2.1 Methane Steam Reforming Kinetic Relationships

Numerous reaction mechanisms were investigated by Froment, and the best reaction scheme is shown in Figure 8 and Table 3 of Reference 1. "Best" was determined by "model discrimination" methods described by Froment. The kinetic expressions used for methane reforming account for adsorption of all the reacting species present (CO, H_2, CH_4, and H_2O), as well as reaction on the surface. The intermediate, adsorbed species concentrations, which are unmeasurable (at least in industrial equipment), are eliminated from the mechanistic steps (at steady-state), and only stable species concentrations (partial pressures) appear in the net reaction stoichiometry and the reaction rate expressions.

Net Reaction Stoichiometry:

$$CO + H2O \longleftrightarrow CO2 + H2$$

$$CH4 + H2O \longleftrightarrow CO + 3H2$$

$$CH4 + 2H2O \longleftrightarrow CO2 + 4H2$$

Reaction Rate Relationships:

$$r1 = \frac{\dfrac{k1}{pH2}\left(pCO * pH2O - \dfrac{pH2 * pCO2}{K1}\right)}{\left(DEN\right)^2}$$

$$r2 = \frac{\dfrac{k2}{pH2^{2.5}}\left(pCH4 * pH2O - \dfrac{pH2^3 * pCO}{K2}\right)}{\left(DEN\right)^2}$$

$$r3 = \frac{\dfrac{k3}{pH2^{3.5}}\left(pCH4 * pH2O^2 - \dfrac{pH2^4 * pCO2}{K3}\right)}{\left(DEN\right)^2}$$

$$DEN = 1 + KCO * pCO + KH2 * pH2 + KCH4 * pCH4 + KH2O * pH2O / pH2$$

The reaction rate "constants", ki's, are functions of temperature, activation energies, and frequency factors, in the classic Arrhenius form:

$$ki = Ai * e^{-Ei / RT}$$

$$i = 1 \text{ to no. of reactions}(\cong 30)$$

Similarly, the adsorption equilibrium constants are functions of temperature, heats of adsorption, and a frequency factor:

$$K_i = A_i * e^{\Delta H_i / RT}$$

$$i = CO, H2, CH4, H2O$$

The reaction equilibrium "constants" are functions of temperature. The equilibrium constants as a function of temperature are listed in References 3 and 4. Those of Reference 3 were used, and have been tabulated and regressed, with results listed in Appendix E.

$$K_i = f(T)$$

$$i = 1 \text{ to no. of reversible reactions}(= 3)$$

The kinetics discussed so far are intrinsic kinetics that are valid in the absence of diffusional or heat transfer resistances. Bulk phase concentrations (partial pressures) and temperatures are not present at the active sites within the catalyst pellets, since the pore structure offers very significant resistance to diffusion, and mild resistance to heat transfer exists between the bulk phase and the pellet surface. Little resistance to heat transfer exists within the pellets (Reference 4, page 69).

The diffusional effects are accounted for using the classic "effectiveness factor".[2, 8, 9, 10, 11]

$$\eta = \frac{\text{actual rate throughout pellet with resistances}}{\text{rate evaluated without resistances}}$$

Evaluating the numerator of this expression requires simultaneous integration of the rate and diffusion relationships. Froment presents the results of that integration in Figure 6 of Reference 2. Rostrup-Nielsen lists results of Haldor-Topsøe's calculations on page 69 of Reference 4. These results are qualitatively similar, but differ in value due to numerous factors. Simplification of the effectiveness factor calculation can be applied for conditions which cause the diffusional resistance to be large (Reference 8, page 431), and that simplification is implied in many of Haldor-Topsøe's equations (Reference 4, pages 37, 69 and Reference 9). The simplification (Reference 8, page 434) is:

$$\eta = \frac{3}{r_p} \sqrt{\frac{K(D_e)}{k(K+1)\rho_p}}$$

Satterfield (Reference 10, page 78) gives further insight to the simplification, when the diffusivities of the forward and reverse reactions are not equal.

Since the kinetics used in the DMCC model are those of Froment, the effectiveness factor profiles in the primary reformer model are based upon his results. Rather than incur the computational burden of calculating the effectiveness factors for each reaction on-line, they are entered as a function of length as constant profiles. Since the effectiveness factors are primarily functions of the catalyst pellet size, pore size, and pore size distribution, and since they are relatively weak functions of operating conditions over the normal range of these conditions, the effect of imposing them as constant profiles is very small. The results of reactor simulations with the effectiveness factor profiles fixed agree very well with measured results from industrial reformers, and with results presented in the literature. Off-line calculations can be done to update the effectiveness factors if catalyst pellet size or pore size are changed significantly, or ultimately these calculations can be added to the on-line model. It is essential that effectiveness factors are used in conjunction with the intrinsic kinetic rate parameters, since the effectiveness factors account for phenomena that attenuate the intrinsic rates by factors on the order of 100. That is, effectiveness factors are on the order of 0.01.

The relationship among actual reaction rate, intrinsic rate, effectiveness factor, and catalyst activity is:

$$r_{i, observed} = \eta_i * \alpha * r_{i, intrinsic}$$

where:

η = effectiveness factor

α = relative catalyst activity

(accounting for aging, sintering, pore closure, etc.)

$r_{i, intrinsic}$ = rate calculated with bulk fluid conditions

The catalyst activity is calculated as a parameter, updated from operating data.

2.2 Naphtha Steam Reforming Kinetic Relationships

Several intrinsic reaction rate relationships have been derived for steam reforming of "higher hydrocarbons" (i.e. higher in carbon number than methane), based on Langmuir-Hinshelwood adsorption and reaction mechanisms (Reference 3, pages 118 and 174, Reference 4, page 54). These reaction rates (as with the reactions associated with the reforming of methane) are greatly attenuated in industrial tubular reactors by mass transfer resistance within the catalyst pellet and, to a lesser degree, by heat transfer resistance from the bulk fluid to the catalyst pellet surface. As with any intrinsic rate relationships, these rate expressions must be used in conjunction with a reactor model that accounts for these resistances for the model to be useful in assessing reformer performance, and predicting performance as independent conditions are manipulated. The relationship among intrinsic rates, observed rates, mass and heat transfer resistances, and catalyst activity has been shown in the Methane Steam Reforming Kinetic Relationships section.

One of the mechanisms (Reference 4, page 54) is based on the following sequence of adsorption, reaction, and desorption:

$$C_nH_m + 2* \xrightarrow{kA} C_nH_z - *_2 + \frac{m-z}{2}H_2$$

$$C_nH_z - *_2 + n* \xrightarrow{kH} C_{n-1}H_{z'} - *_2 + CH_x - *_n$$

$$CH_x - *_n + O - * \xrightarrow{kr} CO + \frac{x}{2}H_2 + (n+1)*$$

$$H_2O + * \xleftrightarrow{KW} O - * + H_2$$

$$H_2 + 2* \xleftrightarrow{KH} 2H - *$$

The "*" symbol represents an active site on the catalyst surface. Concentration of C_nH_z - $*_2$ is generally assumed to be negligible, so the step 2 rate "constant", kH does not appear in the final rate expression. The rate constants kA and kr (adsorption and reforming) do appear in the rate expression. KW and KH are the adsorption equilibrium "constants" for water and hydrogen, respectively.

The mechanisms account for chemisorption of the hydrocarbon and steam, followed by α-scission of the carbon-carbon bonds. The resulting adsorbed C_1 species react with adsorbed steam to form carbon monoxide and hydrogen.

These mechanisms alone would result in no formation of methane, which is of course generated from naphtha feed stocks which are totally free of methane. Methane concentrations of 8 to 10 mole percent (dry basis) are typical at the outlet of industrial naphtha steam reformers. Methane is generated when the hydrogen partial pressure is sufficiently high so that the reverse of the methane reforming reaction:

$$CH_4 + H_2O \xleftrightarrow{} CO + 3H_2$$

is favored. This reaction, along with reforming to carbon dioxide, and the water gas shift reaction are solved simultaneously with the reforming reactions for the higher hydrocarbons. The mechanisms for these reactions are documented in the "Methane Steam Reforming" section.

Competition between methane and the higher hydrocarbons for reactive sites on the catalyst surface is limited to a narrow axial position in the reactor tube, since the higher hydrocarbon species reform quickly, and since methane concentration buildup due to the reverse of the aforementioned methane reforming reaction is only significant after the great majority of the higher hydrocarbons are gone. The hydrogen partial pressure is only high enough to favor the reverse of the methane reforming reaction when the higher hydrocarbon species have essentially disappeared. Therefore the denominator of the higher hydrocarbon rate relationships do not contain a methane adsorption term, nor does the denominator in the methane reforming reaction rate expressions have higher hydrocarbon adsorption terms. Rate expressions can be derived for the region in which the higher hydrocarbon and methane compete for active catalyst sites, however results of simulating a reactor with such a model will be essentially unchanged from the results obtained by applying the rate expressions discussed here.

The concentration profiles that result from solving the methane and the higher hydrocarbon reaction rates simultaneously agree well with the profiles reported by others,[12,13] and the reactor effluent species concentrations agree well with those observed from industrial reformers (Appendix A). The effluent concentrations are fairly insensitive to reaction rates, since industrial reformers operate near equilibrium conditions. Effluent conditions will only be noticeably affected by reaction rates as the catalyst activity declines significantly. The temperature profile, especially in the first on third of the reactor which is furthest from equilibrium, is affected significantly by reaction rates, and therefore is the most affected by the catalyst activity. The case study results in Appendix B illustrate the concentration profiles, and the effects of catalyst activity on temperature profiles.

From Reference 4:

$$r_i = \frac{k_A * p_{CnHm}}{\left(1 + \dfrac{n * k_A}{kr * Kw} * \dfrac{p_{CnHm}}{p_{H2}} * p_{H2} + Kw * \dfrac{p_{H2O}}{p_{H2}} + \sqrt{K_H * p_{H2}}\right)^{2n}}$$

The 2n exponent in the denominator appears to be in error, since similarly derived rate expressions, listed below, have exponents of 2. Also, the 2n exponent leads to very unreasonable results for rates of reaction of higher carbon number hydrocarbons.

From References 3 and 13:

$$r_i = \frac{k_A * p_{CnHm}}{\left(1 + \frac{n * k_A}{kr * Kw} * \frac{p_{CnHm}}{p_{H2}} * p_{H2} + Kw * \frac{p_{H2O}}{p_{H2}}\right)^2}$$

From Reference 3, including Boudart's interpretation of hydrogen and hydrocarbon site differences:

$$r_i = \frac{k_A * p_{CnHm} / p_{H2}^{y/2}}{\left(1 + \frac{n * k_A}{kr * Kw} * \frac{p_{CnHm}}{p_{H2}} * p_{H2}^{(1-y/2)} + Kw * \frac{p_{H2O}}{p_{H2}}\right)^2}$$

The rate expression from "Catalytic Steam Reforming" (1984) can be modified to include Boudart's hydrogen site interpretation, and can be solved in its original form, or the modified form by manipulating the value of "y". A "y" value of zero restores the relationship to its original form. Additionally, that rate expression can be transformed into essentially the expression from "Steam Reforming Catalysts" (1975) by manipulating the value of KH. A small value of KH transforms the 1984 expression into the 1975 one. Generally, the final term in the denominator of the 1984 rate expression (the term including KH) is much smaller than the other terms.

The rate expression from "Catalytic Steam Reforming" (1984), modified to include the Boudart interpretation of the hydrogen site versus hydrocarbon site difference, is the relationship used in the model. The apparent error in the 2n exponent of the denominator has been corrected, and an exponent of 2 is used.

$$r_i = \frac{k_A * p_{CnHm} / p_{H2}^{y/2}}{\left(1 + \frac{n * k_A}{kr * Kw} * \frac{p_{CnHm}}{p_{H2}} * p_{H2}^{(1-y/2)} + Kw * \frac{p_{H2O}}{p_{H2}} + \sqrt{K_H * p_{H2}}\right)^2}$$

Each naphtha species is individually reformed, with its distinct set of rate "constants", kr and kA. Each of these is a function of temperature and has its own pre-Arrhenius factor, as well as an activation energy (for kr), and heat of adsorption (for kA). The same KH and KW is used for each reaction. These too are functions of temperature, with their Arrhenius factors and heats of adsorption taken from Froment's most recent work [1]. The kr and kA values for each reforming reaction for hydrocarbons higher than methane are derived from rate data reported in the literature,[3,4,19] and validated with overall reformer outlet compositions measured in industrial furnaces. Each reaction also has an effectiveness factor associated with it.

Naphtha is characterized as about thirty components, pure or pseudo (several species of similar structure lumped together). The distribution of these components is chosen to best match measured specific gravity, volume average boiling fractions (ASTM-D86 method), and normal paraffinic, branched paraffinic (iso-paraffins), naphthenic, and aromatics (PINA) contents.

Numerous other rate expressions for steam reforming of higher hydrocarbons are listed in the literature (Reference 4, pages 55 & 57). Many of these are simplifications of the relationships shown here, with assumptions imposed on the relative contribution of the various terms. These assumptions may be valid for some operating conditions, or for some catalysts. Many of the rate expressions are simple "power law" relationships that are difficult to relate to mechanistic pathways, and to physical attributes of the catalyst or the reacting species. The rate expression chosen for the model is one that has a broad range of applicability, for different catalyst types, and wide operating conditions.

Specific catalysts are characterized in part by their relative values of kr, KW, and kA. The water adsorption equilibrium constant, KW, is affected by the catalyst support. Alumina supports have lower values of KW than magnesia supports.[13] Alkali content increases KW. Increased alkali concentrations tend to decrease kr, though. Catalysts' shape and size affect pressure drop and effectiveness factors. Their pore structure affects their effectiveness factors as well. Additionally, the support affects the catalyst crush strength, and tendency to hold together under adverse conditions (i.e. wetting) that may be encountered during unplanned plant outages, severe operating conditions, or start-up conditions. Migration of catalyst additives into equipment downstream of the reformers has caused fouling problems in some plants. Catalysts must maintain their activity and strength for several years to be commercially viable.

The following reactions are used in the primary reformer model, for butane and naphtha feeds. The model interprets the "StmReforming" qualifier, and determines the correct stoichiometry. The "User1" and "User2" qualifiers direct the model to use either the Froment reaction rate relationships, or the naphtha reaction rate relationships.

```
Reaction Mechanism = start
!
! USER1 REACTIONS:
!
! Water Gas Shift

    CO          +    H2O    =>   CO2   +    H2                        :User1
```

```
! Methane Reforming to CO

  CH4           +   H2O     =>  CO   +   H2   :StmReforming        :User1

! Methane Reforming to CO2

  CH4           + 2*H2O     =>  CO2  + 4*H2                        :User1
!
! USER2 REACTIONS:
!
! Reforming of Higher Hydrocarbons to CO & H2 (Irreversible)

  Ethane        +   H2O     =>  CO   +   H2   :StmReforming        :User2
  Propane       +   H2O     =>  CO   +   H2   :StmReforming        :User2
  Isobutane     +   H2O     =>  CO   +   H2   :StmReforming        :User2
  N-Butane      +   H2O     =>  CO   +   H2   :StmReforming        :User2
  Isopentane    +   H2O     =>  CO   +   H2   :StmReforming        :User2
  N-Pentane     +   H2O     =>  CO   +   H2   :StmReforming        :User2
  Cyclopentane  +   H2O     =>  CO   +   H2   :StmReforming        :User2
  Iso-C6        +   H2O     =>  CO   +   H2   :StmReforming        :User2
  N-Hexane      +   H2O     =>  CO   +   H2   :StmReforming        :User2
  Cyclo-C6      +   H2O     =>  CO   +   H2   :StmReforming        :User2
  Benzene       +   H2O     =>  CO   +   H2   :StmReforming        :User2
  Iso-C7        +   H2O     =>  CO   +   H2   :StmReforming        :User2
  N-Heptane     +   H2O     =>  CO   +   H2   :StmReforming        :User2
  Cyclo-C7      +   H2O     =>  CO   +   H2   :StmReforming        :User2
  Toluene       +   H2O     =>  CO   +   H2   :StmReforming        :User2
  Iso-C8        +   H2O     =>  CO   +   H2   :StmReforming        :User2
  N-Octane      +   H2O     =>  CO   +   H2   :StmReforming        :User2
  Cyclo-C8      +   H2O     =>  CO   +   H2   :StmReforming        :User2
  Aromatic-C8   +   H2O     =>  CO   +   H2   :StmReforming        :User2
  Iso-C9        +   H2O     =>  CO   +   H2   :StmReforming        :User2
  N-Nonane      +   H2O     =>  CO   +   H2   :StmReforming        :User2
  Cyclo-C9      +   H2O     =>  CO   +   H2   :StmReforming        :User2
  Aromatic-C9   +   H2O     =>  CO   +   H2   :StmReforming        :User2
  Iso-C10       +   H2O     =>  CO   +   H2   :StmReforming        :User2
  Cyclo-C10     +   H2O     =>  CO   +   H2   :StmReforming        :User2
  N-Decane      +   H2O     =>  CO   +   H2   :StmReforming        :User2
  Aromatic-C10  +   H2O     =>  CO   +   H2   :StmReforming        :User2

Reaction Mechanism = end
```

2.3   Coking

Coke deposition is a potential problem in steam reformers, and is favored by low steam to carbon ratio and high temperatures.   High hydrogen composition also lowers the potential for carbon deposition.   The following three reactions can deposit solid carbon, and the thermodynamic equilibrium relationships for these reactions, when carbon is graphite, are well known.[3, 4, 13, 20, 21, 22, 23]

$$CH_4 \longleftrightarrow C + 2 H_2 \qquad \text{Methane Cracking}$$

$$2 CO \longleftrightarrow C + CO_2 \qquad \text{Boudouard Coking}$$

$$CO + H_2 \longleftrightarrow C + H_2O \qquad \text{CO Reduction}$$

Also, higher hydrocarbons can form carbon (3).

$$C_nH_m \longrightarrow \text{polymer} \longrightarrow \text{coke}$$

Equilibrium relationships with carbon of other forms has also been investigated and results reported in the aforementioned references. Whisker or filamental carbon is the form of carbon favored on nickel catalysts, and equilibrium of the aforementioned reactions with this form of carbon is reasonably well understood.

The first three carbon depositing reactions involve only methane and carbon monoxide, and are of particular interest in plants with methane feeds. These reactions can also occur in reformers with higher molecular weight feedstocks, since the reactants exist a short distance into the reformer tube, even when no methane exists in the feed. As mentioned in Reference 23, "The mechanism of thermal cracking (pyrolysis) of higher hydrocarbons is more complicated."   Reference 13 highlights the relative risk of carbon laydown from several paraffinic, aromatic, and olefinic compounds.

Many literature articles imply that carbon deposition can be predicted and avoided by using equilibrium calculations. This implication is not correct, and several references acknowledge this fact.   For methane feeds the equilibrium calculations provide reasonable guidelines, or "rules of thumb", but still are not definitive in predicting whether coke formation will be encountered. For these feeds if the conditions (composition, temperature, and pressure) are constrained so that at equilibrium (with the appropriate carbon form) no carbon exists, then it is likely that none will deposit. As clearly stated in Reference 4, page 82, "The principle (of equilibrated gas) *is no law of nature...*It is merely a rule of thumb, indicating process conditions which are critical for carbon formation". From page 85 of Reference 4, "The principle of equilibrated gas is no law of nature. Rates of carbon formation may be too slow. On the other hand, carbon formation may occur in spite of the principle, if the *actual* gas in the bulk phase, and thus the exterior of the catalyst pellet, shows affinity for carbon formation...Carbon formation is then a question of kinetics and the local approach to the reforming equilibrium." Obviously, the defining relationships which would indicate whether carbon will deposit are the reaction rates associated with the actual gas (not the equilibrated gas), and include the carbon forming as well as the carbon gasifying reactions. These rates have been extensively studied [3,4,20] and are not presently well enough understood to justify including their reactions in the model. Reference 23 states "...the process conditions at which deposition (carbon) occurs can only be determined experimentally for each particular catalyst."

Chapter 5 of Reference 4 presents several mechanisms for coking reactions, for methane as well as higher hydrocarbons. The rate expressions presented cannot be directly and practically used in a reformer reactor model, since many of them are for "without steam present" conditions. Section 2.2 of Chapter 5 presents an empirical approach, which uses a critical steam to carbon ratio for each hydrocarbon species. Critical steam to carbon ratios must be established by experiment. This kind of simple empirical relationship can be added as a model and connected to the reformer model, if the need arises. Presently when applied on-line, operating conditions must be empirically constrained to avoid coke formation. Experience and advice from catalyst vendors can be used to establish the empirical guidelines. Consequently, operating conditions cannot as closely approach coking conditions as would be possible if the coking rates were well defined. Even if these rates were well defined, on-line analysis of reformer conditions (naphtha feed components of high coking potential, and down the tube temperatures) may still dictate that operating conditions be empirically established.

Some confusion may exist pertaining to how catalyst activity affects coke deposition, or the potential for coke deposition. There are three distinct effects of catalyst activity. First, an inactive catalyst can deliver a specified reformer outlet "methane slip" at only slightly higher outlet temperature compared to a catalyst that is significantly more active. Results in Appendix B and Figure 12 of Reference 23 show this significant rise of temperature. For the few furnaces with in-the-tube temperature measurements and the ability to control firing profiles, this rise of temperature can be avoided by automatically adjusting firing profiles. The allowable temperatures along the tube must be empirically determined, as mentioned. Results in Appendix B illustrate this effect as well. Reformers without in-the-tube temperature measurements, or ones with limited capability of controlling firing profiles (i.e. top or floor fired) are at higher risk of coking due to inactive catalyst, since the rise in outlet temperature is subtle, and the rise in the process temperature near the tube inlet many times goes unnoticed. The third effect of catalyst activity on coking is its influence on the coking reaction rates, which as shown are relatively poorly understood. It is clear that the first two effects of catalyst activity are captured well in this model, and that the effect on coking rate is not required to predict, and empirically avoid the higher temperatures near the tube inlet that arise due to inactive catalysts operated at a firing profile similar to that used for active catalysts.

Typical operating conditions in industrial reformers are sufficiently far from coking conditions so that significant benefits can be obtained by applying the model, and constraining operating conditions empirically, as related to coking limits. Many times product values (hydrogen, ammonia or methanol) are sufficiently different from fuel values so that plant-wide optimum operating conditions are at a high steam to carbon ratio (where hydrogen production is favored, but at higher energy usage), and coking limits are not of concern. Plant operators, designers of steam reformers, and catalyst vendors many times focus only on minimum steam to carbon ratios as being "optimum", due to thermodynamic (energy per ton of product), and not economic objective functions.

2.4   Catalyst Poisoning

Catalyst "poisoning" is generally regarded as any process by which the number of catalyst active sites are reduced, and can occur due to sulfur coverage of active nickel sites, sintering of the nickel into fewer active sites, or coke deposition. The model lumps all those effects into the "activity" parameter.

The catalyst activity in the model can be a function of tube length, but typically is specified as uniform along the length. Measured data from an industrial reformer is essentially never available to allow the calculation of the activity profile. Even in reformers with in-the-tube temperature measurements, this information is usually insufficient to determine activity as a function of length. Process gas composition is required as a function of length to establish the activity profile. Clearly, such information can only be obtained in a laboratory environment.

References 23 and 24 address sulfur poisoning in significant detail. Reference 23 illustrates the relative activity of a catalyst as a function of mean sulfur coverage. Also, that reference lists a Temkin-like adsorption isotherm for equilibrium sulfur coverage, as a function of temperature, hydrogen, and hydrogen sulfide mole fraction. As stated, it takes years to establish sulfur equilibrium in a reformer.

A reasonable approach might be to "enhance" the model by imposing an activity profile by incorporating the sulfur equilibrium, or approach to equilibrium isotherm, along the tube length. Without on-line, real time, or at least infrequent laboratory analysis of feed sulfur content, the model would have insufficient information to establish the sulfur isotherm, and subsequent activity profile. Sulfur levels in feed stocks may be very low ($\approx$5 ppb) and difficult to detect. Also, it would be desirable to have feedback of sulfur content on the catalyst as a function of position in the tube, as well as feed sulfur history over the entire catalyst run length to properly validate the model. In view of these issues, and since the model with a uniform activity profile agrees well with measured conditions and outlet compositions, there appears little incentive to extend the modeling functionality in this area at this time.

3.   HEAT TRANSFER RATES AND HEAT BALANCES

Heat transfer rates are calculated to the catalyst tube from the firebox flue gas, through the tube wall, tube fouling, and inside tube film to the bulk fluid, as well as from the bulk fluid to the catalyst surface. As reported in the literature, the catalyst pellets are assumed to have a uniform temperature, since very little resistance to intra-pellet heat transfer exists.

Heat balances are determined for all the reactor models discussed. Therefore the heat duties of the radiant tubes, convection section heaters, waste heat boilers, and associated steam drums are all explicitly calculated and reported. Heat losses are part of the models, as described in the "Heat Losses" section. The enthalpies of all the streams entering and leaving each piece of equipment are calculated and reported. These streams include, for example, the process, fuel, air, and flue gas. Combustion, and all reaction related heat effects, are handled in models using enthalpies of all species including heats of formation based on of the elements at their standard states, so heats of reaction are avoided.

Heat balance terms such as furnace stack heat loss as a fraction of fired fuel heating value and furnace efficiency are not calculated, mainly since these "indices of performance" are not required in an optimization system that has the plant-wide operating profit as an objective function. These indices are remnants of local equipment or design "optimization" approaches. Typically a plant's operation should not be constrained nor its performance judged by these indices. Models can of course be easily used to calculate these indices, and "plant built" to the furnace models to perform these calculations on-line.

Heat transfer rates are calculated to the catalyst tube from the firebox flue gas, through the tube wall, tube fouling, and inside tube film to the bulk fluid, as well as from the bulk fluid to the catalyst surface. As reported in the literature, the catalyst pellets are assumed to have a uniform temperature, since very little resistance to intra-pellet heat transfer exists.

Heat balances are determined for all the reactor models discussed. Therefore the heat duties of the radiant tubes, convection section heaters, waste heat boilers, and associated steam drums are all explicitly calculated and reported. Heat losses are part of the models, as described in the "Heat Losses" section. The enthalpies of all the streams entering and leaving each piece of equipment are calculated and reported. These streams include, for example, the process, fuel, air, and flue gas. Combustion, and all reaction related heat effects, are handled in models using enthalpies of all species including heats of formation based on of the elements at their standard states, so heats of reaction are avoided.

Heat balance terms such as furnace stack heat loss as a fraction of fired fuel heating value and furnace efficiency are not calculated, mainly since these "indices of performance" are not required in an optimization system that has the plant-wide operating profit as an objective function. These indices are remnants of local equipment or design "optimization" approaches. Typically a plant's operation should not be constrained nor its performance judged by these indices. Models can of course be easily used to calculate these indices, and "plant built" to the furnace models to perform these calculations on-line.

Heat transfer rates are calculated to the catalyst tube from the firebox flue gas, through the tube wall, tube fouling, and inside tube film to the bulk fluid, as well as from the bulk fluid to the catalyst surface. As reported in the literature, the catalyst pellets are assumed to have a uniform temperature, since very little resistance to intra-pellet heat transfer exists.

Heat balances are determined for all the reactor models discussed. Therefore the heat duties of the radiant tubes, convection section heaters, waste heat boilers, and associated steam drums are all explicitly calculated and reported. Heat losses are part of the models, as described in the "Heat Losses" section. The enthalpies of all the streams entering and leaving each piece of equipment are calculated and reported. These streams include, for example, the process, fuel, air, and flue gas. Combustion, and all reaction related heat effects, are handled in models using enthalpies of all species including heats of formation based on of the elements at their standard states, so heats of reaction are avoided.

Heat balance terms such as furnace stack heat loss as a fraction of fired fuel heating value and furnace efficiency are not calculated, mainly since these "indices of performance" are not required in an optimization system that has the plant-wide operating profit as an objective function. These indices are remnants of local equipment or design "optimization" approaches. Typically a plant's operation should not be constrained nor its performance judged by these indices. Models can of course be easily used to calculate these indices, and "plant built" to the furnace models to perform these calculations on-line.

3.1   Firebox to Catalyst Tube

Firebox flue gas to catalyst tube heat transfer is primarily governed by radiation, but convection, although proportionally small, is included. The relationships used are derived from Chapter 19 of *Process Heat Transfer* by D. Q. Kern.[14]

$$Q = \sigma * F * \alpha * A_{cp} * (T_g{}^4 - T_{ff}{}^4) + h_c * A_0 * (T_g - T_s)$$

Where:

Q	= total heat transferred
σ	= Stefan-Boltzmann constant
F	= Overall view (exchange) factor (See Kern, Figure 19.15)
α	= Cold plane effectiveness factor (See Kern, Figure 19.11)
A_{cp}	= Cold plane area (of both sides of tube bank, or row)
T_g	= Effective temperature of radiating gas, absolute
T_{ff}	= Temperature of tube front face (surface), absolute
h_c	= Convective heat transfer coefficient
A_0	= Outside area of tube
T_s	= Average outside tube surface temperature, absolute
T_{ff}	= T_s * PMDF
PMDF	= Peripheral maldistribution factor (accounts for circumferential heat flux maldistribution)

This equation is essentially Equation 19.9 from Kern, modified to allow variation of conditions along the tube length. The model integrates this heat transfer relationship along the length of the tube, while simultaneously solving the kinetics, pressure drop relationships, and all other equations related to the reformer (and the rest of the plant).

The effective gas radiating profile as a function of length is related to the burner placement in the firebox, as well as to the relative burner firing rates. The overall effective gas radiating profile is built up from the individual burner row profiles. The individual burner row profiles are similar to normal distributions of temperatures around the centerline location of the burners. The effects of adjacent burner rows overlap. Amplitudes of the temperature distributions are related to the firing rates of the burners. These amplitude to burner firing rate relationships have been derived from the steady-state gains of step response models relating "catalyst" temperatures (i.e. in the tube temperatures) to burner firing rates, and are included in the model. The chain ruling that may seem required to establish the relationship of the effective temperature profile amplitudes to the firing rates is implicitly done, since the steady-state gain relationships are added as constraints, and the amplitude variables then become dependent variables. Furthermore, the steady-state gain relationship intercept values are updated as parameters, for measured values of burner firing rates and measured "catalyst" temperatures. By this means the solution to the Parameter or Reconcile case (calculated prior to each Optimize case) can precisely match the measured down the tube temperatures, including the tube inlet (convection section heater outlet), in-the-tube "catalyst" temperatures, as well as tube outlet temperature. The Appendix A simulation results and discussion illustrate that care must be taken so that the measured catalyst temperatures are not matched with unreasonable effective gas radiating and flux profiles.

Few plants have measured temperatures inside the catalyst tubes. When those temperatures are not available there may be little justification for building models that relate the individual burner profiles to an overall effective gas radiating profile. An overall effective radiating temperature profile is still justified, but without measured feedback, its shape cannot be effectively updated on-line. A simpler gas temperature model can then be employed. Without in the tube temperature measurements it follows that temperature profile control in the tube is not practical. The absolute level of the radiating gas temperature, both in the case when the profile is built up from burner profiles and when it is not, is determined from actual operating conditions (namely measured or specified outlet temperature). The level of the temperature profile is that at which the resulting integrated heat transfer rate supplies the required heat to achieve the specified tube outlet temperature.

An important constraint on the effective gas radiating temperature profile is its relationship to the "bridgewall" temperature, which is the temperature of the flue gas leaving the radiating temperature after it has given up the total heat absorbed in the radiant section tubes and lost (small) from the radiant section walls. This relationship is in many cases handled by a radiant section efficiency parameter. This model does not use that approach, but uses an offset between the length average effective gas radiating temperature and the bridgewall temperature as the analogous parameter. Others, such as Wimpress of C. F. Braun & Co. have used a similar approach.[27] The offset parameter is updated on-line, and is determined by firebox heat balance determined from heat absorbed on the process side, fuel firing rate, and arch oxygen measurements. Heat balances using measured bridgewall temperatures from conventional thermocouples should not be attempted, due to the reasons explained in the "Convection Section" section of this report. The measured fuel flow, for the observed radiant duty, affects this parameter the most. The effective radiating gas temperature profile is not tied to the flue gas temperature at other than the bridgewall position, since the effective temperature (the radiation "source" temperature) is a result of all refractory as well as flue gas temperatures in line-of-sight of the tube front face temperature (radiation "sink" temperature) at any position. If the heat balance with the flue gas were calculated all along the tube, and if that temperature were used as the radiating source temperature, the resulting heat flux profile would not reflect the actual phenomena occurring, and would be significantly in error for most, if not all, industrial furnaces.

The following chart shows a typical net effective gas radiating temperature profile, along with the individual burner row by burner row profiles.

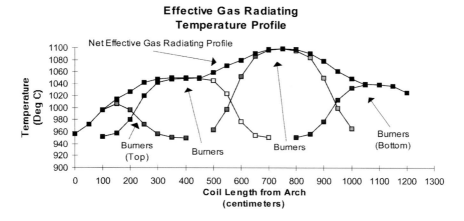

Figure 1. Effective Gas Radiating Temperature Profile

3.2   Conduction Across Tube Wall

Heat transfer across the tube wall occurs by conduction.   The one-dimensional form of Fourier's law of heat conduction (page 245 of Reference 25) is used as the governing relationship between heat flux, thermal conductivity, and temperature gradient. The tube wall thickness is an explicit "discretized" variable in the model.   It therefore can vary with length, although most tubes are of uniform thickness. The thermal conductivity is also a model variable, but is considered constant as a function of length.

3.3   Fouling Resistance

A fouling (or "coke") layer is included in the tube reactor model. Fouling thickness in the model can vary with length. The fouling thermal conductivity is a variable whose value can be specified, and is assumed invariant with length. Theoretically, measured tube metal temperatures at various lengths can be used to "back out" the apparent fouling profile, although inherent errors in pyrometer measurements make this impractical. With little or no fouling the outside tube metal temperature profile approaches the shape of the process (inside the tube) temperature profile, and with appreciate fouling the outside metal temperature profile approaches the shape of the effective radiating gas profile. Hence, the fouling, even if assumed uniform, can be used as a parameter to more closely match tube metal temperatures close to the tube inlet, while the "peripheral maldistribution factor" can be used as a parameter to match the hottest measured tube metal temperature, usually near the tube outlet.

3.4   Inside Tube to Bulk Fluid

The inside tube surface to bulk fluid heat transfer coefficient is calculated using the Sieder-Tate correlation for turbulent flow (similar to Dittus-Boelter), as listed in Reference 15:

$$\frac{h\,D}{k} = 0.023\,(Re^{0.8}\,Pr^{0.3})(\frac{\mu}{\mu_0})^{0.14}$$

Where:

h	= Inside heat transfer coefficient
D	= Inside tube diameter
k	= Process fluid thermal conductivity
Re	= Reynolds Number
Pr	= Prandtl Number
μ	= Bulk fluid viscosity
μ_0	= Fluid viscosity at the tube wall

3.5   Bulk Fluid to Catalyst Pellet

Heat transfer to the pellet is calculated by applying Equation 28 of Chapter 14 in Reference 31:

$$\Delta T\,film = (T_{gas} - T_{pellet\ surface}) = \frac{L*(-rate)*(-\Delta H_r)}{h_s}$$

This relationship assumes that all the reactions occur inside the pellets, and that none occurs in the bulk fluid. This is a very good assumption in a steam reformer. The heat of reaction is backed out of the heat effects, since heats of reaction are not explicitly used elsewhere. Heats of formation are based on elements at their standard reference states, so the overall heat effects (sensible plus reaction) are included in enthalpies.

The bulk fluid to catalyst pellet heat transfer coefficient, h_S, is calculated as a function of fluid properties, catalyst size, and fluid temperature using the heat transfer relationship from Reference 11, page 89, Table 4.2:

$$\frac{h_s\,D_p}{\lambda_f} = 2 + 1.1\,Pr^{1/3}\,Re^{\,0.6}$$

3.6 Within the Catalyst Pellet

Little resistance to heat transfer exists within the catalyst pellet (Reference 4, page 69). Consequently, the catalyst temperature is assumed to be uniform, at the catalyst surface temperature.

3.7 Convection Section

The convection section pre-heaters are modeled using standard heat exchanger models. Considerable differences between measured and calculated temperatures exist on the flue gas side, especially at the higher flue gas temperatures leaving the firebox, or just above the auxiliary burners. These discrepancies can be explained by the shortcomings associated with standard thermocouples employed in high temperature service. The main problem with these thermocouples is that they cannot come close to thermal equilibrium with the very hot flue gas, so their measurements are consistently low. This problem is well understood, and documented in several heat transfer texts.[16,17,18] Reference 16 illustrates the large errors that can arise in high temperature environments, and Reference 17 highlights the futility of using temperatures measured with conventional thermocouples as the bases for heat balances. Reed, in Reference 17, states, "...experience shows that an attempt at heat balance where gas temperatures are taken from fixed thermocouples is largely a pointless exercise in calculation because of inaccuracy of gas temperature measurement and thus, gas heat content". Illustrations on pages 33-17 and 33-18 of Reference 18 show thermocouples that are specially designed for high temperature measurements.

The measured fuel flows, arch oxygen composition, and high pressure steam drum heat balance confirm that the heat duties calculated from the process side (as opposed to the flue gas side) are most accurate, as would be expected. The high pressure steam system and boiler feed water measurements impact significantly on the convection section heat balance since boiler feed water preheat and steam superheat duties make up the majority of the convection section duty. The high pressure steam import flow, and the expected versus measured and calculated synthesis gas compressor steam turbine performance further support that the process side, and not the flue gas side measurements are the most accurate.

3.8   Fuel and Combustion Air System

The fuel-air mixtures are combusted in "extent of reaction" models. Complete combustion to water vapor and carbon dioxide is assumed, with no residual carbon monoxide remaining. The model can be implicitly kept valid by lower bounding the flue gas oxygen content so complete combustion is maintained.

Combustion air preheat is modeled with standard heat exchanger models.

3.9   Heat Losses

The firebox model has a heat loss term that can be calculated as a parameter, however the loss term is essentially a small difference between large numbers (total heat fired minus total heat absorbed in radiant plus convection sections), so it is subject to large relative errors. Additionally, the heat "loss" may be calculated as a heat gain depending on the accuracy of the fired and absorbed heat duties. To keep the heat loss from the firebox reasonable, it is typically set to a small value, on the order of 1% of the heat fired.

There is also a heat loss term associated with the high-pressure steam drum. This term is typically used to understand the whether significant errors exist in boiler feed water flows, steam flows and temperatures, furnace convection section measurements, and fuel flow and composition measurements. During the optimization system commissioning sensitivity of this term to the aforementioned measurements helps those commissioning the system eliminate the worst measurements, and establish the validity of the models.

4.   PRESSURE DROP

Pressure drop calculations in the primary and secondary reformer models are based on the Ergun[5] relationship, as presented in Reference 25. Both the laminar flow and turbulent flow terms are included. The natural parameter that arises from the Ergun equation, which can be updated with measured pressure drop information, is the "TURBULENT_DP_COEF" term in the models of both the primary and secondary reformers. This term affects only the pressure drop, whereas another term, the bed void fraction, which might also have been used as the parameter to update with measure pressure drop, also affects all the reaction rates. The bed void fraction affects the amount of catalyst in a fixed volume reformer tube, and is not an appropriate parameter to use in this case. The void fractions of typical packed beds are shown in Figure 5.70 of Reference (26). Void fractions of 0.4 to 0.6 are typical, and can be determined for specific catalysts sizes and shapes from vendor specification sheets, by measurement, or, with more difficulty, by calculation.

The "TURBULENT_DP_COEF" is the appropriate performance indicator for catalyst pressure drop, and is a better indicator than pressure drop itself, since it is independent of all the known effects of flow rate (both hydrocarbon and steam), gas density, viscosity, catalyst particle diameter, and void fraction. Pressure drop itself is important though, due to the stress it imposes on the catalyst (which raises the potential for crushing) and normally the optimization system has an upper bound on pressure drop. That bound may or may not be active at the solution, depending on the catalyst condition, and whether the solution is maximizing throughput.

The pressure drop profile illustrated is for a primary reformer when the pressure drop across the catalyst was fairly high. The differential pressure indicator showed a 2.9 kg/cm^2 drop across the catalyst tubes.

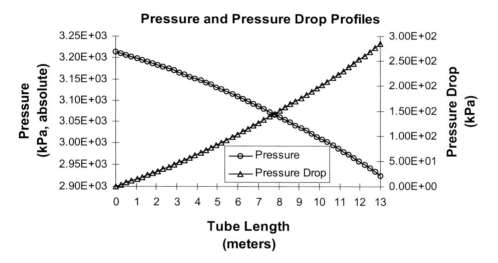

Figure 2. Pressure and Pressure Drop Profile

4.1 Secondary Reformer Reactions and Heat Effects

Most steam reformers are process furnaces, transferring the required endothermic heat from a firebox to the process inside catalyst-packed tubes. Some steam reformers are "auto-thermal", getting their heat from oxidation reactions occurring within the catalyst bed. Secondary reformers in ammonia plants are an example of auto-thermal steam reformers.

There are several categories of reactions and heat effects calculated in the secondary reformer model. First, the primary reformer effluent is mixed with the process air. Then, the effect of consuming the oxygen that is admitted with the process air is determined. This effect was investigated by simulating several possible reaction pathways. The results were insensitive to the pathway chosen, so the simplest, and also the most thermodynamically favored route was used. This assumes that all the oxygen preferentially reacts with primary reformer effluent hydrogen. Sufficient hydrogen exists under all reasonable operating conditions so oxygen is the limiting reactant, therefore no methane is oxidized. The model includes oxidation of methane as a possible pathway, however it is presently not activated. That pathway can be activated without modifying the model by just changing input data (extent of reaction). The adiabatic "flame" temperature is calculated for this oxidation set of reactions, and the outlet from this "extent of reaction" model is the inlet to the secondary reformer packed catalyst bed.

This model assumes that the oxidation is complete before the top of the catalyst bed, and that only methane reforming as described by References 1 and 2 occur in the reactor bed. The same kinetic rate expressions are used as in the primary reformer. The catalyst activity is of course different from the primary reformer activity, and is determined from operating data. An initial estimate of its value, relative to the primary reformer activity, was calculated as the primary reformer activity, times the nickel weight percent ratio, times the density ratio of the catalysts. This estimate assumes that the nickel mass in either catalyst supplies similar number of active sites. The activity calculated in this manner is close to the reconciled activity, based on measured compositions in the primary and secondary outlet, and operating conditions. Effluent compositions are very insensitive to assumptions concerning how the oxygen is reacted. For example, if some of the methane from the primary effluent is oxidized, rather than only hydrogen, the results are essentially unchanged. The high temperatures in the reactor cause the reaction rates to be relatively high, bringing the outlet near equilibrium. This near equilibrium condition is a state, and not a path function, so it is not surprising that various reaction pathways affect the results so little.

The reactor bed is not assumed to be adiabatic, but loses heat from its walls. Presently a constant heat flux from the walls is assumed. A flux profile can easily be imposed, based on heat transfer driving force, but its effect will be negligible. In fact, if the reactor were assumed adiabatic the results would be essentially unchanged. The heat loss from the reactor walls was retained, so that the effects of the heat losses can be easily demonstrated.

4.2   Model Validation

The primary reformer models has been validated by comparing calculated to measured results, and by comparing predicted to observed results. Three cases of calculated versus measured results are presented, one for naphtha feed, and the other for butane feed. One case of predicted versus observed results is presented, for a naphtha feed. The on-line, closed-loop optimization system continuously supplies validation results, by reporting differences (biases) between calculated and measured results on every optimization cycle.

These results, along with calculated parameters, can be monitored by trending their values as a function of time.

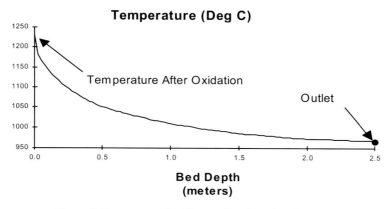

Figure 3. Temperature Profile as a function of Bed Depth

The first two validation cases are Parameter cases, while the third is a Reconcile case. In the first two cases only primary reformer calculations are done, while in the third case both primary and secondary reformer calculations are performed. The Parameter cases manipulate the process steam flow and outlet temperature to match observed outlet CO_2 and CH_4. In the Reconcile case the best values for the primary and secondary reformer catalyst activities are determined as well, using an objective function which minimizes the differences between calculated and observed outlet compositions of both reformers, calculated and observed primary outlet temperature, and calculated and observed process steam flow rate.

4.2.1   Validation Case 1 (Naphtha Feed Parameter Case)

The feed composition used for this validation case was characterized as listed in Table 1.2, based on measured naphtha gravity, ASTM-D86 volume versus boiling point, and PINA (paraffin, isoparaffin, naphthinic, and aromatic content) information, plus recycle desulfurization hydrogen stream measurements. The feed is a mix of the naphtha and desulfurization hydrogen streams. This naphtha feed molecular weight is 65.923.

Table 1.0 Validation Case 1 - Operating Conditions

Feed	Naphtha
Feed rate	17.7 Tonnes/h
Process steam rate	79.0 Tonnes/h
	(Parameterized to 82.92 to match effluent CO_2)
Inlet temperature	470°C
Outlet temperature	760°C (Parameterized to 761.99 to match CH_4)

Table 1.1 Validation Case 1 - Outlet Compositions

Component	Calculated Dry mole %	Observed Dry mole %
HYDROGEN	63.9077	63.6
NITROGEN	0.3675	0.4
ARGON	0.0060	---
CARBON MONOXIDE	9.0188	9.2
CARBON DIOXIDE	16.4000	16.4
METHANE	10.3000	10.3

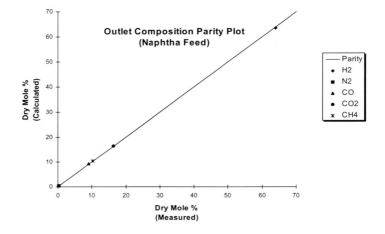

Table 1.2 Validation Case 1 - Feed Composition

Component	Mole%
HYDROGEN	13.807
NITROGEN	4.610
ARGON	0.075
CARBON MONOXIDE	0.000
CARBON DIOXIDE	0.000
METHANE	0.258
ETHANE	0.000
PROPANE	0.000
ISOBUTANE	0.932
n-BUTANE	1.792
ISOPENTANE	17.142
n-PENTANE	24.221
CYCLOPENTANE	2.553
ISO-C6	10.628
n-HEXANE	7.386
CYCLO-C6	5.952
BENZENE	2.064
ISO-C7	2.639
n-HEPTANE	1.952
CYCLO-C7	1.441
TOLUENE	0.391
ISO-C8	0.621
n-OCTANE	0.509

CYCLO-C8	0.314
AROMATIC-C8	0.150
ISO-C9	0.155
n-NONANE	0.103
CYCLO-C9	0.087
AROMATIC-C9	0.066
ISO-C10	0.048
CYCLO-C10	0.027
n-DECANE	0.047
AROMATIC-C10	0.030
WATER	0.000

4.2.2   Validation Case 1a (Naphtha Feed Simulate Case)

This case is simulated at the same conditions as the Case 1 Parameter case, except at an outlet temperature of 778 Deg C, compared to 762 Deg C.

A comprehensive set of plant operating conditions was not available for this case, but the plant indicated a methane slip of about 8.5 % was observed by the plant for this outlet temperature. The calculated outlet methane dry mole percent of 8.8 % is very close to the 8.5 % that was observed, and since all the independent conditions (feed rate, steam to carbon ratio, etc.) are not precisely known for this case, the "discrepancy" may be easily attributed to differences in simulated versus actual conditions.

Table 1.3 Validation Case 1a - Outlet Compositions

Component	Calculated Dry mole %	Observed Dry mole %
HYDROGEN	65.1413	Unavailable
NITROGEN	0.3550	Unavailable
ARGON	0.0058	Unavailable
CARBON MONOXIDE	9.8817	Unavailable
CARBON DIOXIDE	15.8156	Unavailable
METHANE	8.8006	$\cong 8.5$

4.2.3   Validation Case 2 (Butane Feed Parameter Case)

The feed composition used for this validation case was characterized as listed in Table 2.2, based on typical butane feed, plus recycle desulfurization hydrogen stream measurements. The feed is a mix of the butane and desulfurization hydrogen streams.

Table 2.0 Validation Case 2 - Operating Conditions

Feed	Butane
Feed rate	17.7 Tonnes/h
Process steam rate	70.3 Tonnes/h (Parameterized to 70.8 to match effluent CO_2)
Inlet temperature	473°C
Outlet temperature	773°C Average (Range:763 < T < 779)
	(Parameterized to 779 to match CH_4)

Table 2.1 Validation Case 2 - Outlet Compositions

Component	Calculated Dry mole %	Observed Dry mole %
HYDROGEN	63.6562	63.5
NITROGEN	0.2844	0.3
ARGON	0.0046	---
CARBON MONOXIDE	10.3548	10.4
CARBON DIOXIDE	14.7	14.7
METHANE	11.0	11.0

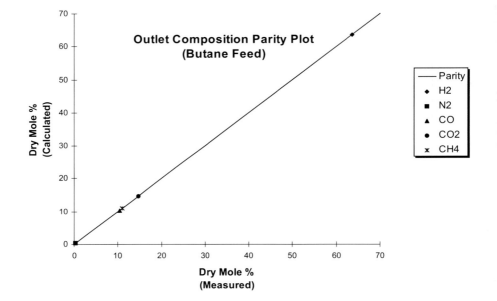

Table 2.2 Validation Case 2 - Feed Composition

Component	Mole %
HYDROGEN	8.100
NITROGEN	2.806
ARGON	0.046
CARBON MONOXIDE	0.000
CARBON DIOXIDE	0.000
METHANE	0.157
ETHANE	0.000
PROPANE	0.000
ISOBUTANE	0.932
n-BUTANE	87.959
ISOPENTANE	0.000
n-PENTANE	0.000

4.2.4 Validation Case 3 (Primary & Secondary Reformer Butane Feed Reconcile Case)

A Reconcile case was executed with the objective function minimizing the sum of the weighted squared differences between the calculated and measured primary outlet compositions, outlet temperature, and secondary reformer outlet compositions. Primary reformer catalyst activity, secondary reformer catalyst activity, primary outlet CO_2 measurement minus calculated value difference, and secondary outlet CH_4 measurement minus calculated value difference were the degrees of freedom (the independent manipulated variables). The degrees of freedom are bounded to assure the results are most reasonable. By posing the problem in this manner the insensitive nature of the catalyst activities (due to near equilibrium conditions) is handled as well as possible. By step bounding the catalyst activities, the errors or "noise" in the composition and process steam flow measurements are not all forced into the catalysts activities, as would be the case if a "square" Parameter case were posed. The best weighting of the terms in the objective function is determined during commissioning when many sets of data are processed in an open-loop mode. Over long time periods (months) the catalyst activities trends downward, since the steps bounds are relative to the activity values from the last solution. Soft bounds (see [DMO]™ Users' Manual) may need to be employed if the data has significant error, causing otherwise (i.e. without soft bounds) infeasible problems to be posed.

This approach allows the relationship between firing duties and catalyst temperatures, if available, (derived from [DMC]™ controller steady-state gains) to be used, with all of the intercepts being updated as parameters, determined primarily by the measured catalyst temperatures. The catalyst temperatures are matched precisely at the solution of each Reconcile case.

The overall approach in determining the catalyst activities relies on reformer outlet compositions which are near equilibrium to update rate parameters (catalyst activities). This kind of analysis is inherently problematic, but can be handled as described. Short term trends of the catalyst activities (hourly, daily, or even over a week or so) will be subject to somewhat erratic, noisy behavior, but will not detract significantly from the ability of the models to predict appropriate outlet compositions and reactor temperature profiles. As the catalysts become less active the aforementioned numerical problems diminish, and the activity trends will become smoother, for the same error in measurements, since the outlet conditions become further (but still not very far) from equilibrium.

The feed composition used for this validation case was characterized as listed in Table 3.2, based on typical butane feed, plus recycle desulfurization hydrogen stream measurements. The feed is a mix of the butane and desulfurization hydrogen streams.

As described, the secondary reformer model was solved along with the primary reformer model for Validation Case 3. The results are listed in Table 3.3.

Table 3.0 Validation Case 3 - Operating Conditions

Feed	Butane
Feed rate	17.451 Tonnes/h
Process steam rate	73.556 Tonnes/h (Reconciled to 74.773)
Inlet temperature	464.2°C
Outlet temperature	772.81°C (Reconciled to 772.92)

Table 3.1 Validation Case 3 - Outlet Compositions

Component	Calculated Dry mole %	Observed Dry mole %
HYDROGEN	64.3261	66.4770
NITROGEN	0.4858	0.2608
ARGON	0.0935	---
CARBON MONOXIDE	9.9671	10.4600
CARBON DIOXIDE	14.8347	14.896
METHANE	10.2928	10.415

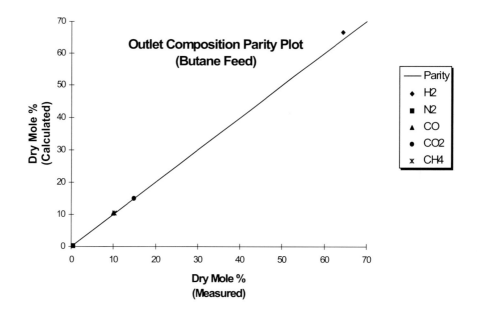

Table 3.2 Validation Case 3 Feed - Composition

Component	Mole %
HYDROGEN	10.8684
NITROGEN	4.5144
ARGON	0.8692
CARBON MONOXIDE	0.0000
CARBON DIOXIDE	0.0000
METHANE	2.9427
ETHANE	0.0000
PROPANE	0.0000
ISOBUTANE	0.8121
n-BUTANE	79.9885
ISOPENTANE	0.0000
n-PENTANE	0.0000

Table 3.3 Validation Case 3 - Secondary Reformer Outlet Conditions

Component	Calculated Dry mole %	Observed Dry mole %
HYDROGEN	52.952	53.553
NITROGEN	22.254	22.254
ARGON	0.327	--------
CARBON MONOXIDE	13.818	13.245
CARBON DIOXIDE	10.164	10.583
METHANE	0.4843	0.4843
Temperature (°C)	966.82	966.52

The outlet methane concentration bias in the objective function was heavily weighted (to drive it toward zero, since the objective function is the sum of the weighted squares biases) and therefore explains the good agreement between calculated and observed values. The process air flow measurement bias (difference between measured and calculated) was a parameter, so the calculated nitrogen composition is precisely equal to the measured value. The air flow measurement bias was 4.4% of the measured value at the solution. The calculated outlet temperature is surprisingly close to the measured value. The heat balance around the high pressure steam drum required only a 1.2% heat loss to close in this case. That balance is of course affected by numerous other measurements, so the calculated secondary reformer outlet enthalpy can only be said to be part of the overall consistent set of information.

5. REFERENCES

1. Xu, J.; Froment, G. F., Methane Steam Reforming, Methanation and Water-Gas Shift: I. Intrinsic Kinetics, *AIChE J.* **1989**, *35 (1)*, 88.
2. Xu, J.; Froment, G. F., Methane Steam Reforming:II. Diffusional Limitations and Reactor Simulation, *AIChE J,* **1989**, *35 (1)*, 97.

3. Rostrup-Nielsen, J. R., Steam Reforming Catalysts, An Investigation of Catalysts for Tubular Steam Reforming of Hydrocarbons, Teknisk Forlag A/S, Copenhagen, 1975.
4. Rostrup-Nielsen, J. R., *Catalytic Steam Reforming*, Springer-Verlag: Berlin/Heidelberg/ New York, 1984.
5. Ergun, S., Fluid Flow through Packed Columns, *Chem. Eng. Prog.* **1952**, *48 (2)*, 89.
6. Elnashaie, S. S. E. H. et al., Digital Simulation of Industrial Steam Reformers for Natural Gas Using Heterogeneous Models, *Can. J. Chem. Eng.* **1992**, *70,* 786.
7. Boudart, M., Two Step Catalytic Reaction, *AIChE J.* **1972**, *18 (3)*, 465.
8. Smith, J.M., *Chemical Engineering Kinetics*, 2nd ed., McGraw-Hill: New York, 1970.
9. Rostrup-Nielsen, J. R.; Wristberg, J., Steam Reforming of Natural Gas at Critical Conditions, Natural Gas Processing and Utilisation Conference, Dublin, Ireland, 1976.
10. Satterfield, C. N.; Sherwood, T. K., *The Role of Diffusion in Catalysis*, Addison-Wesley Publishing Company: Reading, Massachusetts, 1963.
11. Rase, H. F., *Fixed-Bed Reactor Design and Diagnostics Gas-Phase Reactions*, Butterworth Publishers: Boston, 1990.
12. Bridger, G. W., Design of Hydrocarbon Reformers, *Chem. Process Eng.*, January, 1972.
13. Rostrup-Nielsen, J. R., Hydrogen via Steam Reforming of Naphtha, *Chem. Eng. Prog.*, September, 1977.
14. Kern, D. Q., *Process Heat Transfer*, McGraw-Hill: New York, 1950, 1990.
15. Greenkorn, R. A.; Kessler, D. P., *Transfer Operations*, McGraw-Hill: New York, 1972.
16. Krieth, *Principles of Heat Transfer*, International Textsbooks Co.: Scranton, Pennsylvania, 1965.
17. Reed, R. D., *Furnace Operations*, Gulf Publishing Co.: Houston, Texas, 1976.
18. *Steam - Its Generation and Use*, The Babcock & Wilcox Company, 1972.
19. Rostrup-Nielsen, J. R., Activity of Nickel Catalysts for Steam Reforming of Hydrocarbons, *J. Catal.*, **1973**, 31, 173-199.
20. Froment, G. F.; Wagner, E. S., Steam Reforming Analyzed, *Hydrocarbon Process.*, July, 1992.
21. Colton, J. W., Pinpoint Carbon Deposition, *Hydrocarbon Process.*, January, 1991.
22. Rostrup-Nielsen, J. R., Equilibria of Decomposition Reactions of Carbon Monoxide and Methane Over Nickel Catalysts, *J. Catal.*, **1972**, *27*, 343-356.
23. Hansen, J. H. B.; Storgaard, L.; Pedersen, P. S., Aspects of Modern Reforming Technology and Catalysts, AIChE Paper No. 279d, AIChE Ammonia Symposium, *Safety in Ammonia Plants and Related Facilities*, Los Angeles, California, November 17-20, (1991).
24. Christiansen, L. J.; Andersen, S. L., Transient Profiles in Sulphur Poisoning of Steam Reformers, *Chem. Eng. Sci.* **1980**, *35*, 314-321.
25. Bird, R. B.; Stewart, W. E.; Lightfoot, E. N., *Transport Phenomena*, John Wily & Sons, Inc., 1960.
26. Perry, R. H.; Chilton, C. H., *Chemical Engineers' Handbook*, 5th Edition, McGraw-Hill: New York, 1973.
27. Wimpress, N., Generalized Method Predicts Fired-Heater Performance, *Chem. Eng.*, May, 1978.
28. Singh, C. P. P.; Saraf, D. N., Simulation of Side Fired Steam-Hydrocarbon Reformers, *Ind. Eng. Chem. Process Des. Dev.* **1979**, *18(1)*.
29. Rostrup-Nielsen, J. R.; Christiansen, L. J.; Hansen, J. H., Activity of Steam Reforming Catalysts: Role and Assessment, *Appl. Catal.*, **1988**, *43*, 287-303.
30. Phillips. T. R.; Yarwood, T. A.; Mulhall, J.; Turner, G. E., The Kinetics and Mechanism of Reaction Between Steam and Hydrocarbons Over Nickel Catalysts in the Temperature Range 350-500°C, *J. Catal.*, **1970**, *17*, 28-43.
31. Levenspiel, O., *Chemical Reaction Engineering*, John Wily & Sons, Inc., 1962 & 1972.

APPENDIX A   SIMULATION RESULTS

Primary Reformer

Results are shown for naphtha and butane feeds. The naphtha feed case profiles are results from the "Validation Case 1", and the butane feed results are from "Validation Case 3" described in the "Model Validation" section.

The calculated versus outlet composition comparisons have been discussed. This section illustrates the profile results, and compares them to literature sources (for compositions), and to measurements (for temperatures). Only the key component concentrations are plotted, such as methane, hydrogen, carbon monoxide, carbon dioxide, and the main hydrocarbon species. The minor hydrocarbon species, nitrogen, and argon concentration profiles are not shown.

Both the naphtha and butane feed cases show the methane profile rising from very low inlet values to a maximum, and falling to the outlet composition. The hydrocarbon species compositions fall quickly, and are essentially zero at about 4 to 6 meters from the tube inlet. These profiles are in close agreement with the profiles shown in References 4 and 13. For more active catalyst, the hydrocarbon species disappear closer to the tube inlet. The simulated temperature profiles are also in good agreement with profiles in those references, but more importantly, they agree precisely with the measured profiles. Reference 23 shows that the temperature profile in a top fired reformer is significantly different than in a wall-fired furnace.

Only one literature reference[12] was found that illustrated the hydrogen composition profile along the catalyst tube. The simulated results agree well with that reference's profile. At first the dry basis hydrogen composition profile appears strange, since it peaks at a position fairly close to the tube inlet (about 1 to 2 meters), falls, then rises again. On a wet basis the profile is monotonic, and appears reasonable, since no decline occurs. The initial sharp rise in hydrogen concentration is due to the fast rates of reforming of the hydrocarbon species. As carbon dioxide concentration rises, it impedes hydrogen production via the water gas shift reaction, but as shown on the wet basis profile plots, the rate of change is still positive. Since water is a reactant, and not just a dilulent of the hydrocarbon to prevent coking, dry basis profiles are misleading. The Reference 12 dry basis hydrogen profile has the same peak and subsequent rise as the results plotted for naphtha feed. The hydrogen profile for butane feed, for the catalyst activity and firing profile representing late June 1994 operation, shows a much more subtle "peak".

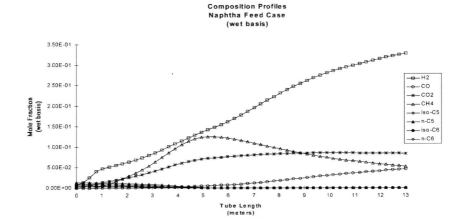

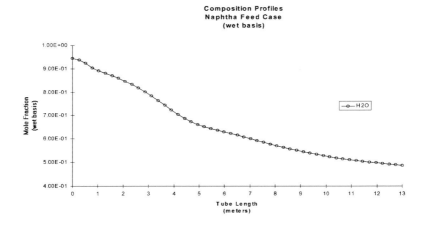

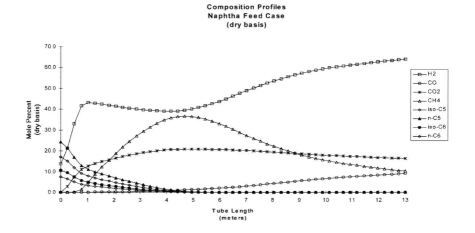

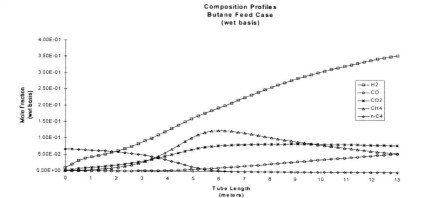

Composition Profiles
Butane Feed Case
(wet basis)

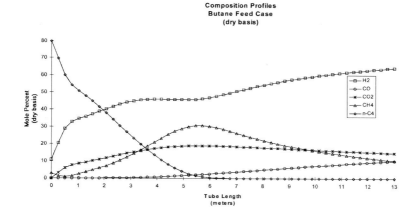

Composition Profiles
Butane Feed Case
(dry basis)

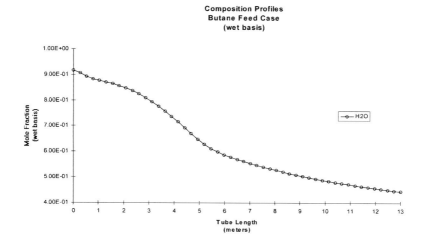

Composition Profiles
Butane Feed Case
(wet basis)

   The flux calculated at the top of the tube is quite small, but must be that small to allow the temperature to drop from the measured convection section feed/steam preheat exchanger outlet temperature to the upper most measured catalyst temperature.   The small flux is insufficient to supply all the net endothermic heat of reaction, so the temperature drops several degrees. Further down the tube the flux is sufficient to supply the reaction plus sensible heat, and the temperature rises. The flux profile is sensitive to the specified catalyst temperatures. Small discrepancies in the catalyst temperatures (or their position) require that the flux profile change significantly, if all the model and measured temperatures are required to match. Reconciliation of the measured temperatures, with shape constraints (rate of change or inflection point restrictions) on the flux profile can be used to assure a reasonable flux profile.

   The measured tube metal temperature (TMT) is precisely matched at the solution of a Parameter or Reconcile case, since the "Peripheral Maldistribution Factor" (See "Firebox to Catalyst Tube" heat transfer section) is used as a parameter to adjust the relationship between the average outside tube metal temperature, calculated assuming uniform heat flux distribution, and the front face tube metal temperature. The front face TMT is higher than the average TMT because heat flux is not uniformly circumferencially distributed. The "Peripheral Maldistribution Factor" is assumed constant in the Simulate or Optimize cases, which is a very good assumption.   As mentioned in the "Fouling Resistance" heat transfer section, the fouling can be used to match a measured TMT near the tube inlet.

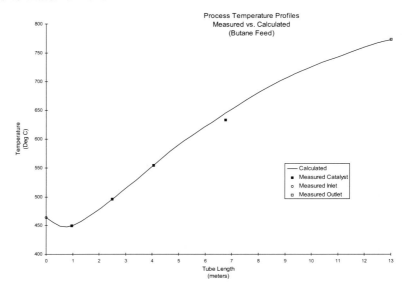

Adiabatic Pre-Reformer

An adiabatic pre-reformer was modeled with the same kinetics used in the previous models. A reactor bed was configured to represent an industrial reactor that was newly commissioned. Consequently the catalyst activity was assumed to be uniform, that is, no poisoned front had time to be established. The catalyst activity was reconciled so that one temperature in the non-equilibrium zone near the front of the reactor was matched. All other simulated temperatures then also matched the remaining measured temperatures, confirming that the kinetics were yielding appropriate heats of reaction, and composition all along the reactor bed. Effluent compositions were directly measured, and are plotted versus simulated compositions in the parity plot, showing excellent agreement.

Good agreement in this adiabatic reactor further confirms the validity of the kinetics. Issues related to heat transfer do not cloud the kinetics validity in this case.

Oxo-Alcohol Synthesis Gas Steam Reformer

A fired tube reactor was configured to match the dimensions and catalyst loading of an existing oxo-alcohol synthesis gas steam reformer. Simulation results at observed conditions (feed gas composition, outlet temperature, steam to carbon ratio, etc.) agree very well with observed results. Catalyst activity is first determined by matching key effluent composition.

APPENDIX B   CASE STUDY OF EFFECTS OF CATALYST ACTIVITY IN A PRIMARY REFORMER

Several cases were simulated to show the effects of catalyst activity. Plots of key operating conditions illustrate many effects that are unnoticed if only outlet compositions are examined. Table C.1 shows how insensitive the outlet conditions are to catalyst activity, over a wide range. The plot of methane composition as a function of tube length for these cases is much more revealing. The methane compositions for all the cases approach similar values near the tube outlet, but the methane composition profiles are significantly different at positions closer to the inlet.

The lower activity cases exhibit operating conditions (temperatures) that are likely to be unacceptable, as shown by the plots of the process temperatures. Both the overall temperature profile, as well as a "close up" of the first 3 meters are shown. As discussed in the "Coking" section, in the paragraph beginning with "Some confusion may exist...", reformers with less active catalyst operated at similar outlet temperature as with active catalyst, **at similar firing profiles**, exhibit higher, and likely unacceptable temperatures within the tube. The plotted profiles are similar in nature to those shown in Figure 12 of Reference 23. The shapes are different due to top firing[23] and wall firing profile differences. The reason the temperature profiles change is because with less active catalyst, less reaction heat sink exists, and more of the heat input raises the sensible heat of the process fluid, and hence the temperature.

Table B.1   Effluent Composition vs. Catalyst Activity, at Constant Inlet & Outlet TemperaturesNaphtha Feed Cases

```
Inlet   Temperature  =   462.0°C
Outlet  Temperature  =   761.8°C
```

All cases have the same firing profile (not level).

Catalyst Activity =>	100%	90%	70%	50%	30%	20%
Component						
HYDROGEN	63.9077	63.9072	63.9057	63.9029	63.8962	63.8872
NITROGEN	0.3675	0.3675	0.3676	0.3676	0.3677	0.3677
ARGON	0.0060	0.0060	0.0060	0.0060	0.0060	0.0060
CARBON MONOXIDE	9.0188	9.0186	9.0181	9.0171	9.0146	9.0109
CARBON DIOXIDE	16.4000	16.4001	16.4004	16.4010	16.4026	16.4049
METHANE	10.3000	10.3006	10.3022	10.3053	10.3130	10.3233

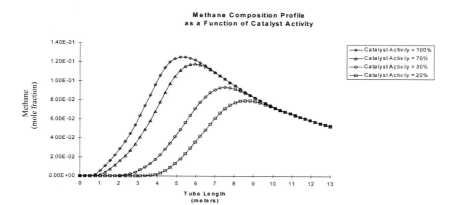

Methane Composition Profile as a Function of Catalyst Activity

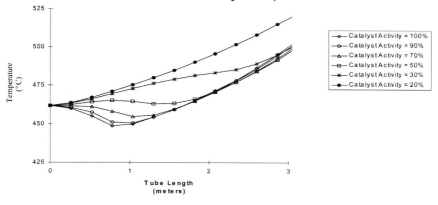

Process Temperature Profile As a Function of Catalyst Activity (at Constant Inlet & Outlet Temperatures, and Firing Profiles)

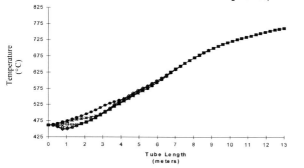

Process Temperature Profile As a Function of Catalyst Activity (at Constant Inlet & Outlet Temperatures, and Firing Profiles)

The unacceptable temperatures can be avoided. Two alternatives of avoiding them are simulated, one resulting in different unacceptable operating conditions (high methane slip), and the other with very acceptable conditions.

The first alternative focuses on the unacceptable temperatures encountered in the first few meters of the reactor, which may cause coking. This alternative keeps the temperature at the approximate one meter from the inlet at the same temperature as the base, 100% catalyst activity, case. This temperature was 449°C. Table C.2 and the accompanying plot shows these results. To keep that temperature at 449°C, and maintain the firing profile, the reactor outlet specification has to be relaxed. The independent operating condition is just moved from the reactor outlet, to the catalyst temperature. As illustrated, under these operating conditions, as the catalyst activity falls from 100% to 90%, and 70%, the methane slip rises dramatically. So does the reactor outlet temperature. As mentioned, this alternative is unacceptable, due to very high methane slip.

Table B.2 Effluent Composition vs. Catalyst Activity, at Constant Upper Bed Catalyst Temperature Naphtha Feed Cases

$$\text{Inlet Temperature} = 462.0 \text{ °C}$$
$$\text{TI211-1 Temperature} = 449 \text{ °C}$$

All cases have the same firing profile (not level).

Catalyst Activity =>	100%	90%	70%
Component			
HYDROGEN	63.9077	63.1242	58.6711
NITROGEN	0.3675	0.3755	0.4209
ARGON	0.0060	0.0061	0.0069
CARBON MONOXIDE	9.0188	8.5117	6.1219
CARBON DIOXIDE	16.4000	16.7405	18.3063
METHANE	10.3000	11.2420	16.4730

The second alternative is to minimize the tube outlet methane slip (dry basis), while modifying the firing profile, keeping the "CAT-1" temperature acceptable (in this case it was upper bounded at 449°C), and while keeping the tube metal temperature (TMT) below acceptable limits. This was posed as an optimization problem. The objective in a plant-wide optimizer is not to minimize the reactor outlet methane content (profit is maximized), but in this sub-plant problem it is the most realistic objective. The ability to modify the firing profile will depend greatly on the control system, burner design, fuel composition, and fire box design. In this case the range over which the optimizer could vary the firing was bounded simply by step bounding the amplitude changes for each burner row. In the overall furnace model the steady-state gain relationship between fuel fired and catalyst temperatures, as well as bounds on the duties that can be fired will constrain the range of firing. The amplitudes of the firing profiles will all be dependent variables, as described in the "Firebox to Catalyst Tube" heat transfer section.

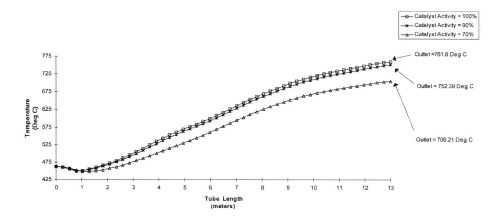

Results of this alternative for a catalyst activity of 90% show that the tube outlet temperature can be raised to 762.43°C, slightly higher than the base (100% activity) case value of 761.8°C, while not violating the "CAT-1" temperature limit, the TMT limit (of 884.92°C, which is the same maximum as in the base case), nor the step bounds on the firing profile. The reason this case is at a slightly higher outlet temperature (and slightly lower methane slip) than even the base (100% activity) case is because the base case is not an optimized (profile) case. Had the base case been an optimized case, its methane slip would have been slightly lower than the optimized 90% activity case.

The optimized profile was reduced (to its lower step bound) for the top burner row, and increased (to the upper step bounds) for the next two burner rows. The bottom row firing was not at its limit, but the TMT in that section of the tube was at its maximum limit (884.92°C). The resulting reactor outlet compositions show that the methane slip is slightly lower, and hydrogen composition slightly higher than the base case. This is to be expected, even for a less active catalyst, for this system which is close to equilibrium, and at a slightly higher outlet temperature. This solution shows that profile control can essentially overcome the otherwise poor operating conditions that would accompany catalyst activity decline. The optimizer increases the heat flux to the part of the tube in which sufficient reaction heat sink exists, and decreases heat flux to the tube where insufficient reaction exists, and where temperature limits would otherwise be violated. Of course, at lower catalyst activities, and with less range of firing control, the debits associated with catalyst activity decline cannot be avoided. For example, as the catalyst activity falls, the ability to reduce firing and maintain the "CAT-1" temperature below reasonable limits, while still maintaining a reasonable reactor outlet temperature will cease to exist.

Table B.3. Effluent Composition Maximize Outlet Temperature Optimize Case Naphtha Feed Case

```
Inlet      Temperature = 462.0°C
TI211-1    Temperature Upper Limit =  449°C
TMT        Temperature Upper Limit =  884.9°C

           Step Bounds on Firing Profile
```

Catalyst Activity =>	90%
Component	
HYDROGEN	63.961
NITROGEN	0.3670
ARGON	0.0060
CARBON MONOXIDE	9.0552
CARBON DIOXIDE	16.3755
METHANE	10.2344

Chapter 25

Hydrogen Production and Supply
Meeting Refiners' Growing Needs

M. Andrew Crews and B. Gregory Shumake
CB&I Process and Technology
Tyler, TX 75701-5013

1. INTRODUCTION

Hydrogen is a key feedstock in many refining operations associated with the production of cleaner gasoline and diesel products. There are several drivers for the increase in hydrogen demand in the refining industry. Crude oil continues to become heavier with additional sulfur and nitrogen species. As the demand for heavy fuel oil diminishes, additional upgrading of the "bottoms" is required to produce a marketable fuel. More stringent environmental regulations have reduced the level of sulfur in both gasoline and diesel products. The combined effect of these trends is that refineries are running short of the necessary hydrogen to meet the increase in hydrotreating and hydrocracking requirements.

There are many routes to the production of hydrogen. In particular, several refinery unit operations produce hydrogen as a by-product. In order to focus the scope of this discussion, this chapter will only cover on purpose hydrogen production. The technologies discussed will be limited to those that can produce hydrogen in sufficient quantities to satisfy typical refinery demands, i.e., from 5 MMSCFD to 200 MMSCFD.

Hydrogen can be produced from a variety of feedstocks and technologies. Feedstock for hydrogen production can range from natural gas to coal and includes all hydrocarbons in-between. The availability of specific feedstock types will limit the number of technology options for the production of hydrogen. The conversion of hydrocarbons to hydrogen requires significant heat input as well as excess (waste) heat resulting from achieving the high temperatures required to obtain sufficient feedstock conversion. This excess heat is typically recovered to produce steam and can also be recovered to improve the overall efficiency of the technology chosen, but with a significant capital cost impact.

2.   THERMODYNAMICS OF HYDROGEN

The production of hydrogen from hydrocarbons can be broken down into three key chemical reactions: (1) steam methane reforming, (2) partial oxidation, and (3) water gas shift.

The use of the term 'reforming' in a refinery environment can be confusing as there are several catalytic 'reformer' unit operations that are used to improve the octane numbers of gasoline. In the context of a hydrogen plant a reformer is a furnace or vessel associated with the steam methane reforming reaction. The steam methane reforming reaction produces hydrogen (H_2), carbon monoxide (CO), carbon dioxide (CO_2), and un-reacted methane (CH_4) from hydrocarbons and water based on the following chemical reaction:

$$C_nH_m \text{ (g)} + nH_2O \text{ (g)} + heat = nCO \text{ (g)} + (m/2+n) H_2 \text{ (g)} \qquad (1)$$

The reforming reaction is endothermic and is favored by high temperature and low pressure. In other words, as the net reaction temperature is increased, the higher the conversion of methane and as the pressure at which the reaction takes place is decreased, the methane conversion is increased. Another variable in the reforming reaction is the steam to carbon ratio. As the amount of steam increases above the stoichiometric requirement, the amount of methane conversion also increases. Conversely as the ratio of steam to carbon decreases, the greater the possibility of forming carbon precursors (coking) through alternative reaction pathways. This is an undesirable condition because the formation of carbon can destroy the reforming catalyst, increase the pressure drop through the reformer, increase the reformer tube wall temperature, and consequently require an unscheduled plant shutdown to repair these problems. See Table 1 for equilibrium constant information on the steam methane reforming reaction.

Partial oxidation refers to a chemical reaction where hydrocarbons react with oxygen in a sub-stoichiometric burn reaction to produce carbon monoxide and hydrogen. Partial oxidation technologies require oxygen as a feedstock. Several other reactions take place in the partial combustion zone that contribute to the overall heat provided by the partial oxidation reaction.

$$C_nH_m\ (g) + (n/2)\ O_2\ (g) = n\ CO\ (g) + (m/2)\ H_2\ (g) + heat \qquad (2)$$
$$C_nH_m\ (g) + (n+m/4)\ O_2\ (g) = n\ CO_2\ (g) + (m/2)\ H_2O(g) + heat$$
$$H_2 + (1/2)\ O_2 = H_2O + heat$$
$$CO + (1/2)\ O_2 = CO_2 + heat$$

Table 1. Reaction and Equilibrium Constants (Used with Permission of Sud-Chemie)

	$CO + H_2O = CO_2 + H_2$		$CH_4 + H_2O = CO + 3H_2$	
Temp., °F	ΔH deg., Btu/lb-mole	Kρ	ΔH deg., Btu/lb-mole	Kρ
200	-17,570	4523	90,021	7.813×10^{-19}
300	-17,410	783.6	91,027	6.839×10^{-15}
400	-17,220	206.8	91,957	7.793×10^{-12}
500	-17,006	72.75	92,804	2.173×10^{-9}
600	-16,777	31.44	93,566	2.186×10^{-7}
700	-16,538	15.89	94,252	1.024×10^{-5}
800	-16,293	9.03	94,863	2.659×10^{-4}
900	-16,044	5.61	95,404	4.338×10^{-3}
1000	-15,787	3.749	95,880	4.900×10^{-2}
1100	-15,544	2.653	96,283	0.4098
1200	-15,299	1.966	96,628	2.679
1300	-15,056	1.512	96,922	0.1426×10^{2}
1400	-14,814	1.202	97,165	0.6343×10^{2}
1500	-14,575	0.9813	97,378	2.426×10^{2}
1600	-14,344	0.8192	97,545	8.166×10^{2}
1700	-14,117	0.697	97,655	2.464×10^{3}
1800	-13,892	0.6037	97,741	6.755×10^{3}
1900	-13,672	0.5305	97,786	1.701×10^{4}
2000	-13,459	0.4712	97,818	3.967×10^{4}
2100	-13,246	0.4233	97,812	8.664×10^{4}
2200	-13,041	0.3843	97,791	1.784×10^{5}

The partial oxidation (POX) reaction is highly exothermic and provides the heat required for the steam methane reforming reaction that occurs following the conversion of hydrocarbons to carbon monoxide and hydrogen. The partial oxidation reaction is favored by high temperature and high pressure. However, the higher the pressure the more likely that alternative reaction pathways will lead to the formation of carbon.

The water gas shift (WGS) reaction is where carbon monoxide is 'shifted' on a mole per mole basis to hydrogen by the flowing reaction:

$$\textbf{CO} + \textbf{H}_2\textbf{O} = \textbf{CO}_2 + \textbf{H}_2 + \textbf{heat} \qquad (3)$$

The water gas shift reaction is favored by low temperature and is mildly exothermic. The WGS reaction is typically the final reaction step that produces hydrogen in a multiple reactor train. Since the WGS reaction is favored by low temperature, it is typically included in the heat recovery train down stream of the high temperature reactors. See Table 2 for thermodynamic and equilibrium data.

Table 2. Equilibrium Constants and Heats of Reaction (Used with Permission of Sud-Chemie)

$$CO = 0.5\ CO_2 + 0.5\ C_{(s)}$$

Temp. in °F	ΔH deg., Btu/lb-mole	Kp
100	-37,130.24	7.5489×10^9
200	-37,244.17	4.7479×10^7
300	-37,313.43	1.1237×10^6
400	-37,345.16	6.3291×10^4
500	-37,346.08	6.4858×10^3
600	-37,321.90	1.0223×10^3
700	-37,277.30	2.2190×10^2
800	-37,216.05	6.1052×10^1
900	-37,141.19	2.0628×10^1
1000	-37,055.16	8.0522
1100	-36,959.90	3.5533
1200	-36,856.96	1.7338
1300	-36,747.60	9.1957×10^{-1}
1400	-36,632.77	5.2304×10^{-1}
1500	-36,513.23	3.1565×10^{-1}
1600	-36,389.57	2.0036×10^{-1}
1700	-36,262.22	1.3284×10^{-1}
1800	-36,131.47	9.1460×10^{-2}
1900	-35,997.52	6.5074×10^{-2}
2000	-35,860.48	4.7655×10^{-2}
2100	-35,720.36	3.5799×10^{-2}
2200	-35,577.17	2.7506×10^{-2}
2300	-35,430.71	2.1563×10^{-2}
2400	-35,280.92	1.7210×10^{-2}
2500	-35,127.58	1.3960×10^{-2}

3.   TECHNOLOGIES FOR PRODUCING HYDROGEN

3.1   Steam Methane Reforming (SMR) Technologies

Steam-methane reformers generate the majority of the world's on-purpose hydrogen. A steam-methane reformer (SMR) is a fired heater with catalyst-filled tubes. Hydrocarbon feedstock and steam react over the catalyst to produce hydrogen. The reacted process gas typically exits the reformer at about 1600°F The flue gas leaving the reformer radiant section is typically about 1900°F.

The flue gas at 1900°F must then be cooled to about 300°F to achieve efficient heat recovery. This heat is recovered in the convection section of the SMR. The process gas at 1600°F must then be cooled to about 100°F before final product purification.

3.1.1   Maximum Steam Export

A maximum export steam plant is defined as a plant that makes the maximum practical quantity of export steam, without auxiliary firing.

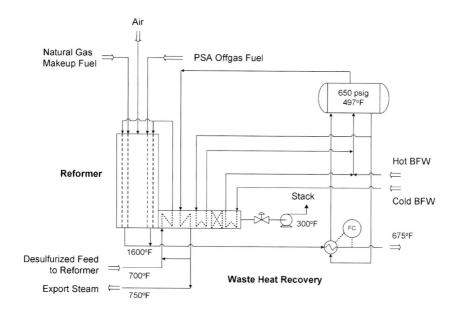

Figure 1. Steam/Methane Reforming with Maximum Steam Export

Table 3 gives a typical feed and utility summary for a maximum export steam plant, as defined above.

Table 3. Maximum Export Steam Feed and Utility Summary

	Units	Units/M SCF H$_2$ Product
Natural Gas	MM Btu LHV	0.450
Export Steam	Lbs	86.8
Treated Water	lbs	117.0
Power	kWh	0.55
Cooling Water Circ	gal	12.7

3.1.2 Limited Steam Export

A limited export steam plant is defined as a plant that makes some export steam but significantly less than the maximum. This is typically achieved by adding a combustion air preheat unit (CAP). This unit consists of a modular heat exchanger that heats the combustion air to the SMR by heat exchange with the flue gas from the SMR. The hot combustion air reduces the fuel requirement for the SMR, which in turn reduces the steam production.

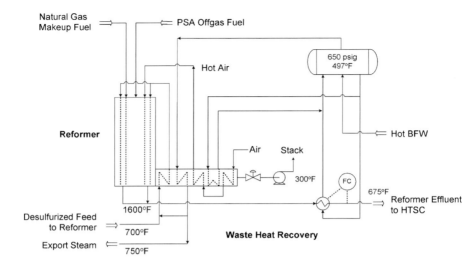

Figure 2. Steam/Methane Reforming with Limited Steam Export

A typical combustion air preheat temperature for this type of plant is approximately 750°F. This permits use of carbon steel ducting in the air preheat distribution system. Table 4 gives a typical feed and utility summary for a limited export steam plant, as defined above.

Table 4. Limited Export Steam Feed and Utility Summary

	Units	Units/M SCF H$_2$ Product
Natural Gas	MM Btu LHV	0.386
Export Steam	lbs	38.0
Treated Water	lbs	67.7
Power	kWh	0.58
Cooling Water Circ	gal	13.1

Note that the limited export steam plant makes less than half as much export steam as the maximum steam export case, but also uses less natural gas. At the same time, the limited steam export plant capital cost is higher, because of the addition of the air preheat unit.

3.1.3   Steam vs. Fuel

The steam value and fuel price typically determine which of these two cases is the most economic. A high steam value (relative to fuel price) favors the Maximum Steam Export Case. A high fuel price and a low steam value (relative to fuel price) favor the Limited Steam Export Case.

Figure 3 is a plot of natural gas price versus steam value. For a given natural gas price, if the steam value is above the curve, then maximum steam export is favored. But if the steam value is below the curve, then CAP (limited steam export) is favored. For example, for a natural gas price of 3.00 $/MM Btu LHV, Figure 3 indicates that if the steam value is above 3.30 $/M lbs, then steam generation is favored. But if the steam value is below 3.30 $/M lbs, then CAP (limited steam export) is favored.

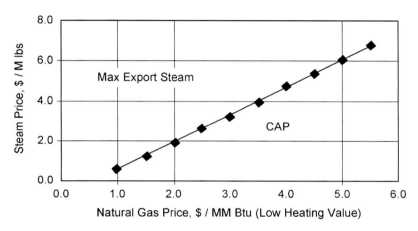

Figure 3. Steam price versus natural gas price

3.1.4   Minimum Export Steam

A minimum export steam plant is defined as a plant that optimizes heat recovery in the plant to the maximum extent possible. This is often done when the value of steam is essentially zero or the price of the feedstock and fuel are exceptionally high. Minimum steam export is often achieved by first increasing the SMR process gas inlet temperature and then the combustion air preheat temperature, both of which reduce the fired duty of the SMR. Typical temperatures are 1150°F for the process gas and 900°F for the air preheat. These changes reduce the export steam to a low level, but typically not completely to zero. Additional modifications are required to reduce the export steam to an absolute minimum.

Adding reaction steps to the process can further reduce the amount of feed and fuel required by the hydrogen plant. The first reaction step that can be added is a prereformer. The prereformer is an adiabatic reactor where the feed is heated to approximately 900°F and the gas is partially reformed over a catalyst bed and hydrocarbon components of ethane and heavier are converted to methane. The conversion of ethane plus components to methane is exothermic. If the feedstock is essentially methane (i.e., natural gas), the amount of heat required to drive the steam methane reforming reaction to equilibrium will produce a temperature decrease across the reactor. However, if the feedstock is butane or heavier, the methane conversion reaction produces more heat than is required by the steam methane reforming reaction and there is a net temperature increase across the reactor.

The prereformer is an excellent way to produce a consistent feed to a steam methane reformer. Heavier feedstocks to the SMR increase the potential for carbon formation. The ultimate in energy recovery can be achieved by utilizing a heat recuperative reforming step in the process. Since the steam methane reforming occurs at high temperature, the reformed gas is typically cooled by generating steam. However, if the heat can be recovered by utilizing it in an additional reforming step, the net heat input required to complete the reaction is reduced and, therefore, the quantity of fuel required to operate the furnace is reduced. The reformer effluent gas is passed through the shell side of a tubular heat exchanger with catalyst filled tubes.

3.2   Oxygen Based Technologies

Steam-methane reforming (SMR) has been the conventional route for hydrogen and carbon monoxide production from natural gas feedstocks. However, several alternative technologies are currently finding favor for an increasing number of applications.

These technologies are:

SMR/O$_2$R: Steam-Methane Reforming combined with Oxygen Secondary Reforming.
ATR: Autothermal Reforming.
POX: Thermal Partial Oxidation.

Each of these alternative technologies uses oxygen as a feedstock. Accordingly, if low cost oxygen is available, they can be an attractive alternate to an SMR for natural gas feedstocks. Low cost oxygen is now available in many large industrial sites where a large air separation plant already exists or can be economically installed to serve the needs of the area. (Note that unlike ammonia plant technology, air cannot be used in place of oxygen since the contained nitrogen would dilute the hydrogen/carbon monoxide product.)

A brief description of each technology follows.

3.2.1   SMR/O$_2$R

An SMR/O$_2$R essentially consists of an SMR followed by an oxygen secondary reformer (Figure 4). The oxygen reformer is a refractory lined vessel containing catalyst and a burner. The reaction mixture from the SMR is fed to the top of the oxygen reformer where it is mixed with oxygen fed through the burner. Partial oxidation reactions occur in a combustion zone just below the burner. The mixture then passes through the catalyst bed where reforming reactions occur. The gas exits at about 1900°F.

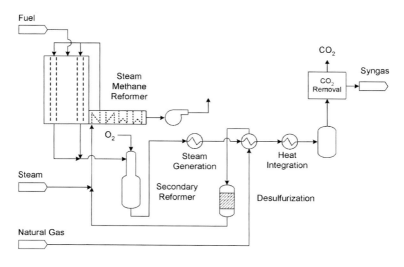

Figure 4. SMR/O2R Plant

3.2.2 ATR

An ATR is similar to an oxygen secondary reformer except that it does not receive feed from an SMR. Instead, it is fed directly with a natural gas/steam mixture, which is mixed directly with oxygen from a burner located near the top of the vessel (Figure 5). Again, partial oxidation reactions occur in a combustion zone just below the burner. The mixture then passes through a catalyst bed where reforming reactions occur. The gas exits at about 1900°F.

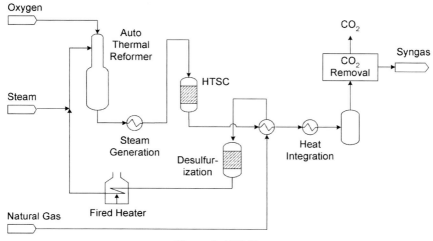

Figure 5. ATR Plant

3.2.3   POX

A POX is similar to an ATR except that it does not contain a catalyst and does not require steam in the feed. It is fed directly with a natural gas stream, which is mixed directly with oxygen from a burner located near the top of the vessel. Partial oxidation and reforming reactions occur in a combustion zone below the burner. The gas exits at about 2500°F.

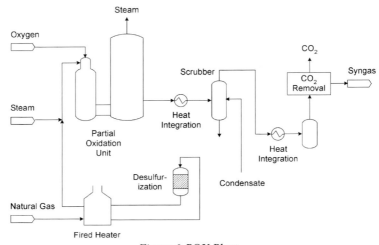

Figure 6. POX Plant

3.2.4   Products

The exit gas from each of the above reactors consists of hydrogen, carbon monoxide, carbon dioxide, steam, and residual methane. Small quantities of nitrogen and argon from the original feedstocks may also be present. This gas mixture is then typically processed to yield one or more of the following products:
(1) High Purity Hydrogen
(2) High Purity Carbon Monoxide
(3) A Hydrogen/Carbon Monoxide Gas Mixture

If only hydrogen is required, the plant becomes a hydrogen plant. If only CO is required, the plant becomes a carbon monoxide plant. If both hydrogen and carbon monoxide are required as separate streams, the plant is typically known as a HYCO plant. If only a hydrogen/carbon monoxide mixture is required, the plant is typically known as a synthesis gas (or syngas) plant. If all three products are required, the plant is considered a combination (hydrogen/carbon monoxide/syngas) plant.

3.2.5 H_2/CO Ratio

A key variable in the design of these plants is the product H_2/CO ratio.

For HYCO and syngas plants, the product H_2/CO ratio typically varies from 1.0 to about 3.0. But for hydrogen plants, the ratio is almost infinite. And for carbon monoxide plants, the ratio is essentially zero.

3.2.6   Natural Ratio Range

Each of the above technologies makes products with inherently different H_2/CO ratios. The natural ratio range is defined herein as the range of ratios that can be achieved when varying only the CO_2 recycle. The natural ratio ranges for each technology are shown in Table 5, as applicable to natural gas feedstocks. (Other feedstocks will produce a different set of natural ratios).

Table 5. Hydrogen to CO Ratio Summary

	Membrane Or Import CO_2	Total CO_2 Recycle	No CO_2 Recycle	Increase Steam	Add Shift Converter
SMR	<3.0	3.0	5.0	>5.0	infinity
SMR/O_2R	<2.5	2.5	4.0	>4.0	>5.0
ATR	<1.6	1.6	2.7	>2.7	>3.0
POX	<1.6	1.6	1.8	>1.8	>2.0

As seen in Table 5, the natural ranges for natural gas are as follows: For the SMR: 3.0 to 5.0. For the SMR/O_2R: 2.5 to 4.0. For the ATR: 1.6 to 2.65. And for the POX: 1.6 to 1.8. Note that some ranges are broader than others, and that some of the ranges do not overlap.

The technology that offers a natural ratio that spans the required product ratio is considered to have an inherent advantage and merits careful consideration.

3.2.7 CO_2 Recycle

CO_2 recycle is the typical means of tailoring the natural ratio span to meet the desired H_2/CO product requirement.

As noted earlier, carbon dioxide is present in all the reactor effluents. Since CO_2 is considered an impurity, it must typically be eliminated. In most designs, the CO_2 is removed from the process gas by selective adsorption in a suitable solvent such as an amine solution. The CO_2 is then stripped from the solvent as a separate stream.

(Note: synthesis gas for methanol plants typically retains some or all of the carbon dioxide. This is a special case, which is outside the scope of this discussion).

The separate CO_2 stream can be removed from the system or it can be recycled back to the reactor. If it is recycled, the CO_2 can be converted to CO in the reactor by the reverse water gas shift reaction. This reduces the final H_2/CO ratio because the carbon atoms in the original CO_2 are converted to CO. On the other hand, if the CO_2 is not recycled, the H_2/CO ratio is increased, because the carbon atoms in the original CO_2 are not converted to CO. (CO_2 that is not recycled is typically vented to the atmosphere; however, sometimes it is sold as a byproduct).

The natural H_2/CO range for each technology represents the spread between full CO_2 recycle and no CO_2 recycle. See Figure 7. By operating with partial recycle, any H_2/CO ratio within the natural range can be achieved.

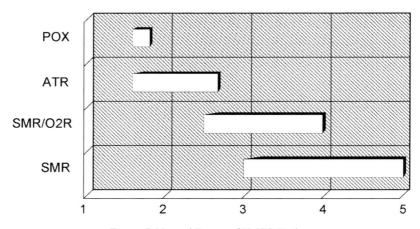

Figure 7. Natural Range of H_2/CO Ratio

But as noted earlier, some of the natural ratio ranges do not overlap; hence for a given product H_2/CO ratio, at least some of the technologies will require additional ratio adjustment to meet the product requirements. Fortunately, a number of process techniques are available to do this. The viability of the technology will depend on the economics of using these techniques.

To decrease the ratio below the natural range, common techniques include importing CO_2; use of a membrane; or use of a cold box. To increase the ratio, common techniques include increasing the quantity of steam in the reactor feed or adding a shift converter. These techniques, along with CO_2 recycle, are summarized in Table 6.

Table 6. Techniques for H_2/CO Ratio Adjustment

	Decreases Ratio	Increases Ratio
Recycle CO_2	X	
Import CO_2	X	
Use Membrane	X	
Remove CO_2		X
Increase Steam		X
Add Shift Converter		X
Use Cold Box	See Note	

Note: A cold box decreases the ratio with respect to CO and increases the ratio with respect to H_2.

3.2.8   Import CO_2

The H_2/CO ratio can be decreased by importing CO_2 as feed to the reactor. The import CO_2 is added to the CO_2 already being recycled. The carbon atoms in the import CO_2 can then be converted to CO by the reverse water gas shift reaction.

3.2.9   Membrane

The H_2/CO ratio can also be decreased by use of a membrane unit. Hydrogen is removed from the membrane as a permeate stream, which reduces the H_2/CO ratio of the main process stream.

3.2.10   Cold Box

A cold box is often used to separate the hydrogen and carbon monoxide into a high purity CO stream and a H_2-rich stream. This reduces the ratio to near zero with respect to the CO stream. The H_2-rich stream may undergo further purification (typically in a PSA unit) to yield a high purity H_2 stream. This increases the ratio to near infinity with respect to the H_2 stream.

3.2.11   Steam

Adding steam to the reactor converts CO to H_2 by the water gas shift reaction. This increases the H_2/CO ratio.

3.2.12   Shift Converter

A shift converter is a catalytic reactor which catalyzes the conversion of CO to H_2 by the water gas shift reaction. This step can be used to increase the H_2/CO ratio to a very high value. It is widely used in hydrogen plants.

Table 6 summarizes the ranges achievable by each technology by inclusion of these techniques. Typically, the viability of the technology decreases with the magnitude of the added steps necessary to achieve the required product H_2/CO ratio.

3.2.13   Other Considerations

As stated above, these ratios are based on natural gas feedstocks. If heavier feedstocks are used, the ratios will be lower because the hydrogen to carbon ratio of the feed will be lower.

The above ratios are based on typical reformer outlet temperatures and pressures for each technology. The reformer outlet temperature is typically set by the residual methane requirement. The reformer outlet pressure is typically set by compression requirements. Some design flexibility often exists between these limits. However, within the normal range of flexibility, the H_2/CO ratio does not change appreciably with the reformer outlet temperature or pressure. Hence, for a given design, it is normally impractical to achieve a significant change in H_2/CO ratio by varying the reformer outlet temperature or pressure. Instead, the techniques discussed above should be utilized.

If it is impractical for a given technology to achieve the desired ratio by the techniques discussed above, then excess hydrogen or carbon monoxide containing gas can be burned as fuel in the SMR (for SMR and SMR/O_2R plants) or in the Process Heater (for ATR or POX plants). If necessary, the Process Heater can be expanded to include a second radiant section to permit burning of excess product by generating steam. Alternatively, a package boiler can be used to burn excess gas and generate steam.

Thus far, considerable discussion has centered on the H_2/CO ratio, since as noted above, the technology that can most easily achieve the required product H_2/CO ratio has an inherent initial advantage. But there are other considerations that could over-ride this advantage. These are outlined below.

3.3   Technology Comparison

Table 7 is a comparison of some key process advantages and disadvantages for the technologies under consideration for hydrogen/carbon

Table 7. Technology Comparison

SMR	ATR
*Advantages	Advantages
– Most commercial experience	– Natural H_2/CO ratio often favorable
– Does not require oxygen	– Lower reforming temperature than POX
– Reforming temperature relatively low	– Low residual CH_4
– Best natural H_2/CO ratio for hydrogen product	_ Tailor residual CH_4
	*Disadvantages
*Disadvantages	– Limited commercial experience
– Natural ratio high for CO product	– CO_2 cycle higher than POX
– Highest atmospheric emissions	– Require oxygen
	– Makes carbon (soot)
	– Complex heat recovery section
	– Licensed process
	– Significant third party involvement
SMR/O$_2$R	**POX**
*Advantages	*Advantages
– Smaller than SMR	– Does not require feedstock desulfurization
– Low residual CH_4	– Does not require oxygen
– Tailor residual CH_4	– Low residual CH_4
*Disadvantages	– Low natural H_2/CO ratio
– Two-step reforming	*Disadvantages
– Higher reforming temperature	– Narrow H_2/CO ratio
– Require oxygen	_ Cannot tailor residual CH_4
	– Require oxygen

monoxide plants with natural gas feedstock. Please note that while the oxygen requirement is shown as a disadvantage, it may not be a disadvantage in cases where low cost oxygen is available.

3.3.1 Process Parameters

Key process parameters and feedstock requirements for each technology include the following:

PRESSURE - Most SMR plants run at reformer outlet pressures between 150 and 400 psig. The maximum pressure is about 550 psig, due to metallurgical considerations in the SMR outlet piping. The same applies to SMR/O_2R plants.

This limits the final product gas pressure to about 500 psig. Above this, product gas compression is required.

ATR and POX plants can run at much higher pressures. For example, some plants require synthesis gas at about 800 psig; this can be supplied directly from an ATR or POX without synthesis gas compression.

TEMPERATURE - SMR plants typically run at reformer outlet temperatures of 1550 to 1700°F.

SMR/O_2R and ATR plants typically run at outlet temperatures of 1750 to 1900°F.

POX plants typically run at about 2500°F. This high temperature is needed to maintain low residual methane without a catalyst.

STEAM/CARBON - Steam is typically required to prevent carbon formation on the reformer catalyst. The steam requirement is generally expressed as the steam/hydrocarbon-carbon ratio (steam/hcc). This is the ratio of the moles of steam to moles of hydrocarbon-carbon in the feed (the carbon in any CO_2 recycle is not included in computation of the ratio).

Both the SMR and SMR/O_2R plants typically use a 2.5 to 5 steam/hcc ratio for no CO_2 recycle, and a 2/1 steam/hcc ratio when all the CO_2 is recycled. (A lower ratio can be used with CO_2 recycle since the CO_2 makes some H_2O by the water gas shift reaction). For partial recycle, a ratio between 3/1 and 2/1 is typically used.

ATR plants can operate at a lower ratio, often about 1 to 1.5 with or without CO_2 recycle.

POX plants require no steam. Some soot is formed, but since there is no catalyst, there is obviously no possibility of catalyst bed plugging. With proper design and operation, the soot can be processed and removed in the downstream equipment without affecting the plant on-stream time.

METHANE SLIP - The exit gas from each reactor contains some residual methane, which corresponds to unreacted natural gas feed. This is typically referred to as the methane slip. A low methane slip increases the purity of a synthesis gas product. It also reduces the natural gas requirement.

An SMR typically has a relatively high methane slip (3 to 8 mol pct, dry basis). This is primarily because the SMR operates at a lower outlet temperature. The other technologies operate at higher outlet temperatures and typically have low slips (0.3 to 0.5 mol pct, dry basis).

If a methane-wash cold box is used, typically a methane slip of at least 1 to 2 mol percent is required to maintain the methane level in the box. This is not a problem for an SMR design because its methane slip is considerably more than this. It is also easily achieved by the SMR/O_2R and ATR designs by simply reducing the reactor outlet temperature. However, this methane level is more difficult to achieve with a POX, since in the absence of a catalyst, a high temperature and corresponding low methane slip should be maintained to ensure consistent performance.

RAW MATERIALS - Natural gas and oxygen are the two major raw materials. For each technology, the quantities required must be considered on a case-by-case basis, since they depend on the H_2/CO requirement for each application.

For example, the product H_2/CO ratio may require that CO_2 be either recycled or vented, depending on the ratio and the technology.

If the required H_2/CO product ratio is 2.0, all the CO_2 would be recycled in an SMR or SMR/O_2R design; part of the CO_2 would be recycled in an ATR design; and no CO_2 would be recycled in a POX design.

The extent of CO_2 recycle changes the raw material requirements for a given technology, because if CO_2 is vented, its contained carbon and oxygen is not recovered.

To provide a basis of comparison, a design can be considered wherein all the CO_2 is recycled for each technology. This would normally occur for designs for a H_2/CO product ratio less than 1.6.

A typical design meeting this requirement would be a carbon monoxide only plant (H_2/CO ratio of zero). For such a design, we have compared all four technologies for a proposed plant on the US Gulf Coast.

For this plant, the natural gas and oxygen requirements are given in Table 8. The natural gas requirement shown is the net requirement on a heating value basis and includes credit for excess hydrogen at fuel value. (For this plant, all the technologies produce excess hydrogen since CO is the only required product and there is no requirement for product hydrogen).

Table 8. Feedstock Requirements

	Units per 100 SCF CO Product			
	SMR	SMR/O2R	ATR	POX
Net Natural Gas* (MM Btu LHV)	0.729	0.724	0.531	0.509
Oxygen Import, lb	0	14.2	56.5	58.7

* includes credit for excess hydrogen at fuel value.
All cases based on full CO_2 recycle.

For the following discussion, the SMR and SMR/O_2R are called SMR-based technologies because the SMR is the dominant component. And the ATR and POX are called oxygen-based technologies because the partial oxidation reactor is the dominant component.

From Table 8, note that the natural gas requirements for the SMR-based technologies are about the same. Note also that the natural gas requirements for the oxygen-based technologies are about the same. But also note that the oxygen-based technologies require only about 70 percent as much natural gas as the SMR-based technologies.

The natural gas requirement is lower for the oxygen-based technologies because most of the oxygen in the feed reacts with carbon in the natural gas to form carbon monoxide. The oxidation reaction is therefore more efficient for the formation of carbon monoxide, which is the desired product for the referenced plant.

Another reason for the lower natural gas requirement is that for the oxygen-based technologies, the heat required for the reaction is applied directly to the process gas in the reactor vessel. This is more energy efficient than the SMR-based technologies, which require a temperature driving force between the combustion gas and the intube process gas.

With respect to oxygen consumption, the oxygen requirement for the SMR is obviously zero. For the SMR/O_2R, the oxygen requirement is about 25 percent of that for the ATR or POX.

The ATR and POX oxygen requirements are about the same. This is to be expected, since both technologies use the same direct combustion process, and for the referenced plant, both technologies recycle all the CO_2 and produce essentially the same H_2/CO product ratio.

3.3.2 Export Steam

For the referenced plant, the export steam is relatively low since the excess hydrogen is not burned within the plant to make export steam. Accordingly, the export steam for the SMR, SMR/O_2R, and POX averaged only about 150 lbs per 1000 SCF of carbon monoxide and any differences were not considered significant. However, the ATR export steam was less than half this value because the energy requirement for the CO_2 removal system is larger.

3.3.3   Economic Considerations

The economics of each technology depend on the conditions and requirements of each project and must be studied on a case-by-case basis. Some of the most important considerations are as follows

3.3.4   Oxygen Availability

The availability of oxygen is obviously a key parameter. Since the SMR requires no oxygen, it is the obvious choice if oxygen is unavailable or prohibitively expensive. Studies indicate that the oxygen-based technologies can be attractive if oxygen is available at less than about $25 per short ton. If oxygen is greater than about $30 per ton, the SMR is favored.

The oxygen price can be less than $25 per ton in large industrial sites where a large air separation plant already exists or can be economically installed to serve the needs of the area. Hence, in such cases the oxygen-based technologies merit serious consideration.

3.3.5   Hydrocarbon Feedstock

This section is based on natural gas feedstock. With natural gas, all technologies are technically viable. However, if the feedstock is a heavy hydrocarbon (such as fuel oil or vacuum bottoms), the POX is the only viable technology. This is because the other technologies cannot process feedstocks heavier than naphtha (to avoid carbon deposition on the catalyst).

3.3.6 H_2/CO Ratio

The importance of the product H_2/CO ratio has already been discussed extensively. As previously stated, the technology that offers a natural ratio that spans the required product ratio is considered to have an inherent advantage and must be seriously considered.

3.3.7   Natural Gas Price

Since the SMR-based technologies directionally require more natural gas, they are favored by low natural gas prices.

3.3.8   Capital Cost

The SMR-based technologies typically have a somewhat higher capital cost because the SMR furnace with its high alloy tubes and large flue gas heat recovery section is inherently more expensive than the ATR or POX technologies which are refractory lined carbon steel vessels without external flue gas heat recovery.

An SMR has a lower capital cost than an SMR/O_2R when the operating pressure is relatively low. Since low pressure favors the reforming reaction, under such conditions most of the reforming occurs in the SMR, and the O_2R does not significantly reduce the size of the SMR. But at higher pressures, an SMR/O_2R can cost less, because more of the reforming is done in the O_2R, and the size of the SMR can often be significantly reduced.

The ATR and POX technologies have offsetting capital costs, depending on the extent of CO_2 recycle.

For full CO_2 recycle, the CO_2 removal system is much larger for the ATR than the POX, and the added capital cost associated with this difference tends to make the ATR more expensive overall.

However, for no CO_2 recycle, the ATR tends to cost less than the POX because its reaction and heat recovery section is inherently less expensive and it does not carry any third party royalty.

3.3.9   Conclusions

Based on the above, the following conclusions can be drawn.
1) The SMR/O_2R, ATR, and POX technologies can be attractive if low cost oxygen is available.
2) For competing technologies, the H_2/CO product ratio is typically the most important process parameter.
3) For low methane slip, the SMR/O_2R, ATR, and POX technologies are favored.
4) For full CO_2 recycle, the POX is typically better than the ATR.
5) Relative to the POX, the ATR is a non-licensed technology which avoids third-party involvement.
6) The economics of each technology is dependent on the conditions and requirements for each project and must be evaluated on a case-by-case basis.
7) The technology with the most favorable H_2/CO ratio for producing hydrogen is the SMR.

3.4   Hydrogen Purification

3.4.1   Old Style

Many refiners are still operating hydrogen plants that were designed and built 20 or more years ago. These older plants were generally designed with the best available technology of the time. These typical plants consist of a steam-methane reformer (SMR) to convert the hydrocarbon feed to a syngas mixture followed by a high temperature shift converter (HTSC) and low temperature shift converter (LTSC) to shift most of the CO to hydrogen. This hydrogen-rich gas is purified by a CO_2 removal unit where a hot potassium carbonate or MEA solution removes the CO_2 and then followed by a methanator to convert the remaining CO and CO_2 to methane and water. The final product gas is typically about 95-97% hydrogen. Figure 8 shows a typical layout for an Old Style hydrogen plant.

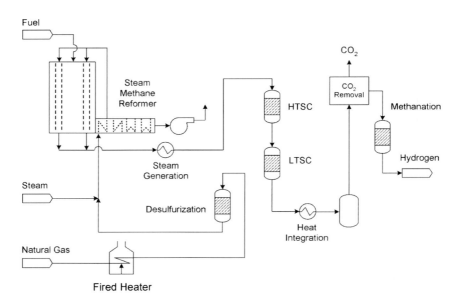

Figure 8. Old Style Hydrogen Plant

3.4.2   Modern

The main process difference between a Modern hydrogen plant and an Old Style hydrogen plant is the hydrogen purification technology. A Pressure Swing Adsorption (PSA) unit replaces the CO_2 removal unit and methanator and allows the Modern plant to produce hydrogen product with a much higher purity. In addition, many significant improvements in technology and design allow the Modern plant to operate at much higher efficiency and with substantially lower operating costs.

A Modern hydrogen plant includes an SMR followed by an HTSC. A PSA unit purifies the syngas effluent from the HTSC. The final product gas is typically 99.99% hydrogen. Since the PSA unit removes the impurities from the syngas, an LTSC is not required to further reduce the CO content.

The PSA unit produces an offgas stream that can be used by the SMR as the primary fuel source. In addition, the increase in product purity from the PSA unit has potential benefits in downstream units. For example, the higher purity hydrogen makeup to a hydrotreater increases reactor hydrogen partial pressure, lowers the recycle flow, and potentially reduces compression costs and/or increases Hydrotreater performance and catalyst run life. Figure 9 shows a typical layout for a Modern hydrogen plant.

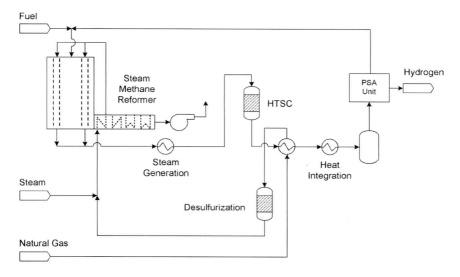

Figure 9. Modern Hydrogen Plant

4. DESIGN PARAMETERS FOR SMR'S

Steam reformers are used in hydrogen generation plants throughout the refining and petrochemical industries and are expected to remain the most cost effective approach for on-purpose hydrogen production. The steam reformer for hydrogen plants consists of a fired heater containing catalyst-filled tubes. This article discusses process design considerations for a modern reformer of this type.

4.1 Function

The primary function of a steam reformer is to produce hydrogen. Hydrocarbon feed gas is mixed with steam and passed through catalyst-filled tubes. Hydrogen and carbon oxides are produced by the following reactions.

$$C_nH_m \text{ (g)} + nH_2O + heat = nCO \text{ (g)} + (m/2+n) H_2 \text{ (g)} \quad (1)$$

$$CO \text{ (g)} + H_2O \text{ (g)} = CO_2 \text{ (g)} + H_2 \text{ (g)} + heat \quad (2)$$

The first reaction is the reforming reaction, and the second is the shift reaction. Both reactions produce hydrogen. Both reactions are limited by thermodynamic equilibrium. The net reaction is endothermic. These reactions take place under carefully controlled external firing, with heat transfer from the combustion gas in the firebox to the process gas within the catalyst-filled tubes.

The carbon monoxide in the above product gas is subsequently shifted almost completely to hydrogen in a downstream catalytic reactor by further utilization of reaction (eqn. 2).

4.2   Feedstocks

Typical hydrocarbon feedstocks for the reformer include natural gas, refinery gas, propane, LPG, and butane. Naphtha feedstocks with final boiling points up to about 430°F can also be processed.

4.3   Fuels

Typical fuels for the reformer are light hydrocarbons such as natural gas and refinery gas, although distillate fuels are sometimes used. Residual fuels are not utilized since the contained metals can damage the reformer tubes.

In most hydrogen plants, a pressure swing adsorption (PSA) system is used for hydrogen purification. In such plants, a major portion of the reformer fuel is PSA offgas, with the makeup fuel being a hydrocarbon stream as discussed above.

4.4   Design

Reformers are fired to maintain a required process gas outlet temperature. Most modern reformers are top-fired. In this design, the burners are located in the top of the furnace, and fire downward. The process gas flows downward through the catalyst-filled tubes in the same direction.

Accordingly, the top-fired design features co-current flow of process gas and flue gas. The co-current flow permits use of (1) the highest flue gas temperature when the intube process gas temperature is lowest, and (2) the lowest flue gas temperature when the intube process gas temperature is highest. This in turn provides tubewall temperatures that are quite uniform over the length of the tube. Because of this uniformity, the average tubewall temperature is lower. And lower tubewall temperatures reduce tube cost and increase tube life.

Another advantage of co-current flow is that as the flue gas cools, it sinks in the same direction as its normal flow. This provides an inherently stable furnace operation. In particular, it avoids flue gas backmixing that inherently occurs in alternative designs. Since there is no backmixing, the flue gas outlet temperature tends to more closely approach the process gas outlet temperature than in alternative designs. This improves the furnace efficiency.

Note that the highest tubewall temperature occurs at the outlet, which is quite common for this type of furnace. Also note that at the tube outlet, the tubewall temperature is very close to the required process gas outlet temperature. This is because the heat flux at the outlet is quite low, having fallen steadily for the last two-thirds of the tube.

Therefore, the tubewall temperature is minimized for the required process gas outlet temperature, which in turn minimizes the required tubewall thickness.

In addition, note that the flue gas temperature falls off steadily for the last two-thirds of the tube, and is at a minimum at the tube outlet. Since the furnace must be fired to provide the necessary flue gas outlet temperature, the fact that the flue gas outlet temperature is relatively low means that the fuel requirement is minimized.

The burners are housed at the top of the reformer by an enclosure commonly referred to as a penthouse. This housing provides a convenient shelter for the burners and for the inlet piping and valving.

The flue gas is collected at the bottom of the reformer in horizontal fire-brick ducts, often referred to as "tunnels." The flue gas then exits horizontally into a waste heat recovery (WHR) unit. The combustion gas is drawn through the WHR unit by an induced draft fan and the gas is then discharged to the atmosphere through a stack. The WHR unit and fan are essentially at grade, which facilitates operation and maintenance.

When the above features of the top-fired design are considered and weighed against alternative arrangements, the consensus of the industry favors the top-fired design.

The remainder of this section will therefore focus on the top-fired design.

The process considerations discussed below focus on steam reformer parameters specifically related to hydrogen plant design. However, many of these parameters also apply to steam reformers for other types of plants (such as syngas, ammonia, and methanol plants).

4.5 Pressure

The shift reaction equilibrium is independent of pressure. However, the reforming reaction equilibrium is favored by low pressure. Therefore for the overall reaction, the lower the pressure, the higher the conversion of hydrocarbon to hydrogen. Accordingly, from the reaction standpoint, it is best to operate at low pressure.

However, there are other considerations. For example, modern hydrogen plants typically use a pressure swing adsorption (PSA) unit for hydrogen purification. PSA units are more efficient at higher pressure. The minimum pressure for acceptable operation is typically 150 to 200 psig. The optimum pressure may be 300 to 450 psig. Therefore, to accommodate the PSA unit, the reformer pressure will need to be adjusted accordingly.

Another consideration is product hydrogen pressure requirements. In many applications, the hydrogen product from the hydrogen plant requires subsequent compression to a much higher pressure, for example for refinery hydrotreating applications. In such cases, the compression cost can be reduced considerably if the hydrogen from the hydrogen plant is produced at a higher pressure.

As discussed below, hydrogen plant reformers typically operate at process gas outlet temperatures up to 1700°F. At these temperatures, metallurgical considerations for the outlet piping typically limit the reformer outlet pressure to about 550 psig. This corresponds to a hydrogen product pressure of about 500 psig.

Many hydrogen plant reformers are designed consistent with a hydrogen product pressure of about 300-350 psig. Although this decreases the reforming reaction conversion, the overall plant economics often favors the higher pressure.

On the other hand, if the product pressure requirement is low, then the PSA unit will typically determine the minimum pressure.

Therefore, for hydrogen plants, the reformer outlet pressure typically runs between 150 and 550 psig.

4.6   Exit Temperature

The reforming reaction equilibrium is favored by high temperature. At the pressure levels used in hydrogen plants, the reformer process gas exit temperature typically runs between 1500 and 1700°F. Lower temperatures give insufficient conversion. Higher temperatures increase metallurgical requirements, tubewall thickness, and fuel consumption.

4.7   Inlet Temperature

The reforming reaction rate becomes significant at about 1000°F, so it is usually advantageous to design for an inlet temperature near this value. This is typically achieved by preheating the reformer feed against the hot flue gas in the WHR section of the reformer.

A higher reformer inlet temperature decreases the absorbed duty requirement and therefore decreases the number of tubes, the size of the furnace, and the fuel requirement. It also decreases the steam generation from the waste heat recovery unit.

If steam has a high value relative to fuel, it may be economical to reduce the reformer inlet temperature somewhat in order to maximize steam generation. In many such cases, the optimum inlet temperature is about 1050°F. This is low enough to maximize steam generation, but high enough to keep the furnace size down.

If steam does not have a high value, the optimum reformer inlet temperature is often about 1100°F. Above this, metallurgical considerations with the inlet piping become a factor.

Up to 1050°F, 2-1/4 Cr -1 Mo material is satisfactory for the inlet piping. At higher temperatures, stainless steel would ordinarily be used, but stainless steel is generally considered unacceptable because of the possibility of chloride stress cracking. Such cracking can occur if the steam in the feed has chloride-containing water droplets carried over from the steam drum. Incoloy is considered the next step up in metallurgy and is not subject to chloride stress cracking, but its cost typically prohibits its use. Accordingly, for metallurgical reasons the maximum inlet temperature should remain at about 1050°F.

4.8   Steam/Carbon Ratio

The hydrocarbon feed must contain sufficient steam to eliminate carbon formation. The relationship between steam and hydrocarbon is typically expressed as the steam-to-carbon ratio. This corresponds to the moles of steam per mol of carbon in the hydrocarbon. The design steam-to-carbon ratio is typically about 3.0 for all hydrocarbon feedstocks. Lower values (down to about 2.5) can be used for some feedstocks, but there is a higher risk from potential operating upsets that might occur which could drop the ratio down to where carbon formation could occur.

For heavier feedstocks, carbon formation is more likely, and an alkali-based catalyst must be used to suppress carbon formation. Such catalysts are readily available, and it is understood that the 3.0 steam-to-carbon is applicable only if the proper catalyst is chosen. Otherwise, higher steam-to-carbon ratios are required.

4.9   Heat Flux

The reformer heat flux is typically defined as the heat input per unit of time per square unit of inside tube surface. For a given absorbed duty, the heat flux is therefore determined by the amount of tube surface.

A low heat flux provides extra catalyst volume and lower tubewall temperatures. This provides several advantages: The extra catalyst volume increases the reforming reaction conversion. The lower tubewall temperature reduces the tubewall thickness, which in turn reduces the cost per tube and increases the tube life. The lower tubewall temperature also reduces the fuel requirement.

A high heat flux has the opposite effect, but has the advantage of reducing the number of tubes.

Because of these trade-offs, commercial heat fluxes typically vary from about 20,000 Btu/hr-ft2 to 28,000 Btu/hr-ft2. A conservative design will typically run from 20,000 to about 25,000 Btu/hr-ft2.

The above fluxes are average values for the entire furnace. The point flux is highest in the zone of maximum heat release, and then falls off to a relatively low value at the tube outlet. The maximum point flux is often 35,000 to 40,000 Btu/hr-ft2.

4.10   Pressure Drop

The reformer pressure drop depends primarily on the number of tubes, the tube diameter, and the catalyst selection. Typically, the overall design pressure drop will range from 40 to 60 psi.

4.11   Catalyst

Reforming catalysts are typically manufactured in a ring or a modified ring form. The modified ring forms have increased surface area and a higher activity for about the same pressure drop, but they cost more.

Reforming catalysts use nickel as the principal catalytic agent. For heavier feedstocks, an alkali promoter is typically used to suppress carbon formation. For heat fluxes above about 25,000 Btu/hr-ft2, modified ring shapes are needed to maintain the reforming reaction conversion.

The reaction conversion is typically measured in terms of "approach to equilibrium." A typical design approach to reforming equilibrium is 20°F. This means that the reforming reaction (equation 1 above) is at equilibrium corresponding to a temperature 20°F lower than the actual temperature. This corresponds to a typical end-of-life catalyst condition. Fresh catalyst typically runs at essentially a 0°F approach to reforming equilibrium (that is, for fresh catalyst the reforming reaction is essentially at equilibrium at the actual temperature). The shift reaction (equation 2 above) is rapid and is considered in equilibrium at all times.

Typical catalyst life is 4 to 5 years. To obtain the performance discussed above over this time frame, the appropriate catalyst must be selected, consistent with the design heat flux and allowable pressure drop.

For higher flux reformers, a dual charge of catalyst is typically used. The top half of the tube is loaded with a high activity catalyst that prevents carbon formation in the zone of maximum flux. The bottom half of the tube can be a more conventional, less expensive catalyst.

Catalyst pressure drop is also an important consideration. Fortunately, the modified ring catalysts can provide higher activity for about the same pressure drop. If significantly higher activity is needed, a higher pressure drop must be accepted.

4.12   Tubes

Reformer tubes typically operate at maximum temperatures of 1600 to 1700°F and are designed for a minimum stress-to-rupture life of 100,000 operating hours. Today's preferred metallurgy is a 35/25 Ni/Cr alloy modified with niobium and microalloyed with trace elements including titanium and zirconium.

Smaller tube diameters provide better intube heat transfer and cooler walls. The cooler walls reduce tubewall thickness, which reduces tube cost and increases tube life. The cooler walls also decrease fuel consumption. However, more tubes are required and the pressure drop is higher. Based on these parameters, the optimum inside tube diameter is typically 4 to 5 inches.

Thin walls increase tube life because secondary stresses are minimized during thermal cycling on startups, shutdowns, and upsets. Accordingly, the tubewall thickness should be minimized consistent with meeting the tensile strength requirement. For many applications, the minimum sound wall (MSW) can be as low as 0.25 inches.

Longer tubes provide a longer co-current heat transfer path in the reformer, and thereby reduce the flue gas exit temperature, which conserves fuel. Longer tubes also reduce the number of tubes. However, the pressure drop is higher. The optimum tube length is typically 40 to 45 feet.

Increasing the tube pitch (the center-to-center spacing) reduces the shielding effect between tubes and lowers the peak temperature around the tube circumference. This improvement is significant for short tube pitches, but falls off at longer pitches.

If the pitch is too short, the design must be modified to avoid overlapping the tube inlet flanges. The optimum pitch is typically 2 to 3 tube diameters.

The lane spacing between tube rows must be sufficient to avoid flame impingement from the burners. Typical spacing ranges from 6 to 8 feet.

4.13   Burners

The burners are located between the tube rows. Increasing the number of burners reduces the heat release per burner, which permits a smaller flame diameter and reduced lane spacing. The number of burners varies with the design, but a ratio of one burner for every 2 to 2.5 tubes will provide a very uniform heat release pattern and is considered good design practice.

For most hydrogen plants, the burners are a dual-fired design, which will fire both PSA offgas and the supplemental makeup gas. Lo-NOx burners are typically used to meet modern environmental requirements. In some burner designs, the makeup gas is used to induce flue gas into the flame, thereby reducing the flame temperature and the NOx level. With a properly designed burner, NOx levels of as low as 0.025 lbs/MM Btu LHV of heat release can be expected.

4.14   Flow Distribution

It is important to obtain good flow distribution for all reformer streams. The piping should be designed such that the variation in gas flow to the reformer tubes and to the burners does not exceed plus or minus 2.5 percent. Otherwise, the tubewall temperatures may not be sufficiently uniform.

Special consideration must be given to the PSA offgas flow since it is typically available to the burners at only about 3 psig. Even greater consideration must be given for preheated combustion air (if used), since the differential air pressure across each burner is typically no more than 2 inches H_2O.

To help ensure good distribution, the piping should be as symmetrical as possible, and detailed pressure drop computations should be made and analyzed.

It is especially important that the flue gas tunnels be properly designed. The tunnels are rectangular fire-brick structures located at the bottom of the reformer which serve as horizontal ducts for flue gas removal. Selected bricks are removed along the bottom of the tunnels to provide openings for the flue gas. To ensure plug flow of flue gas down the box, these openings must be designed for uniform flow distribution of the flue gas from the box into the tunnels. To ensure proper design, detailed hydraulic calculations should be made and analyzed

4.15   Heat Recovery

The flue gas typically exits the radiant box at 1800 to 1900°F. A waste heat recovery (WHR) unit is provided to recover heat from this gas. Typically, this consists of a package unit containing a reformer feed preheat coil, followed by a steam superheat coil (if applicable), followed by a steam generation coil, followed by a boiler feedwater preheat coil. If combustion air preheat is used, the air preheat unit typically replaces the boiler feedwater coil. The flue gas typically exits the WHR unit at about 300°F.

On this basis, and with a typical heat loss of 3 percent of the absorbed duty, the overall efficiency of the reformer (radiant plus WHR) is about 91 percent on an LHV basis.

Steam is also generated in a process steam generator, which extracts heat from the reformer outlet process gas. The WHR unit and the process steam generator typically share a common steam drum.

Steam generation pressure must be sufficient to provide steam to the reformer. Typically, the minimum required pressure is 100-150 psig above the hydrogen product pressure, depending on the plant pressure drop. Significantly higher steam pressures can easily be accommodated, and reformers generating 1500 psig steam are not uncommon in some industries.

5.   ENVIRONMENTAL ISSUES

The federal government's passage of the Clean Air Act in 1970 set national limits on industrial emissions to reduce air pollution and acid rain. Since 1970, the Clean Air Act has been amended and some states, such as California, have set even stricter limits on several of these emissions. Therefore, a major point of emphasis when designing a hydrogen plant is the reduction of plant emissions. A typical hydrogen plant has three sources of emissions: 1) Flue gas from the combustion chamber of the reformer; 2) Condensate from the process; and 3) Wastewater from the steam generation system.

5.1   Flue Gas Emissions

There are five primary pollutants found in the flue gas: nitrogen oxides, carbon monoxide, sulfur oxides, unburned hydrocarbons, and particulates. These components are formed during the combustion process.

The majority of the flue gas emissions consist of the nitrogen oxides (NOx). NOx is an environmental concern because it can cause photochemical smog and acid rain. The nitrogen oxides from a hydrogen plant primarily consist of nitrogen oxide (NO) and nitrogen dioxide (NO_2). There are three types of NOx formation: prompt NOx, fuel NOx, and thermal NOx. Prompt NOx is formed when fragments of hydrocarbons in the fuel combine with nitrogen in the combustion air. This form of NOx is considered to be negligible as compared to thermal NOx in hydrogen plant applications. Fuel NOx is formed when nitrogen-containing hydrocarbons in the fuel are burned. However, this is typically not a concern for hydrogen plants because the makeup fuel is usually natural gas. Thermally produced NOx represents the largest contributor to NOx formation in a hydrogen plant. Thermal NOx is produced by N_2 and O_2 in the combustion air reacting in the hottest part of the burner flame. The rate of thermal NOx formation is based on the burner flame temperature, the amount of combustion air, and residence time. As the flame temperature increases, the amount of NOx formed increases. Issues such as type of fuel gas, amount of excess air and combustion air temperature affect the amount of thermal NOx formation. Therefore, to reduce the flame temperature, many burner manufacturers have developed burner designs that incorporate staged combustion and flue gas recirculation. The recirculation of the flue gas cools the flame thus lowering the amount of thermal NOx formation. The recirculation will also lower the amount of CO exiting the burner, as a portion of it will be recycled and combusted.

A NOx removal system may be required to meet a site's emission requirements. The typical system used for this case is a Selective Catalytic Reduction unit (SCR). The SCR unit reduces NOx by reacting ammonia with the flue gas over a catalyst, which yields nitrogen and water vapor. An SCR unit can reduce the NOx emission from a hydrogen plant by approximately 90%. There is a small slipstream of ammonia, approximately 10 ppmvd that will exit with the flue gas. This unit adds additional capital costs as well as operating costs and ammonia handling issues but is sometimes required as the best available technology (BACT) for NOx removal.

The CO, combustion produced particulates, and the unburned hydrocarbons are all related to the amount of excess air, type of fuel, and the amount of mixing of the fuels within the burner. Insufficient air and inadequate mixing can result in incomplete combustion thus raising the amount of CO and unburned hydrocarbons in the flue gas. Particulates formed in the combustion process are very low for gaseous fuels but will increase with the use of liquid fuels.

The sulfur oxides that are formed are directly related to the amount of sulfur found in the fuel. These emissions are typically low when PSA offgas is burned with natural gas. The PSA offgas is sulfur-free due to feedstock pretreatment to protect catalyst beds within the hydrogen plant. Natural gas also typically contains very low levels of sulfur. However, if refinery fuel gases are used as a makeup fuel to the reformer, then the sulfur emissions can increase dramatically as these streams often contain large amounts of sulfur.

Burner manufacturers will have typical values of these emissions for their burners based on actual tests of their burner design. The emission limits of the CO, sulfur oxides, unburned hydrocarbons, and particulates are generally met since light gaseous fuels are typically used as the fuels for the reformer. Table 9 below contains some typical emission values for a standard hydrogen plant utilizing Low-NOx burners when burning PSA offgas with natural gas.

Table 9. Typical emission values for a standard hydrogen plant

Emission	Units	Value
NOx	lb/MM Btu LHV	0.03
CO	ppmv (3% O_2 dry basis)	25
Unburned Hydrocarbons	ppmv (3% O_2 dry basis)	5
Particulates	lb/MM Btu LHV	0.005

5.2   Process Condensate (Methanol and Ammonia)

The process condensate from a standard hydrogen plant contains impurities such as dissolved gases, ammonia, methanol, and traces of other organic compounds. The dissolved gases that are found in the process condensate consist of the components in the syngas (CO, CO_2, H_2, CH_4, and N_2). Henry's Law can be used to calculate the amounts of these gases that are in the condensate.

Henry's Law: $P = H x$

where   P is Partial Pressure of a Component

H is Henry's Law Constant
X is Mole Fraction of Component in Solution

The CO_2 represents the largest amount of dissolved gas in the condensate, typically in the range of several 1000 ppmw. The dissolved H_2, CO, and CH_4 are typically in lower concentrations of less than a hundred ppmw each.

In a hydrogen plant, ammonia is produced in the reformer. The steam methane reformer's operating conditions are similar to that of an ammonia synthesis reactor. The amount of ammonia produced is based on the amount of nitrogen in the feed. Feedstocks with negligible amounts of nitrogen will not produce an appreciable amount of ammonia. Ammonia formation depends on the reaction equilibrium and the residence time inside the reformer. The ammonia equilibrium reaction is given below.

$$[NH_3] = Kp_{NH3} * [N2]^{0.5} [H2]^{1.5} * P_T$$

where [ ] is Mole Fractions
Kp_{NH3} – Equilibrium Constant
P_T – Total Pressure, atm

Ammonia formation is lessened due to the fact that the residence time of the steam methane reformer is designed for hydrogen production not ammonia production with recycle. In an NH_3 plant, the ammonia is considered to be at equilibrium in a secondary reformer that is operating approximately 300°F to 500°F higher than a traditional steam methane reformer. Therefore for calculations, the worst case for ammonia production in a hydrogen plant can be assumed to be equilibrium at the reformer outlet temperature plus an additional 300°F.

Methanol is produced as a by-product in the shift converters within the hydrogen plant. In High Temperature Shift Converters (HTSC), the formation of methanol is an equilibrium reaction similar to that of ammonia formation in the reformer. The equilibrium reaction for methanol is given below.

$$[CH_3OH] = Kp_{CH3OH} [CO_2][H_2]^3 P_T^2 / [H_2O]$$

where [ ] is Mole Fractions
Kp_{CH3OH} – Equilibrium Constant
P_T – Total Pressure

The typical concentrations of methanol in an HTSC application are approximately 100 to 300 ppmw. In a Low Temperature Shift Converter (LTSC) application, the methanol production is greater than that of an HTSC application. The formation of methanol is not just related to equilibrium for an LTSC but also by the catalyst characteristics and kinetics. Therefore, the catalyst vendor should be contacted in reference to calculating the expected amount of methanol from an LTSC application.

In the majority of hydrogen plants today, the process condensate is used as makeup water to the steam generation system of the plant. Typically, boiler feedwater makeup is mixed with the process condensate and sent to a deaerator. The deaerator uses steam to strip the dissolved gases, namely O_2 and CO_2, and the other contaminants from the boiler feedwater. These contaminants can be harmful to downstream equipment and boiler operation. These contaminants are then emitted to the atmosphere. Since there are large amounts of CO_2, the type of deaerator for hydrogen plant use typically has a vertical stripping section consisting of either trays or packing located on top of the deaerator. To calculate the amount of each contaminant leaving the deaerator's stripping section, one can use the equations below.

Mole Fraction Remaining:

$$(Lx_{OUT}) / (Lx_{IN}) = (S-1) / (S^{n+1} - 1)$$

and Mole Fraction Stripped:

$$(Gy_{OUT}) / (Lx_{IN}) = S (S^n - 1)/(S^{n+1} - 1)$$

where   G & L are Molar Flow Rates
        S – Stripping Factor (KG / L)
        n – Number of Equilibrium Stages
        K – Vapor-liquid distribution coefficient, $(y / x = H / P_T)$

As methanol emissions continue to be monitored more closely, there are some methods of reducing the methanol in the deaerator vent. The vent stream could be condensed and sent to the reformer or the steam system. Catalytic combustion could be used to reduce the methanol. A scrubber system could be added to remove the methanol. In some instances a condensate stripper is added instead of the deaerator to remove the ammonia, methanol, and other contaminants from the condensate. This system recycles the vent stream to the reformer as process steam and the bottoms are mixed with the incoming boiler feedwater makeup. However, this system adds considerable capital cost to a project.

5.3   Wastewater

The wastewater from a hydrogen plant typically consists of only the blowdown from the boiler system. The boiler feedwater that feeds the steam generation system has small amounts of impurities such as sodium, chlorides, silica, and organic carbons. These impurities will accumulate within the boiler system and create sludge, scaling of the boiler tubes, and possible carryover of solids into the process steam. Blowdown of the boiler water is performed to prevent these issues from affecting the operation of the steam system. The blowdown is typically sent to the sewer or the on-site waste treatment plant for treatment and disposal.

6.   MONITORING PLANT PERFORMANCE

There are several areas of focus when monitoring the performance of a hydrogen plant. The first area of focus is the performance of the catalyst beds employed in the hydrogen plant. The Hydrotreater, which converts sulfur compounds to H_2S, should be checked periodically for pressure drop through the bed. The hydrotreating catalyst life is approximately 3 years. If the catalyst bed's pressure drop is exceeding design then it may be time to change the catalyst out due to catalyst degradation and activity loss.

The next typical catalyst bed in a hydrogen plant is the desulfurizer bed. The desulfurizer removes the H_2S from the feed gas. The desulfurizer catalyst bed design is based on the loading of the sulfur compounds. Therefore, the exiting gas from the desulfurizer bed should be checked to ensure that the sulfur levels are below 0.1 ppmv. If the original design life of the catalyst bed is known, then periodic feedstock analysis can be used to forecast when the bed may require change out.

Since the reforming reaction constitutes the majority of the hydrogen production in the plant, it is important to monitor the reforming catalyst. The reforming catalyst is typically designed for a 5-year life. Pressure drop measurements should be taken across the catalyst filled tubes in the reformer. If the pressure drop increases as time goes by, this could be an indication that catalyst attrition or possible carbon formation on the catalyst could be taking place. The activity of the catalyst can be checked by comparing the outlet composition of the reformer to the expected composition at the design approach to equilibrium conditions. If the pressure drop increases or the activity has decreased, then it is probably time to change out the reforming catalyst.

The examination of the high temperature shift catalyst is similar to that of the reforming catalyst. The catalyst life for the high temperature shift catalyst is approximately 5 years. Pressure drop readings should be taken to check for possible catalyst attrition. The outlet composition should be validated with the expected composition with the design approach to equilibrium conditions.

The pressure swing adsorption system (PSA) has a catalyst life of approximately 20 years or equal to the expected plant life. However, the hydrogen purity and recovery should be recorded periodically to watch for possible catalyst poisoning. The amount of nitrogen in the feedstock also has an effect on these two operating parameters as well and should be checked. A higher nitrogen content in the feedgas than design could lower the hydrogen recovery in the PSA unit.

The reformer tubes are one of the most important pieces of equipment in a hydrogen plant. These tubes are built from micro-alloyed materials in order to handle the extreme environment in which they are exposed to. The industry standard for reformer tube design is for 100,000 hour life. To ensure that the tubes will last the designed 100,000 hour life, the reformer tubewall temperature readings should be taken on a regular basis. An optical pyrometer is the instrument of choice for this task. In order to obtain the correct tubewall temperature reading, the tubewall and the furnace background should be shot for each tube reading. The background temperature is required to correct for the background radiation. The equation below can be used to correct the tubewall temperature readings. The emissivity on the pyrometer should be set to 1 before taking the readings.

$$Tt = [(T_{MT}^4 - (1 - e) * T_{MB}^4) / e]^{1/2}$$

where   Tt is True Tube Temperature, °R
         T_{MT} is Measured Tube Temperature, °R
         T_{MB} is Measured Background Temperature, °R
         e is Average furnace emissivity (typical = 0.82)

The maximum allowable tube stress will need to be calculated in order to calculate the actual reformer tube life. This is accomplished by using the Mean Diameter Formula given below.

$$S = (P * D_M) / (2 * t_{MIN})$$

where   S is Tube stress, psi max
         P is Tube inlet pressure, psig
         D_M is Tube Mean Diameter = $(D_O + D_I)/2$
         D_O is Tube OD, as cast, inches
         D_I is Tube ID = $D_O - 2 * t_{MIN} - 2 * CA$
         CA is Casting allowance, (typically 1/32" on outside wall, 0" on inside wall)
         t_{MIN} is Minimum sound wall, inches

The Larson-Miller equation correlates the stress-temperature to life of the tubes. The equations below are for Manaurite XM material tubes.

$$S = 0.145 * B * 10^{(-0.0062*P^2 + 0.2955*P - 1.5426)}$$
$$P = T * (22.96 + LOG(t))10^{-3}$$

where   S is Minimum stress to rupture, ksi
         B is Ratio of minimum to average stress (Typical temperature range, B = 0.85)
         P is Larson-Miller parameter, dimensionless
         T is Tube wall temperature, °K
         t is tube life, hours

If the measured temperature is less than design, then the reformer tubes should last their expected life. However, if the measured temperature is greater than the design temperature, then the reformer tube life will be shortened. If the actual life of the reformer tubes is below the expected life, then the operating conditions of the reformer should be further investigated.

The steam system is an area that requires constant attention to ensure proper operation of the hydrogen plant. If the steam quality decreases, it can lead to solids carryover from the steam drum. These solids will plate out in the feed preheat coil of the convection section and subsequently lead to an equipment failure and plant shutdown. To reduce the probability of an upset, the steam drum should be manually blowndown on a regular basis to reduce the amount of dissolved solids and other impurities in the steam drum. Next, the boiler feedwater should be analyzed on a regular basis to ensure that the treatment it is receiving is adequate for the desired steam generation conditions.

Cooling water is generally used for the final trim cooling of the process gas before the hydrogen purification step. In most instances, the cooling water is considered to be a dirty medium. Therefore, it is recommended to develop a temperature profile of the cooling water side of the exchanger based on design data. Additional temperature profiles should be developed based on data from the operation of the plant and compared to the original profile. This data can be used as a means to check for fouling of the cooling water side of the exchanger and improper heat transfer through the exchanger.

Another means of monitoring the performance of the hydrogen plant is to check the pressure drop through the entire plant. If the pressure drop is higher than design, then individual pieces of equipment should be checked for excess pressure drop. This may signify several things such as catalyst attrition, excessive exchanger fouling, or equipment failure.

The final method of checking the performance of the unit is to calculate the overall utility consumption of the plant and compare it to the design summary or previous summaries. The feed and fuel flows are typically converted to energy since their utility cost are typically on an energy basis (for example: MM Btu/hr LHV). These numbers as well as the other utility quantities are divided by the hydrogen product flow. This method allows for calculating the total price of hydrogen on a $/M SCF H_2 basis. Once the summary has been created, comparisons to earlier summaries and analysis of major cost areas can be investigated. The effects of process improvements can be evaluated on this basis.

7. PLANT PERFORMANCE IMPROVEMENTS

The most common method of improving the plant performance is to increase the capacity of the hydrogen plant. The plant design should be evaluated as to its capability to handle an increased load before this is attempted. The process design specifications should be retrieved on all of the relevant plant equipment and compared to the vendor's design specifications. If there are any significant differences between the two sets of specifications, investigation into the cause of the difference is warranted. From this comparison, a final set of equipment design data should be developed for each piece of equipment.

The operating data of the equipment should then be compared to the design capabilities of the equipment in order to set the available capacity increase for each piece of equipment. Compressors should be checked for capacity versus design flow, spillback design, inspection of valves, and motor requirements. The fans should be checked for operating versus design capacity, vane inspection, and motor requirements. The pumps should be checked for operating versus design capacity, inspection of screens and impellers, and motor requirements.

The next area of evaluation consists of the pressure profile of the plant. The operating pressure profile should be compared to the design pressure profile. Any significant differences should be investigated and possibly corrected. An acceptable capacity increase for the equipment should be used to set a proposed capacity increase for the plant. The pressure drop for this raised capacity case should be calculated and compared to design pressure profile. Depending on the results of this calculation, investigation into the available feed pressure may be warranted. Relief settings for the raised capacity case should be evaluated to ensure proper operation. Some pieces of equipment may have to be checked to determine if they can be re-rated for a higher operating pressure than design.

The reformer tubes will need to be evaluated as to their capability of handling an increase in capacity. Temperature readings should be taken at several plant capacity increments (50%, 75%, 100%). This data should be compared with the maximum design tubewall temperature. If the tubewall temperature is approaching the maximum design tubewall temperature, there are several things that can be considered. The catalyst vendor should be contacted about the possibility of a more active reforming catalyst. The burners could be revamped or replaced. The burner pressure drop should be monitored and compared to design data. The excess air used for the combustion should be checked to ensure proper operation. If the burners were to be replaced, it would affect firing capacity, flame pattern, and NOx emissions. The reformer tube metallurgy should be checked to see if it could be upgraded. An upgrade in tube metallurgy would affect the tube ID and the design tubewall temperature.

Most hydrogen plants utilize a PSA system for their hydrogen purification. This system should be checked for H_2 purity, and a corresponding H_2 recovery should be calculated. This data should be compared to the original design data of the unit. The cycle times should be compared to the design cycle time. To check the available capacity increase for the PSA unit, the adsorption time should be increased in small increments until the maximum permissible impurity breakthrough occurs. The cycle time should be adjusted and the H_2 recovery calculated. The purge time in this cycle should be compared with the design purge time. The additional purge time is a measure of the additional capacity that is available in the PSA unit.

If the PSA unit is capacity limiting, then the following options should be considered: reduction of offgas drum back pressure, improvement of feed conditions, relaxation of purity specifications, setting unit on automatic purity control, change cycle (equalizations), adsorbent change out, or add additional vessels.

The reduction of the utility consumption of the plant is another method of improving the plant performance. It is important to make certain that the plant is not using an excessive amount of utilities. The reformer is a key target area for this investigation. The burners should be checked to ensure that they are using the correct amount of excess air (typically 10%, 20% for heavier fuels). For example, if the burners were using 20% versus 10% excess air, this would translate to an approximate increase in reformer firing of 10%. Therefore, using the correct amount of excess air does not waste fuel. The bridgewall temperature should also be checked. For every 20°F above the required bridgewall temperature, there is a 1.5% increase in reformer firing. The bridgewall temperature should be set at the temperature required to obtain the process outlet temperature that yields the proper CH_4 slip through the reformer. As discussed above, the PSA recovery should be maximized. For every 1% increase in recovery, there is an approximate savings of 0.5% of feed and fuel. The deaerator vent and steam system blowdowns should be checked. Excessive blowdown adds additional treating chemicals and boiler feedwater flow while reducing the export steam quantity.

Another option is to debottleneck an existing hydrogen plant by revamping or upgrading portions of it. However, these cases should be evaluated on a case-by-case basis as different sites have different utility and/or plot considerations. Some of the upgrades are discussed in Section 8:

8.   ECONOMICS OF HYDROGEN PRODUCTION

There is a growing focus in the refining industry on hydrogen capacity. Hydrogen is generally required for deep sulfur removal from hydrocarbon products. As sulfur restrictions on gasoline and diesel become increasingly stringent, the refining demand for hydrogen continues to grow.

By evaluating the hydrogen utilization in their facilities, refiners are coming to the realization that they need additional hydrogen supply. There are a number of options available to address this need. Refiners may be able to meet the increased demand by improving operations of their existing hydrogen plants. They may choose to separate hydrogen from waste or off-gas streams or even purchase hydrogen from 3rd parties. As an alternative, after careful technical and economic evaluation, they may conclude that the best solution to optimize the economic benefit to them over the longer term is to build a new hydrogen plant.

In recent years, many advances have been made in hydrogen plant technology. Substantial improvements have been incorporated into hydrogen plant design to significantly improve overall life cycle costs. Based on experience with hydrogen plant benchmarking, it has become clear that the optimum economic solution in some cases may be to replace an existing hydrogen plant with a new modern hydrogen plant.

There are a number of options available to refiners to meet the increase in hydrogen demand. Before deciding to proceed with any option, refiners should conduct a comprehensive technical and economic evaluation of their existing operations and evaluate the technical and economic benefits of the options available to them. The option that provides the optimum economic and operations benefits will be different for each situation and will depend on such things as the existing steam balance, the cost and availability of utilities, plot limitations, and the condition of existing hydrogen plants.

The first option for refiners who are operating Old Style (See Section 3.4) hydrogen plants is to consider ways to increase the capacity of these plants. Following an evaluation of the condition and efficiency of the existing plant, they may be able to effectively increase the capacity by either tightening up on operations or selectively upgrading portions of the old plant.

In many refineries, hydrogen is treated like a utility. There may not be much of a focus on the details of the actual production and the plants are sometimes operated "loose." In this case, simple operational changes could significantly increase production and efficiency. Refiners could also debottleneck an existing hydrogen plant by revamping or upgrading portions of it. Debottlenecking options can also have a positive efficiency impact. Some of these potential upgrades may include:

Replacing reformer tubes with upgraded metallurgy and thinner walls will allow for more throughput and a higher heat flux, which would increase capacity.

- Adding a Pre-Reformer would unload the primary reformer so capacity can be increased.
- Adding a Secondary Reformer would increase methane conversion, which increases capacity.
- Adding Combustion Air Preheat would lower the fuel requirement and potentially unload the waste heat recovery unit and fluegas fan, resulting in a capacity increase.
- Upgrading the CO_2 removal unit would minimize hydrogen loss in the Methanator, resulting in a potential capacity increase.
- Adding a PSA unit would decrease hydrogen production, but would typically produce more cost effective and higher purity hydrogen.

A second option is to separate hydrogen from a waste stream or an offgas stream that is currently being sent to fuel. This would typically require the addition of separation equipment and possibly some compression. In addition, separating hydrogen out of the fuel system will usually result in addtional makeup fuel. This could change the heating value of the refinery fuel system and possibly have an impact on other fuel burning equipment.

A third option is to buy hydrogen from a 3rd party. Various industrial gas suppliers are willing to sell hydrogen to refiners either by pipeline or, depending on location, by a stand-alone plant. This option requires a minimal capital investment by the refiner but the delivered hydrogen will probably be more expensive per unit than if self-produced.

The last option is to build a new hydrogen plant. Building a Modern hydrogen plant is typically the most capital intensive of the options available; however, the capital investment could pay off if there is a significant gain in efficiency. Below we will analyze and compare the typical overall production cost of hydrogen between a Modern and an Old Style hydrogen plant.

8.1   Overall Hydrogen Production Cost

The most significant economic factor in evaluating options to increase hydrogen capacity is the overall production cost of hydrogen. The overall production cost can be estimated over the life of the hydrogen plant by using the different cost parameters of constructing, operating, and maintaining the hydrogen plant. This overall production cost reflects a complete picture of the hydrogen plant economics over the life of the plant.

The efficiency of producing hydrogen and by-products are very important in minimizing the production cost of hydrogen. Utility costs are the major operating cost in hydrogen production. Utilities typically include usage of feed, fuel, boiler feed water, power, and cooling water, and generation of export steam (steam is typically a by-product of the hydrogen production process). Of these, feed and fuel make up the largest portion of the utility costs. In addition, the credit for export steam can have a significant impact on utility costs, especially when refinery utility costs are favorable for steam production. The remainder of the utilities combined typically make up less than 10% of the total operating cost.

These utility costs, together with other economic parameters applicable to the plant being evaluated, can be incorporated into a cash flow model and the overall production costs of hydrogen can be evaluated. The other economic parameters include such things as capital cost, start-up cost, other operating costs (including catalyst replacement and tube replacement), and maintenance costs. From this model, the internal rate of return (IRR), net present value at various rates of return (NPV), net cash flow, and a generated income statement can also be developed.

8.2   Overall Production Cost Comparison

Building a new hydrogen plant is typically not the most appealing alternative to refiners. A new hydrogen plant requires a significant capital investment, and although hydrogen is required to support many of the refinery unit operations, it is generally not viewed as a direct "money maker."

However, once all factors are taken into account and a total production cost of hydrogen over the life of the plant is determined, the best economic solution may be replacing an Old Style hydrogen plant with a new Modern hydrogen plant. The following evaluation illustrates this potential.

8.3   Evaluation Basis

To demonstrate the economics of building a Modern hydrogen plant versus continuing to operate an Old Style plant, two representative plants each producing 90 MM SCFD of contained hydrogen from a natural gas feed will be compared. Natural gas will also be used for fuel, and both plants will export 600 psig superheated steam. The Old Style plant will consist of the major processing units described above and will produce hydrogen with a purity of 95%. Other parameters for the Old Style plant will be based on typical observed values.   The Modern plant will consist of the major processing units described above and will produce hydrogen with a purity of 99.99%. The Modern plant design will be based on producing maximum export steam.

This evaluation could be done based on a variety of other plant configurations, but for demonstration purposes, this evaluation is limited to the plant types described. For comparison purposes, the cost of utilities will be based on the following:

Natural Gas	$4.00 per MM Btu
HP Steam	$5.00 per 1000 lbs
Boiler Feedwater	$0.50 per 1000 lbs
Power	$0.05 per kWh
Cooling Water	$0.10 per 1000 gals

8.4   Utilities

As previously discussed, the utility costs of a hydrogen plant are among the most important economic factors in determining the overall production cost of hydrogen. Simulation models were built for both the Old Style and Modern plants to calculate utility costs. The Old Style plant utilities are based on a simulation model built to reflect typical plant performance. The Modern plant utilities are based on a simulation model for the design of a typical new plant. Table 10 shows the utilities and utility cost of hydrogen for each plant.

Table 10 clearly illustrates a lower total utility cost for producing hydrogen in the Modern Plant than in the Old Style plant.   The Modern plant produces hydrogen at rate of $1.409 per M SCF of contained hydrogen, while the Old Style plant produces at a rate $1.908 per M SCF of contained hydrogen. For a plant producing 90 MM SCFD of contained hydrogen, this

Table 10. Utiltiy Cost of Hydrogen Production

Utilities per 100 SCF Contained Hydrogen		
	Old Style	**Modern**
Natural Gas Feed, MM Btu LHV	0.275	0.317
Natural Gas Fuel, MM Btu LHV	0.200	0.126
Total Feed,+ Fuel, MM Btu LHV	0.475	0.443
HP Export Steam, lbs	20	90
Boiler Feedwater, lbs	45	120
Power, kWh	0.65	0.52
Cooling Water, gal	530	8
	Old Style	**Modern**
Natural Gas Feed, MM Btu LHV	1.100	1.268
Natural Gas Fuel, MM Btu LHV	0.800	0.504
Total Feed,+ Fuel, MM Btu LHV	1.900	1.772
HP Export Steam, lbs	-0.100	-0.450
Boiler Feedwater, lbs	0.023	0.060
Power, kWh	0.033	0.026
Cooling Water, gal	0.053	0.001
Total Utility Cost, $	**1.908**	**1.409**

difference results in an annual utility savings for a Modern plant of $16.4 MM. Of course, the utility cost alone does not complete the picture of the overall production cost of hydrogen. We must also evaluate a number of other economic and operating factors.

8.5 Capital Cost

Capital must be invested to build the new plant in order to capture the utility savings of the Modern plant. A total capital investment of about $55 MM would be expected for a new typical 90 MM SCFD hydrogen plant. This approximate installed sales price is based on inside battery limits, natural gas feed, no compression, no buildings, no water treatment units, no SCR's or MCC's, industry engineering standards, and delivery to the US gulf coast. For purposes of comparison, no capital investment is considered for the Old Style plant.

8.6 "Life of the Plant" Economics

A cash flow economic model can be generated using the utility costs developed for each plant, the capital cost required for the Modern plant, as well as a number of other economic factors. These other economic factors, along with their associated values are listed below:
- New Plant construction length – 2 years
- Escalation Rate 1.5% for all feed, product and utilities

- Labor – 2 operators per shift
- Overhead – 50% of labor
- Maintenance – 2% of plant cost per year
- Other Misc – 1% of plant cost per year
- Catalyst costs accrued in year of change out
- Reformer tube replacement – every 10 years
- On Stream Time – 98.5%
- Working Capital – 45 days
- Debt Level – Borrow 75% for New Plant
- Cost of Capital – 8%
- Debt Length – 7 years
- Depreciation Life (of Capital Cost) – 10 years
- Tax Rate – 34%
- Project Life – 25 years (includes 2 years of construction)
- Internal Rate of Return – 0% to obtain actual cost of Hydrogen production

This resulting average overall production cost of hydrogen, calculated for the life of the plant, is shown in Table 11.

Table 11. Overall Production Cost of Hydrogen

	Old Style	Modern
Plant Capacity, MM SCFD (Contained)	90	90
Capital Cost, MM$	-	55
Average H2 Production Cost, $ per 100 SCF	1.996	1602
Average H2 Production Cost, $ per year	65,568,600	52,625,700
Average Annual Production Cost Savings, $	-	**12,942,900**

The evaluation shows that taking all relevant economic factors into consideration, the overall production cost of hydrogen is lower for the Modern plant. The Modern plant produces hydrogen at a rate of $1.602 per M SCF of contained hydrogen while the Old Style plant produces hydrogen at $1.996 per M SCF of contained hydrogen. This lower overall production cost results in an annual savings for the Modern plant of about $12.9 MM per year. This evaluation proves that for the cases analyzed, it is economically feasible and potentially advantageous to build a new, more efficient hydrogen plant.

8.7   Sensitivity to Economic Variables

Economic parameters for each refinery are different. The major parameters that can significantly alter these results are the efficiency of the existing plant, the feed and fuel price, and the export steam credit. The efficiency of the existing plant can span a wide range and is specific to each plant. For the remaining evaluations, the Old Style plant utilities will be held constant. The overall production cost of hydrogen will be analyzed as a function of the other two major factors, feed and fuel price and export steam credit.

8.8   Feed and Fuel Price

The cost of feed and fuel is typically the largest component of the overall production cost of hydrogen. Feed and fuel usually account for more than 80% of the total before the steam credit is taken. The overall operating cost changes significantly as the natural gas price varies. Figure 10 shows the effect of varying the natural gas price. For this evaluation, the export steam credit to natural gas price ratio was held constant.

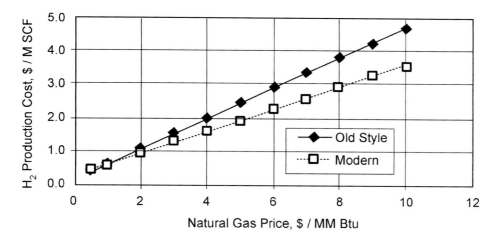

Figure 10. H2 Production vs Natural Gas Price

As the price of natural gas increases, the Modern plant becomes more favorable. This is due to the overall feed and fuel efficiency advantage of the Modern plant. For example, if the natural gas price were changed from $4 to $8 per MM Btu (double the base case credit), the overall production cost would increase by $1.326 per M SCF of hydrogen for the Modern plant and $1.804 for the Old Style plant. The higher natural gas price would increase the average annual savings for the Modern plant from $12.9 MM to $28.6 MM.

8.9   Export Steam Credit

The export steam credit also has a significant effect on the overall production cost. The value a refinery places on steam depends on utility factors and the existing steam balance in the refinery. For example, during the winter, steam tracing is generally used more heavily and the value of steam could be higher than average. Conversely, in the summer, when less steam tracing may be utilized, steam may have a lower than average value. Modern hydrogen plants typically export much more steam due to the fact that they are more efficient and have a CO_2 removal regeneration requirement. Figure 11 shows the effect of varying the export steam credit. For this evaluation, the natural gas price remained the same.

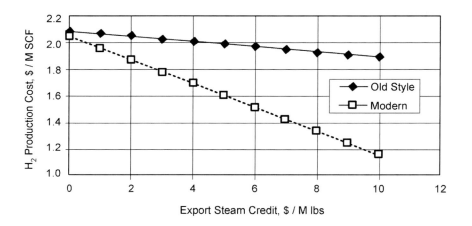

Figure 11. H2 Production Cost Vs Steam Credit

As the export steam credit increases, the economics for the Modern plant become more favorable. For example, if the export steam credit were changed from $5 to $10 per M lbs (double the base case credit), the overall production cost would drop by $0.450 per M SCF of hydrogen for the Modern plant and $0.100 for the Old Style plant. The higher steam credit would increase the average annual savings for the Modern plant from $12.9 MM to $24.4 MM.

9.   CONCLUSION

Refiners have a number of different options to consider in addressing the demand for additional hydrogen. A comprehensive technical and economic evaluation of existing operations is required to determine the optimum solution for each specific application. The optimum solution for each refinery will be different due to site-specific items such as the cost and availability of utilities, plot limitations, refinery steam balance, emission requirements, and the condition of any existing hydrogen plants. The rising cost of utilities, and particularly natural gas, has placed a premium on the overall plant efficiency. Therefore, the overall plant efficiency has not only become a key factor in determining the economics of what option to pursue, but also in the performance enhancement of existing hydrogen plants.

Today's Modern hydrogen plants take advantage of numerous technological improvements and offer a much more efficient plant with lower overall life cycle cost. Once technical and economic evaluations are performed, it should not be surprising that the most economically attractive and most feasible approach may be to build a new hydrogen plant.

10. ADDITIONAL READING

1. Tindall, B.M.; King, D.L., Design Steam Reformers For Hydrogen Production, *Hydrocarbon Process.*, July **1994**.
2. Tindall, B.M.; Crews, M.A., Alternative Technologies to Steam-Methane Reforming, *Hydrocarbon Process.*, p. 80, August, **1995**.
3. King, D.L.; Bochow, C.E., What Should and Owner/Operator Know When Choosing and SMR/PSA Plant, *Hydrocarbon Process.*, May **2000**.
4. Hoitsma, K.; Snelgrove, P., Effective Reformer Design and Erection, *World Refining*, p. 24 July/August **2002**.
5. Tindall, B.M.; Crews, M.A., Economics of export steam for hydrogen plants, *Hydrocarbon Engineering*, 39-41, February **2003**.
6. Boyce, C.; Ritter, R.; Crews, M.A., Is It Time For A New Hydrogen Plant, *Hydrocarbon Engineering*, February **2004**.
7. Ernst, W.S.; Venables, S.; Christensen, P.S.; Berthelsen, A.C., Push Syngas Production Limits, *Hydrocarbon Process.*, 100-C through 100-J, March **2000**.
8. de Wet, H.; Shaw, G., Ammonia Plant Safety & Related Facilities, American Institute of Chemical Engineers, Vol. 35, 315-335.
9. Haldor Topsoe A/S, Autothermal Reforming, 10-1 through 10-11, June **1995**.
10. Christensen, T.S.; Primdahl, I.L., Improve Syngas Production Using Autothermal Reforming, *Hydrocarbon Process.*, 39-46, March **1994**.
11. Lowson, A.; Primdahl, I.; Smith, D.; Wang, Shoou-I, Carbon Monoxide Production by Oxygen Reforming, Air Products and Chemicals, Inc., 2-21 1475H, March 1990.
12. Jahnke, F.C.; Ishikawa; Rathbone, T., High Efficiency IGCC Using Advanced Turbine, Air Separation Unit, and Gasification Technology, Presentation at the 1998 Gasification Technologies Conference in San Francisco, CA., 1-8, October 1998.
13. Wallace, P.S.; Anderson, M.K.; Rodarte, A.I.; Preston, W.E., Heavy Oil Upgrading by the Separation and Gasification of Asphaltenes, Presented at the 1998 Gasification Technologies Conference in San Francisco, CA., 1-11, October 1998.
14. Liebner, W.; Hauser, N., The Shell Oil Gasification Process (SGP) and its Application for Production of Clean Power, Lurgi and Shell, 1990's.
15. Texaco Development Corporation, Texaco Gasification Process for Gaseous or Liquid Feedstocks, Texaco Image Services Tulsa, January 1993.

16. Heaven, D.L., Gasification Converts A Variety Of Problem Feedstocks And Wastes, *Oil Gas J.*, 49-54, May 27, **1996**.

17. Fong, W.F.; O'Keefe, L.F., Syngas Generation from Natural Gas Utilizing the Texaco Gasification Process, Presented at the 1996 NPRA Annual Meeting in San Antonio, TX., AM-96-68 Pages 1-12, March 17-19, 1996.

18. Stellaccio, R.J., Partial Oxidation Process for Slurries of Solid Fuel, U.S. Patent No. 4,443,230, April 17, 1984.

19. Hauser, N.; Higman, C., Heavy Crude and Tar Sands – Fueling for a Clean and Safe Environment, 6th UNITAR Int'l. Confereence on Heavy Crude & Tar Sands – Houston, TX., February 12 – 17, 1995.

20. Lurgi Corporation, Integrated Gasification Combined Cycle Process (IGCC), June 1995.

21. Lurgi, Perspectives and Experience with Partial Oxidation of Heavy Residues, Presented at the Tale Ronde – Association Francaise des Techniciens du Petrole – Paris, June 28, 1994.

22. Reed, C.L.; Kuhre, C.J., Make Syn Gas By Partial Oxidation, *Hydrocarbon Process.*, 191-194, September **1979**.

23. Abbott, J., Advanced Steam Reforming Technology in GTL Flowsheets, Synetix (ICI), 1999.

24. Hicks, T., A Decade of Gas Heated Reforming, FINDS, Volume XI, Number 3 Third Quarter 1996.

25. Abbishaw, J.B.; Cromarty, B.J., The Use of Advanced Steam Reforming Technology for Hydrogen Production, Presented at the 1996 NPRA Annual Meeting – San Antonio, TX., AM-96-62 Pages 1-15, March 17-19, 1996.

26. Farnell, P.W., Commissioning and Operation of ICI Katalco's Leading Concept Methanol Process, Prepared for presentation at the AIChE Ammonia Safet Symposium – Tucson, Arizona, September 18-20, 1995.

27. Farnell, P.W., Commissioning/Operation of Leading Concept Methanol Process, Ammonia Technical Manual, 268-277, 1996.

28. Goudappel, E.; Herfkens, A.H.; Beishuizen, T., Gas Turbines for crude oil heating and cogeneration, PTQ Spring 2000, 99-109.

29. Shahani, G.; Garodz, L.; Baade, B.; Murphy, K.; Sharma, P., Hydrogen and Utility Supply Optimization, *Hydrocarbon Process.*, 143-150, September **1998**.

30. Hairston, D., Hail Hydrogen, Chemical Engineering, 59-62, February 1996.

31. Farnell, P.W., Investigation and Resolution of a Secondary Reformer Burner Failure, Prepared for Presentation at the 45th Annual Safety In Ammonia Plants and Related Facilities Symposium – Tuscon, Arizona – Paper no. 1D, September 11-14, 2000.

32. Shahani, G.H.; Gunardson, H.H.; Easterbrook, N.C., Consider Oxygen for Hydrocarbon Oxidations, *Chem. Eng. Prog.*, 66-71, November **1996**.

33. Jung, C.S.; Noh, K.K.; Yi, S.; Kim, J.S.; Song, H.K.; Hyun, J.C., MPC Improves Reformer Control, *Hydrocarbon Process.*, 115-122, April **1995**.

34. Baade, W.F.; Patel, N.; Jordan, R.; Sekhri, S., Integrated Hydrogen Supply-Extend Your Refinery's Enterprise, AM-01-19 Presented at the NPRA 2001 Annual Meeting – New Orleans, LA., March 18-20, 2001.

35. Grant, M.D.; Udengaard, N.R.; Vannby, R.; Cavote, C.P., New Advanced Hydrogen Plant in California Refinery by the Topsoe Process, AM-01-27 Presented at the NPRA 2001 Annual Meeting – New Orleans, LA., March 18-20, 2001.

36. Howard, Weil, Labouisse, Friedrichs Inc., GAS-TO-LIQUIDS, SOLIDS-TO-LIQUIDS, LIQUIDS-TO-LIQUIDS, Fischer-Tropsch Technology, 1-54, December 18, 1998.

37. Sanfilippo, D.; Micheli, E.; Miracca, I.; Tagliabue, L., Oxygenated Synfuels from Natural Gas, PTQ Spring 1998, 87-95.

38. Barba, J.J.; Hemmings, J.; Bailey, T.C.; Horne, N., Advances in Hydrogen Production Technology: the Options Available, *Hydrocarbon Eng.*, 48-54, December/January **1997/98**.
39. Dybkjaer, I.; Madsen, S.W., Advanced Reforming Technologies for Hydrogen Production, *Hydrocarbon Eng.*, 56-65, December/January **1997/98**.
40. DaPrato, P.L.; Gunardson, H.H., Selection of Optimum Technology for CO Rich Syngas, *Hydrocarbon Eng.*, 34-40, September/October **1996**.
41. Hutchings, G.J.; Ross, J.R.H.; Kunchal, S.K., *Methane Conversion*, 1-81, 1994-97.
42. ICI Group, ICI Catalysts for Steam Reforming Naphtha, ICI Group 72W/033/1/CAT46, 1-19, 46 Series.
43. Brown, F.C., Alternative Uses for Methanol Plants, PTQ Spring 2000, 83-91.
44. Mii, T.; Hirotani, K., Economics of a World Class Methanol Plant, PTQ Spring 2000, 127-133.
45. Carstensen, J., Reduce Methanol Formation in Your Hydrogen Plant, *Hydrocarbon Process.*, 100-C – 100-D, March **1998**.
46. LeBlanc, J.R.; Schneider III, R.V.; Strait, R.B., *Production of Methanol*, Marcel Dekker, Inc., 51-132, Copyright 1994 by Marcel Dekker, Inc.
47. Parks, S.B.; Schillmoller, C.M., Improve Alloy Selection for Ammonia Furnaces, *Hydrocarbon Process.*, 93-98, October **1997**.
48. Cromarty, B., Effective Steam Reforming of Mixed and Heavy Hydrocarbon Feedstocks for Production of Hydrogen, Presented at the NPRA 1995 Annual Meeting – San Francisco, CA., March 1995.
49. Martin, R.P., Nitrogen Oxide Formation and Reduction in Steam Reformers, Paper Presented at the International Fertilizer Industry Association, October 2-6, 1994, Amman, Jordan.
50. Clark, L., Controlling NOx, Posted on www.JohnstonBurner.com, April 1, 2002.
51. Garg, A., Reducing NOx from Fired Heaters and Boilers, Presented at Chemical Engineering Expo 2000 at Houston, TX, 2000.
52. Kunz, R.G.; Smith, D.D.; Patel, N.M.; Thompson, G.P.; Patrick, G.S., Air Products and Chemicals, Inc., Control NOx from Gas-Fired Hydrogen Reformer Furnaces, Paper presented, AM-92-56.
53. Yokogawa, Boiler Blowdown, Analytical-SIC4900-02, Posted on www.Yokogawa.com.
54. Harrell, G., Boiler Blowdown Energy Recovery, Energy Matters Newsletter, Winter 2003.

Chapter 26

HYDROGEN: UNDER NEW MANAGEMENT

Nick Hallale,[1] Ian Moore,[1] and Dennis Vauk[2]
1. AspenTech UK Limited, Warrington, UK
2. Air Liquide America L.P., Houston, Texas

1.   INTRODUCTION

These days, you can't be involved in the oil refining industry without coming across something on hydrogen management. However, most articles and presentations follow the same route, namely discussing the stricter fuel specs on sulfur and aromatics as well as the changing product markets. They then move on to say that more hydrotreating and hydroprocessing will be required as a result. The conclusion? Refineries are going to need more hydrogen. But are they telling the refiners anything that they didn't already know? Of course, some authors go a step further and subtly (or not so subtly) remind the refiner that he or she will need to buy more hydrogen from them or perhaps buy a hydrogen plant from them in the future. As if they really need to have it rubbed in.

This chapter aims to be different. Instead of focusing on the problems, we will look at the opportunities. The intention is to get refiners thinking of their hydrogen as an asset not a liability.

At present, refineries generally fit into one of three situations with respect to hydrogen:

Type 1: The refinery has an excess of hydrogen, which is routed to the fuel gas system. Refinery operations are not constrained by hydrogen. The price of hydrogen is based on its fuel gas value. Direct pressure-control letdowns to fuel gas have no economic penalty. In process units, high-pressure purge rates can be increased to increase hydrogen partial pressure in reactors without penalty. In these cases, catalytic reformer hydrogen is normally the only source of hydrogen supply. Typical hydrogen price (depending on marginal fuel price) is €350/tonne (January 2003 prices).

Type 2: The refinery is often "short" of hydrogen, with the catalytic reformer acting as the "swing" hydrogen producer. Refinery operations and profitability are constrained by hydrogen. Key process units "compete" for hydrogen, which is priced in the refinery LP model based on plant-wide profitability (e.g. gasoline over-production). Direct pressure-control letdowns have a high economic penalty. In process units, high-pressure purges are minimized, reducing hydrogen partial pressure in the reactors. In these cases, catalytic reformer hydrogen is normally still the only source of hydrogen supply, although Type 2 also applies when on-purpose generation or import capacity is limited and operating at maximum. Under these conditions, hydrogen value can be three times higher – €1000/tonne.

Type 3: The refinery can meet hydrogen demand through "on-purpose" hydrogen production (e.g., with an SMR) or through external import. These are the refinery "swing" hydrogen producers. Refinery operations are not constrained by hydrogen. Hydrogen is priced based on its marginal production cost or import value. Direct pressure control letdowns have an economic penalty based on the value of hydrogen as a reactant in process units relative to its value as fuel. In process units, high-pressure purge rates are optimized to balance the value of higher hydrogen partial pressures (better yields, increased capacity, longer catalyst life) against the cost of purge losses. The value of hydrogen depends on whether or not the cost of capital is included in the price. For a refinery that has pre-invested in a hydrogen plant, the marginal production cost could be €500/tonne. If hydrogen is imported from an external supplier, the cost is more likely to approach €900/tonne.

Most refiners are on a painful journey from the fondly remembered days of Type 1 to a current situation of Type 2. Larger and more profitable refineries are developing and implementing plans that take them to Type 3, while a small number may choose to stay within Type 2 because major new investment simply cannot be justified.

2.   ASSETS AND LIABILITIES

In his book, *Rich Dad, Poor Dad,*[1] best-selling author Robert Kiyosaki explains that the reason most people never become wealthy is because they don't know the difference between assets and liabilities. He keeps away from textbook accountancy and uses a definition of assets and liabilities simple enough for even chemical engineers to understand: *an asset is something that puts money into your pocket while a liability is something that takes money out of your pocket.* Many people view expensive cars, wide-screen TVs and of course their homes as assets, but do these put any money in their pocket? In fact these are all liabilities. True assets are those investments which make money for their owners, such as real estate, businesses, stocks and shares. Recognising the difference between the two is vital.

What does all this have to do with hydrogen? Well, as we will show, refiners will never get the most value from their hydrogen unless they have the correct view of it. For the most part, refiners tend to view hydrogen as a utility that has to be supplied in order for them to operate. It is a necessary evil whose cost must simply be borne, just like fuel, electricity and water. Supplying hydrogen takes money out of their pockets and so it is a liability. Right?

Wrong! As this article will show, hydrogen - if properly managed - can be an asset, something that *makes* money for the refinery. Just as there are good and bad investments, there are good and bad ways to use hydrogen in a refinery. The secret is in finding the good ways. To do this requires a willingness to question conventional wisdom and to take a wider view of the issues. This chapter will discuss some of the important tools that will help to do this. It will also discuss some of the lessons learned from industrial projects carried out by AspenTech and Air Liquide as part of their alliance called PRO-EN Services.[2]

3.   IT'S ALL ABOUT BALANCE

Accountants talk about balance sheets. Chemical engineers talk about mass balances. The principle is the same: what goes in must be accounted for. When we are dealing with hydrogen systems, the total amount of hydrogen produced and/or supplied must equal the total hydrogen that is chemically consumed, exported, burned as fuel or flared. Unfortunately, it is very rare to find a refinery where all hydrogen is accounted for. There is usually a poor hydrogen balance to begin with. Stream flow rates and compositions are often not measured and when they are, the data may contain significant inconsistencies. In many cases, large imbalances are often accepted and attributed to "leaks," "distribution losses," "meter error" and "unaccounted flow." If actual currency were involved, the accountants would probably use less euphemistic terms, such as "embezzlement," "fraud," "misappropriation" and "theft." If we take the view that hydrogen is a valuable asset, not having a decent hydrogen balance is tantamount to throwing money away or letting someone steal it. It has been our experience that significant benefits - hundreds of thousands of dollars or more per year - can be achieved by examining the hydrogen balance and finding no-cost "housekeeping" improvements. Simple as this may sound, it can only be realized by understanding the overall hydrogen system and this in turn needs systematic analysis and modelling capabilities.

To give an example, we recently performed a hydrogen system study for a U.S. refinery. The refinery supplemented the hydrogen generated by its two catalytic reformers with purchased hydrogen. Flow rate and composition data were collected from the refinery and used to set up a flow sheet balance with surprising results. We found that hydrogen worth $2 million per year was being lost to flare and fuel gas through three specific areas, one of which was a leaky valve. We also found other opportunities, including reducing the amount of hydrogen purged for pressure control and avoiding the needless re-compression of rather large gas streams.

So what is needed to set up a hydrogen balance? Firstly, flow rates, compositions and pressures must be determined at certain key points in the hydrogen system. Flow diagrams of the hydrogen-consuming processes are also needed so that reactor and separator configurations as well as recycle locations can be determined.

Secondly, models of the consumers are required. These models need not be totally rigorous ones because this would take far too long in the early stages of a project. However, they should be sufficiently detailed to capture the important operating features of the units as well as to predict their performance. AspenTech has been developing such models using fit-for-purpose simplified models for reactors, flash drums and separation columns, which can be used as building blocks for the entire hydrogen network.

Thirdly, data correction and reconciliation tools are required. During this process, many refiners are surprised to discover that their hydrogen-system flow meter readings, if properly corrected, aren't so bad after all. Think of currency exchange rates. No semi-decent accountant would try to convert South Africa Rands into U.S. Dollars at the exchange rate used in the 1970s. However, it is surprising to see how often people are happy to trust flow meters calibrated with molecular weights that are now quite different from present operation. Because the molecular weight of hydrogen is so low, small changes in stream composition can affect the stream molecular weight significantly. For example, a mixture consisting of 99% hydrogen and only 1% ethane has a molecular weight 14% greater than that of pure hydrogen. Refiners typically correct flow meter readings for temperature and pressure, but not for composition. In other words, if hydrogen were currency, they would be using outdated exchange rates. The software mentioned above can automatically correct measured stream flow rates for deviations from the calibration point. It can also carry out a data reconciliation whereby the user can enter whatever data are available and then specify how much confidence he or she has in each value. For example, a flow rate can be an accurate measurement (within 5%), an accurate estimate (within 20%), a rough estimate (within 50%) or a guess (within 500%). The software will then reconcile the values to achieve a balance while staying within the confidence bounds. *Figure 1* shows what a completed hydrogen system balance might look like.

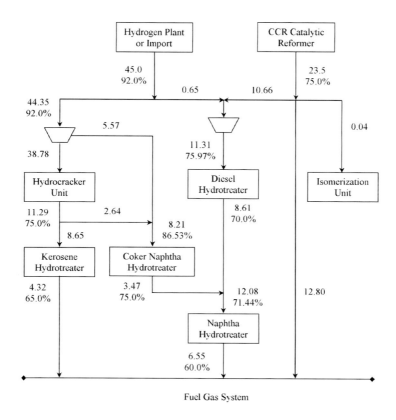

Figure 1. A typical refinery H_2 balance (flows in MMCFD, H_2 purity in mol%)

4. PUT NEEDS AHEAD OF WANTS

In *Rich Dad, Poor Dad*, Kiyosaki advises that people should not spend money on unnecessary items while trying to build a foundation for wealth. He argues that by delaying gratification now, people will be able to use their money to buy assets which build up enough real wealth so that they can afford anything they want later. The trick is to put needs ahead of wants.

The same principle applies when trying to make more money from hydrogen in refineries. Don't worry, we are not suggesting that refinery operations should have to make do without hydrogen for a few years!

However, we are suggesting that refiners question whether certain units actually need to be fed with hydrogen of very high purity. If there is a supply of hydrogen at 99% purity available, chances are that engineers or operators will claim that their unit "needs" to have 99% pure hydrogen. It is this mindset that needs to be challenged. Once we break away from this, the scope for re-using and recycling hydrogen becomes a lot greater and benefits can be substantial.

One European refinery used a significant flow rate of very pure hydrogen for drying purposes. When questioned about this, the refinery engineers replied that this was the way it had always been done. That made sense because until recently, the refinery had a large surplus of hydrogen. Now, however, that hydrogen is worth several hundred thousand dollars per year. It would be far more sensible to use the offgas from another consumer, accomplishing the same job for free!

Admittedly, most refineries do not have such obvious savings. Most hydrogen consumers require a certain flow rate of gas and hydrogen purity in order to operate properly. The flow rate is needed to maintain a gas to oil ratio high enough to prevent coking, while the purity is required to maintain the required hydrogen partial pressure for effective kinetics. With fixed flow rate and purity demands, it may not look like there is any scope for improvement. However, the opportunities are still there if we know where to look for them.

The secret is simple: look at reactor inlets, not make-up streams. *Figure 2* and *Figure 3* illustrate this point. *Figure 2* is a diagram of a hydrogen consumer showing the make-up hydrogen and the high pressure and low pressure purges. Looking at it in this way, it is very easy to be misled into thinking that the unit requires 10,000 sm^3/hr of hydrogen at 99% purity. However, this is wrong. In order to find the real requirement, one needs to look at the internal workings of the consumer as shown in *Figure 3*.

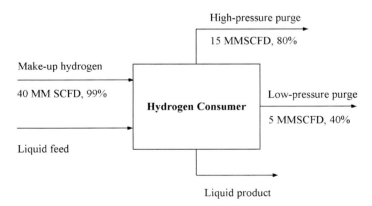

Figure 2. A black-box view can mislead us to think the consumer needs 99% pure hydroge

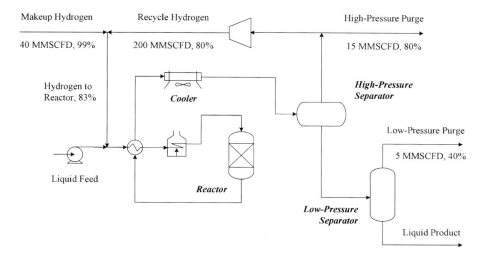

Figure 3. The true reactor inlet purity requirement is only 83%

As can be seen, the hydrogen make-up stream is mixed with hydrogen recycle before being fed to the reactor inlet. Therefore, the purity that the reactor actually sees is only 83%. Recognising this immediately opens up the scope for re-using other gases in the make-up stream and using less of the 99% hydrogen. As long as the flow rate and purity of the hydrogen going into the reactor do not change, the make-up purity is not that important. Put another way, the consumer may "want" 99% purity in the make-up, but it really needs 83% purity at the reactor inlet.

A simple example is shown in *Figure 4* and *Figure 5*. Two consumers are both taking make-up hydrogen from an external supplier at 99% purity, with a total demand of 20,000 sm³/hour. If we make the mistake of saying that the make-up purity must be fixed, there is clearly no scope at all for re-using hydrogen (*Figure 4*).

However, if we do the correct thing and focus on the reactor inlet, *Figure 5* shows that it is possible to re-use the purge from Consumer A as part of the make-up of Consumer B. This allows the demand from the external supplier to be reduced by 1710 sm³/hr or 8.7%. With typical hydrogen costs, this could be worth between half a million and a million U.S. dollars per year. The flow rates being compressed by B's make-up and recycle compressors are lower, giving power-cost savings too. Pretty good value for the cost of a pipe! Notice that Consumer B still has exactly the same flow rate and hydrogen purity at the reactor inlet as before. All that has changed are the make-up and recycle flow rates … and of course our way of thinking!

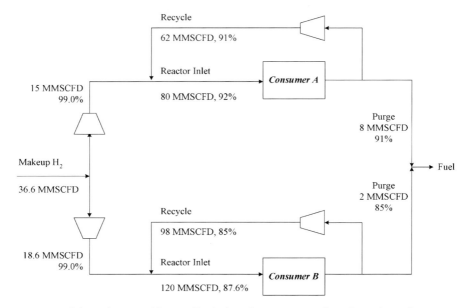

Figure 4. If the makeup purities are fixed, there is no way to reduce the makeup flow rate

Right about now you may be thinking that's all fine and well when there are only two consumers, but real refineries have a lot more than that. Should we use hydrocracker off-gas to feed the diesel hydrotreater, or should we send it to the naphtha hydrotreater? Maybe we should use cat reformer hydrogen instead of imported hydrogen, but should we use it directly or blend it with hydrogen from elsewhere? Maybe that hydrocracker off-gas should be used as fuel instead? Or maybe we could purify it in a PSA. Or why not a membrane? Since we're thinking of PSAs and membranes, maybe we should purify the naphtha hydrotreater off-gas as well? Where should we send the purified product?

At this point, many people might just throw up their hands and forget about the whole thing. What we have right now works, so why mess with it? And if we run out of hydrogen in the future we'll just buy more from a third party and be done with it. But, would you take this attitude with money? Would you be content to let someone steal or gamble away your cash each month while you keep borrowing from the bank in order to replace it? Probably not. So why do it with other assets?

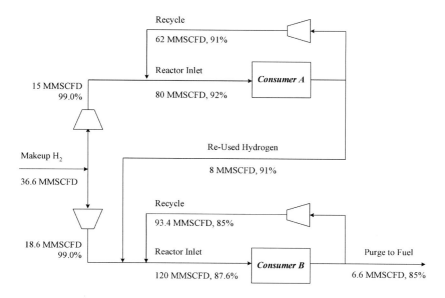

Figure 5. Focusing on reactor purity reduces required H_2 by 8.7%

Wouldn't it be neat if you had a quick and systematic way of cutting out all the fuss and knowing immediately what the maximum hydrogen recovery achievable is? Imagine being able to say to your boss that you know how to recover 15% of the refinery's hydrogen through re-piping, *before* spending any time analyzing all the options? Or perhaps you'll be able to tell him or her that only 1 or 2% can be recovered, so there is no point wasting time looking for a better configuration.

Hydrogen pinch analysis lets you do just that. The method is similar to the well-known energy pinch analysis used for designing heat exchanger networks.[3] Instead of looking at enthalpy and temperature, though, we concern ourselves with gas flow rate and hydrogen purity. The method in its original form aims at maximising the in-plant re-use and recycling of hydrogen to minimise "on-purpose" or "utility" hydrogen production.[4] Later in this article we will discuss whether this is, in fact, the correct thing to do in all cases.

The first step in the Pinch analysis is to plot hydrogen composite curves (*Figure 6*). These are plots of hydrogen purity versus flow rate for all the sources and all the demands for hydrogen in the refinery. Sources are streams containing hydrogen that potentially could be used. They include catalytic reformer hydrogen, "on-purpose" hydrogen as well as the overhead gases from high and low pressure separators in the various consuming units. When plotting hydrogen demands, remember the lesson on needs versus wants. Don't base your calculations on make-up purities!

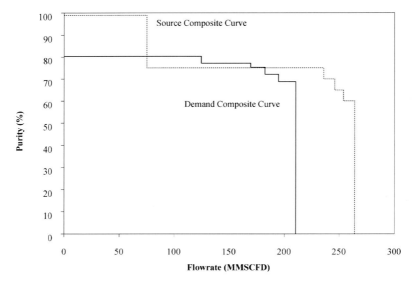

Figure 6. Composite curve showing hydrogen flow and purity

By plotting hydrogen purity versus the area enclosed between the source and sink composites, a hydrogen surplus diagram is constructed (*Figure 7*). This diagram is analogous to the grand composite curve in heat exchanger network synthesis and shows the excess surplus hydrogen available at each purity level. If the hydrogen surplus is positive throughout the diagram (as is the case in *Figure 7*), there is some slack in the system.

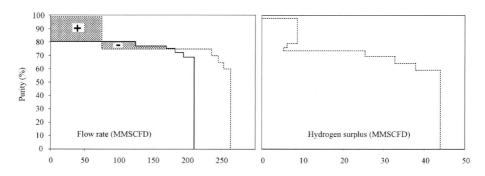

Figure 7. Hydrogen surplus diagram

The hydrogen generation flow rate can be reduced until the surplus is zero (*Figure 8*). The purity at which this occurs is termed the "hydrogen pinch," and it is the theoretical limit on how much hydrogen can be recovered from the sources into the sinks. It is analogous to the heat recovery pinch.[3] The "on-purpose" hydrogen flow rate that produces a pinch is the minimum target and is determined *before* any network design analysis. With the appropriate software, a task that would have taken days, weeks or months can be accomplished in hours. If the gap between the target and the current use is large, it is worthwhile spending time reconfiguring the hydrogen network. The key is to avoid cross-pinch hydrogen transfer and to be especially careful about the above-pinch purity, as this is the region that is most tightly constrained. Hydrogen streams with purities greater than the pinch purity should not be used to feed consumers that can use hydrogen below the pinch purity. Also, hydrogen streams above the pinch purity should not be sent to the fuel system or flared. If the gap is not large, your time would be better spent looking at other improvement options, such as purification.

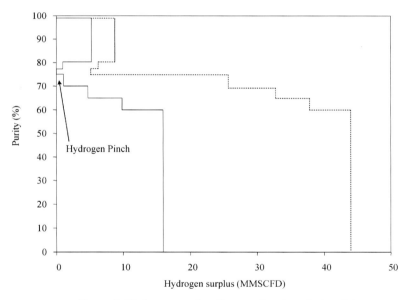

Figure 8. Hydrogen surplus diagram showing pinch point

There are three main options for purification. In case you are wondering, these are *not* PSAs, membranes and cryogenic systems. Those processes all do fundamentally the same thing, which is to split a feed stream into a product stream with a high purity and residue stream with a low hydrogen purity. The options we are talking about relate to the placement of the purifier relative to the pinch. As *Figure 9* shows, there are three possible placements: above the pinch, below the pinch and across the pinch.

Placing the purifier below the pinch is not a wise idea. This simply takes hydrogen from a region where it is in excess and purifies it before sending it back to the same region. In essence, you would be buying an expensive unit to purify hydrogen for burning!

The best option is to place the purifier across the pinch. This moves hydrogen from a region of excess to a region that is tightly constrained on hydrogen. It frees up hydrogen from "on-purpose" sources, and any hydrogen lost to residue would have ended up in the fuel system anyway.

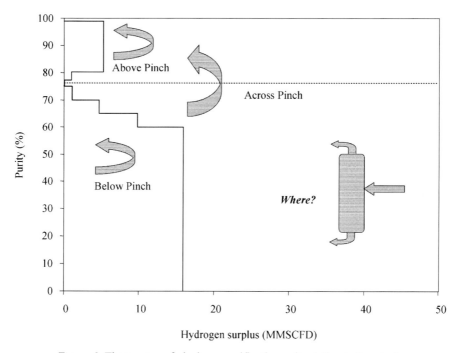

Figure 9. Three ways of placing a purification unit relative to the pinch

5. BEYOND PINCH

The pinch analysis approach discussed above is good for getting an immediate overview of the system, setting targets, and even doing some initial screening of ideas. However, in our project work, we have found a number of limitations.

Firstly, it does not fully cope with all the complexities of network design. The two-dimensional representation only considers flow rate and purity, but does not incorporate other important practical constraints such as pressure, layout, safety, piping, operability and of course capital cost.

One of the more important constraints is stream pressure. The targeting method assumes that any stream containing hydrogen can be sent to any consumer, regardless of the stream pressure. In reality, this is only true if the stream has a sufficiently high pressure. Thus, Pinch analysis targets usually are too optimistic. They can encourage refiners to waste time developing projects that are infeasible or far too expensive due to the need to install compressors. Compressors are among the most expensive capital items in any chemical or refining process. Therefore, a sound retrofit design should make best use of existing compression equipment. Often, the economic feasibility of a hydrogen recovery project is determined by bottlenecks in existing compressors and not by purity and flow rate constraints alone.

Obviously, direct re-use of hydrogen between consumers is only possible if the pressure is sufficient. However, it is certainly possible to re-use a low-pressure hydrogen stream indirectly, i.e. by routing it through an existing compressor, if certain conditions are met. Firstly, there has to be sufficient capacity in a compressor to accommodate the stream; re-using hydrogen will change the make-up and recycle flow rates throughout the system, so capacity might be available in one or more other compressors. Also, the pressure of the re-used stream must satisfy the inlet-pressure requirements of the target compressor. In addition, the discharge pressure of the compressor should be high enough for use in the target consumer.

To account for pressure and other important constraints, a mathematical programming or optimization approach is required. Work on this was started at the University of Manchester Institute of Science and Technology (UMIST), and AspenTech has taken the lead in further development.[5]

5.1 Multi-Component Methodology

At this point we should be able to begin process engineering of the proposed re-use project. However, if we were to do so we would find that the simulated benefits are very much lower than those predicted by the optimiser. This is due to another major limitation of the hydrogen pinch approach, its assumption of a binary mixture.

Consider two hydrogen streams, each of 85 mol% purity. The first is ethylene plant export, containing almost 15% methane. The second is catalytic reformer export, containing roughly equal amounts of methane, ethane and propane, plus small amounts of heavier material. Hydrogen pinch techniques cannot differentiate between these streams, and would identify no penalty or benefit from switching between them as a source of make-up gas. Yet in reality the ethylene plant gas would require operation with a much higher purge rate, due to tendency of methane to build-up in recycle loops. In fact, in certain circumstances it is more efficient to substitute a make-up supply with a lower purity hydrogen source (but lower methane content), while the hydrogen pinch methodology would lead you to do the opposite.

To meet this challenge, we developed a multiple-component optimization methodology that fully accounts for the behaviour of individual components within the process reactors, separators and the recycle gas loop. While in the binary pinch approach the composition and flows of reactor feed and separator gas are fixed, our multiple-component approach allows these compositions to float, so long as constraints such as minimum hydrogen partial pressure and minimum gas-to-oil ratio are met. Simulation models for reactors, high-pressure and low-pressure separators are used to correctly model overall process behaviour. We also developed a network simulation tool based on AspenTech's Aspen Custom Modeler™ (ACM) software.

By extending our analysis to include components other than hydrogen, we can simultaneously optimise downstream amine scrubbers, LPG recovery systems and the entire fuel gas system. Our project experience shows that the benefits from increased LPG recovery can outweigh the value of hydrogen savings.

5.2   Hydrogen Network Optimization

While preparing the optimization software, we first set up a superstructure that connects every sink with every source, provided that the source pressure is greater than or equal to the sink pressure. Compressors are included as both sources and sinks. We formulate basic constraints such as balances on total flow rate and hydrogen flow rate, as well as any compressor limitations such as maximum power or maximum throughput. A whole host of other constraints can also be incorporated, such as limited space, no new compressors allowed, and the old favorite: "I don't want to spend any capital!"

Next, we subjected the superstructure to mathematical programming that eliminates undesirable features while satisfying an objective function, which could be minimum hydrogen generation. There is actually no need to limit ourselves to minimising hydrogen generation. For example, we could choose to minimise operating cost or total annual cost. All relevant costs can be considered, including hydrogen cost, compressor power cost, fuel gas credit and capital cost of new equipment. We will not bore you with all the details, as these can be found elsewhere.[5] Suffice it to say that the optimization tool uses a combination of linear and non-linear programming.

The ability to add constraints at will means that all practical considerations can be built in. The cost of adding a constraint can also be determined. For example, an engineer may say he or she doesn't like the idea of running a long pipe across a road to connect two units. This option can be banned and the optimization carried out again. The difference in cost between the two solutions indicates how much his or her "not liking" the connection will cost. If it costs a million dollars per year, the engineer may decide that he or she doesn't actually mind it *that* much after all!

The other major limitation of pinch analysis is that it gives fundamental guidelines about purification placement, but does not always help with selecting which streams to purify or whether to use a PSA, membrane, cryogenic or other purification process. This is where know-how becomes vital. Purifier experts can rapidly assess different technologies for candidate feed streams, simulate their performance and give quick cost estimates. These experts can weigh the benefits of different options, recognising that each application will be unique. For example, in a recent European study, a refinery was facing a large increase in hydrogen demand to meet upcoming sulfur-in-fuel specifications. In this case, hydrogen recovery was more important than product purity, so Air Liquide experts determined that, for a certain stream, a membrane would be a better choice than a PSA unit.

Another issue with purifiers is that different technologies have very different pressure requirements. A PSA unit gives a product pressure very close to the feed pressure, but the PSA residue pressure is extremely low. On the other hand, a membrane requires a large pressure drop in order to perform properly, the product pressure is much lower than the feed pressure, and the residue pressure is almost the same as the feed pressure. These issues need to be considered in the specific context of the refinery and can be handled easily using the approach described above. Ultimately, economics should determine the optimum trade-off between product purity, product pressure and hydrogen recovery.

Purification expertise also saves time. Hydrogen network analysis experts work in parallel with purification experts to rapidly assess options and develop a project flow sheet. This is much more time-effective than having one company generate a small number of options using Pinch analysis and then sending the options to another company for cost quotes. With the parallel approach, more options are evaluated in less time, and both the network analysis engineers and the purification engineers see the entire picture.

The following case study is representative of a real refinery system, but certain data have been disguised to maintain customer confidentiality. The existing hydrogen system is shown in Figure 10. The objective of the study was to retrofit the network to minimise operating costs. Process and cost data are given elsewhere.[56]

Several constraints were imposed by the refinery:
1. The existing compressors have 5% spare capacity.

2. There is only space on the site for one new compressor and one new purification unit.
3. A payback of more than 2 years will not be acceptable

The network retrofit was designed by setting the objective function to be minimum operating cost while constraining the payback time (capital cost divided by annual operating cost savings) to be two years or less. Figure 10 shows the resulting design. Dotted lines indicate new equipment.

In the recommended project, both a new compressor and a PSA are used and substantial re-piping is required. Notice that the new compressor accommodates the increased recycle requirement for the NHT as well as the need to compress one of the feeds to the PSA so that its product can be used in the hydrocracker.

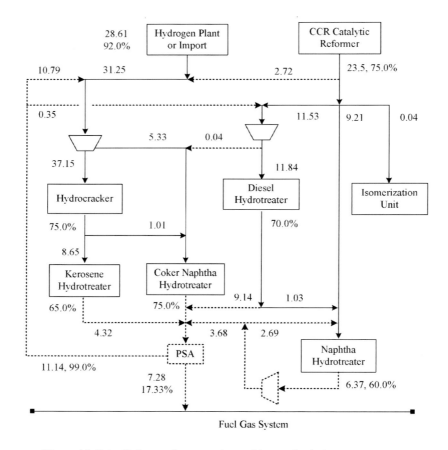

Figure 10. Retrofit for maximum savings with a payback time ≤ two-years

The total capital investment of the retrofit is $9.8 million and the operating-cost savings are $6 million per year. The payback period is therefore 1.6 years, which is better than required.

Now, refineries often have limited capital budgets, so even modification with a good payback might be too expensive. It would be valuable to know the maximum savings achievable with a fixed capital budget, say US$5 million. Adding capital expenditure as an additional constraint and re-optimizing gives the solution shown in Figure 11. The best investment is a PSA unit requiring no new compressor. Notice too that fewer new pipes are installed. The operating cost savings are smaller (only $3.5 million per year) but this is to be expected.

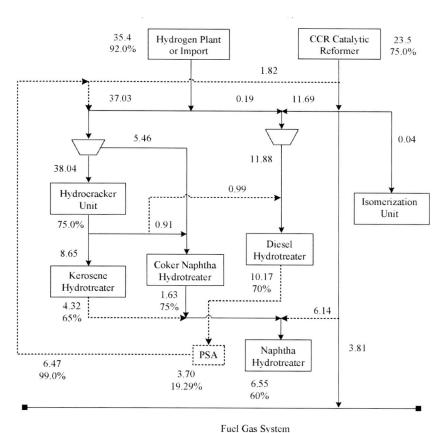

Fuel Gas System

Figure 11. If capital expenditure is limited to US$5 million, the design changes

6.   YOU DON'T GET RICH BY SAVING

Up to now, this article has addressed the issue of minimising hydrogen supply, in other words, saving hydrogen. This is certainly a very real problem for refiners, and many of our clients have specified this as their objective. After all, hydrogen is expensive, and a few percent saved can mean millions of dollars per year. However, this is often not the most profitable course of action.

As *Rich Dad, Poor Dad* puts it, nobody has ever become rich by saving money. How many people have become millionaires by opening a savings account? If you put your money in the bank, all you can expect is a modest increase from the interest it earns. When you account for time and inflation, money in the bank actually loses its value gradually. According to Robert Kiyosaki, the rich have their money work for them by investing it in assets that generate passive income. Such investments include real estate, which generates rental income, capital gains, and businesses that generate revenue and dividends.

If you agree that hydrogen is money, let's go a step further and look at hydrogen consumers as investments. Instead of merely *saving* hydrogen, why not consider re-investing it where it will *make* money for you?

How do we re-invest hydrogen? By feeding more and/or purer hydrogen into the appropriate reactors. For example, the hydrogen freed up using the network design methods can be used to process cheaper feed stocks. It could also be used to boost partial pressures to enhance conversion, throughput, yields and catalyst life.

The important question is how do we know which units will be the most profitable ones to invest hydrogen in? When analyzing money investments, there are certain techniques that wise investors use. These include performing a due diligence and analyzing the financial statements and cash flows of the investment. A good return on investment is the goal. Nobody wants an investment that loses money, and likewise a refiner should use hydrogen where it will provide a better return than simply saving it would give. We have our own techniques for analyzing hydrogen investments. They are reactor modelling and refinery LP modelling.

Rigorous kinetic modelling is used to obtain a good understanding of process operation under different feed hydrogen conditions. AspenTech has developed rigorous tools for modelling fixed-bed hydroprocessing units, such as hydrocrackers, reformers, FCC pretreaters, and desulfurization units. They model kinetics for denitrogenation, desulfurization, saturation, and cracking, and are capable of accurately predicting yields, hydrogen consumption and product properties for widely different feeds and operating conditions. These reactor models can easily be connected to rigorous fractionation models, creating a fully integrated model of the entire hydrotreater, reformer or hydrocracker complex.

The reactor models include a unique catalyst deactivation tool, which allows refiners to minimize catalyst life giveaway and to calculate future conversions, yields, and product properties. Hydrocracking models can be used to optimize tradeoffs between feed rate, conversion, catalyst cycle life, feedstock severity, operating conditions and costs. For recycle hydrocracking units, there are tradeoffs between fresh feed rate and conversion-per-pass in single-stage units, or between 1st stage and 2nd stage conversion in two-stage units.

It is not necessary to model every reactor and determine the benefits in a trial and error way. A combination of the LP model, understanding of current process operations and constraints, and hydrogen network analysis can be used to develop a short list of key processes and potential changes to those processes. For example, the LP model will show the bottlenecks to increasing refinery profit. If increasing hydrogen partial pressure can eliminate one of these bottlenecks, then this unit will be a candidate for applying rigorous reactor modelling and subsequent process analysis. Hydrogen network analysis can also give insights into process operations. Above a threshold hydrogen purity, hydrotreater operation is insensitive to hydrogen purity. If reducing the hydrogen partial pressure to a hydrotreating reactor gives a large saving in the overall refinery hydrogen target, this unit can be also be selected for rigorous modelling to determine the true impact on reactor operation.

All process changes need to be modelled, the impact on the hydrogen network evaluated, and the final benefit established through the LP model. The reactor models can be linked to Aspen PIMS, a leading PC-based LP software package used by the petroleum and petrochemical industries. By providing automatic updates to Aspen PIMS vectors, the models can enhance detailed operations planning, economic evaluation and scheduling activities. In particular, they can help a refiner decide how best to apportion intermediate heavy distillate streams between different conversion units.

As an illustration, *Figure 12* shows case-study results from a rigorous model for a two-reactor hydrocracker with partial recycle of unconverted oil.[7] The unit runs at about 60% conversion-per-pass. A fixed flow of unconverted bottoms is recycled. The remainder is exported to the FCC unit or heavy diesel blending. At present, the "export" comprises about half of the total unconverted oil.

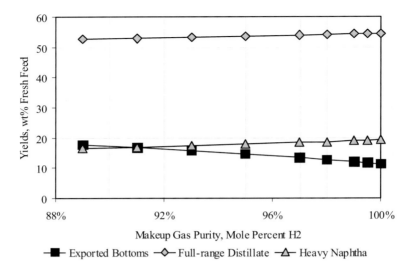

Figure 12. Hydrocracker performance improves with increased hydrogen purity

For the case study, the purity of the make-up hydrogen was varied from the current value (89%) to 100%. The feed rate, recycle oil flow rate, and weighted average reactor temperature (WART) for both reactors was held constant. Under these conditions, the following effects were noted:

1. Full-range distillate yield increases from 52.8% to 54.6%
2. Heavy naphtha yield increases from 16.6% to 19.4%
3. Exported bottoms decreases from 17.8% to 11.3%.

According to the model, the net benefits of increasing the makeup-gas purity form 89% to 100% were on the order of $2 million per year. But how much would it cost to make available the extra hydrogen needed? Using a simple marginal cost of hydrogen is not the way to answer this.

Figure 13 shows how our network design and analysis tools can fit into an overall profit optimization study. They identify quickly how much additional hydrogen can be made available for certain levels of cost. Typically, a few percent more hydrogen can be squeezed out with simple modifications requiring low or no cost, for example piping modifications. Then there will be a step-change where getting any more hydrogen will require a purification system. Finally, there will be another step-change where purification reaches its limit and any more hydrogen will have to be supplied from external sources, such as imports or a hydrogen plant. Knowing the true cost of providing additional hydrogen makes it easy to weigh it against the benefits that are suggested by process modeling and LP optimization.

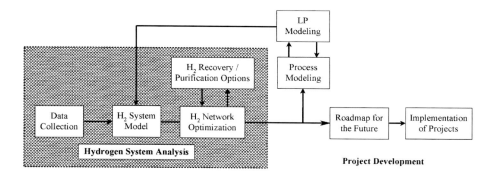

Figure 13. Network design tools compare costs and benefits for project options

Figure 14 shows a typical road-map that can be drawn up, showing where the refinery is now and where it wants to be at different points in the future (e.g. one year, five years, ten years). All of the tools described in this article give the refiner an easier job of planning how to tackle the future. They are used to systematically choose the best set of projects that achieve immediate objectives but also fit in with long-term goals. This avoids the "Regret Capital" syndrome.

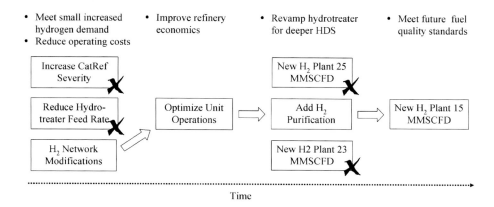

Figure 14. Systematic tools are used to construct a roadmap for the future

7. CONCLUSIONS

In conclusion, several hidden opportunities typically are not addressed by refiners when developing plans for Clean Fuels. In our experience:

Time spent in investigating the current balance pays back. This can identify substantial operating savings, and it eliminates "unconscious over-design" in future investments.

Make the best use of hydrogen purity. There is scope to reduce both capital investment and future operating costs by maximizing the benefits from high-purity hydrogen sources. Reactor modelling is an important part of evaluating these benefits and optimizing hydrogen use.

Hydrogen is not the only consideration. A proper methodology for optimization must extend beyond the hydrogen system to include low pressure purges, amine treating, LPG recovery and the fuel gas system. Purification upstream of LPG recovery units can debottleneck those units and increase their overall recovery efficiency, adding significantly to the value of the hydrogen-recovery project. Reducing hydrogen loss to fuel can have both positive implications (improved calorific value) and negative implications (increased fuel oil firing and hence increased SOx emissions).

New model-based network optimisation methodologies allow the identification of no- and low-CAPEX projects to improve current operations, particularly for refineries of Types 2 and 3. These savings are often in excess of €1 million/year, with payback times measured in months.

If there is one conclusion you should take away with you, it should be that hydrogen is money. Stop thinking about it as merely an unglamorous utility and start looking at ways to make more money from it. And who knows? By applying these ideas to your personal finances you well may end up becoming a rich dad (or mum) too. We are still working on it!

8. REFERENCES

1. Kiyosaki, R.T. and Lechter, S. L. *Rich Dad, Poor Dad*, Warner Books: New York (2000)
2. Cassidy, R.T.; Petela, E. "Life Cycle Utilities Management," NPRA Annual Meeting, Paper No. AM-01-63 (2001)
3. Linnhoff, B. "Use Pinch Analysis to Knock Down Capital Costs and Emissions," *Chem. Eng. Progress*, 90 (8), pp. 33-57 (August 1994)
4. Alves, J. "Analysis and Design of Refinery Hydrogen Distribution Systems," PhD Thesis, University of Manchester Institute of Science and Technology (1999)
5. Hallale, N.; Liu, F. "Refinery Hydrogen Management for Clean Fuels Production," Advances in Environmental Research, 6, 81-98 (2001)
6. Robinson, P.R.; Thiessen, J.M.; Hanratty, P.J.; Mudt, D.R.; Pedersen, C.C. "Plan For Clean Fuel Specifications With Rigorous Models," NPRA Computer Conference, Paper No. CC-00-143 (2000)

Chapter 27

IMPROVING REFINERY PLANNING THROUGH BETTER CRUDE QUALITY CONTROL

J. L. Peña-Díez
Technology Centre
Repsol-YPF
P.O. Box 300
28930 Móstoles - Madrid (Spain)

1.   INTRODUCTION

Near-term refining trends are demanding continuous improvements to current practices to ensure the profitability to survive in highly competitive global markets. To the common complexity of refinery processes, and the effect of crude oil market volatility and low and cyclic business margins, increasing environmental regulations have made companies face to the need to continuously operate assets at or near the economic optimum.

Refining companies are successfully implementing supply chain integrated solutions to solve these problems and improve refinery operation and profitability. However, these applications depend on the availability of accurate and live data. Some points of the supply chain—in particular planning and operation areas—are greatly influenced by the availability of reliable and updated crude oil libraries reflecting the information of the crude oil and the products that will be obtained in the different processes.

Although the consistency of estimating crude oil relative values with gravity and sulfur-based valuations is still defended[1], there has been a clear tendency in the oil industry towards product-based valuations, considering the value of the products than can be obtained from a crude oil in a particular refinery. The relative value for the same crude oil may differ from one refinery to another, and LP models are commonly used to maximize crude oil value optimizing the product distribution.[2]

LP crude oil valuation is specific to a refinery and its facilities, markets and constraints.[3] The combination of LP methodologies and refinery simulation models provides the required process and economic data to adequately evaluate crude oil feedstocks. Consistency between the quality of received crude oils and the assay data used with planning LP models is critical to guarantee the success in the optimization process. If crude oil quality suffers significant variations, a continuous assay update is required to take advantage of the benefits offered by supply-chain technologies.

This chapter will briefly review some of the new techniques proposed for crude oil quality control, and an alternative new approach will be presented.

2.   CRUDE OIL QUALITY CONTROL

Crude oil quality changes with time, although the effect of these variations in planning uncertainties may be significantly different. To illustrate the effect of these quality changes, Figure 1 shows the historical API variation of several representative crude oils.

- The first group of crude oils (A1 and A2) shows a relatively constant quality with time, that makes them desirable from the planning point of view.
- The most common situation is shown in crude oils B1 and B2, with a clear trend in quality with time. This trend will force a periodical update of the assay in the database.
- The third group (crude oils C1 and C2) exhibit significant variations in quality with time without any trend, making planning a very difficult task due to the uncertainty associated to the crude oil quality to be received in the next cargo.

Assays of the crude types B and C will need frequent revisions in the crude oil assay libraries, but the nature will be different. With crude oils from group B there is a continuous need to update the crude assay in the databank, which can be planned following two possible criteria: on a fixed delta time basis or on a delta quality basis, which is the usual procedure.

Crude oils from group C present a more serious problem. A considerable effort in updating crude assays can be made without any improvement, due to the random quality variations. It is very difficult to define the appropriate moment to update the assay, and most of the times different qualities are included in the crude oil library to represent different scenarios

The problem is that these quality variations affect both the planning and operation processes. If planning and operation tools are not updated to reflect the changes in crude oil quality in the form of LP delta vectors for each crude, the model may converge in a solution different from the optimal, and transfer this uncertainty to the operation in the refinery. Increasing differences are expected to appear between planning and operation.

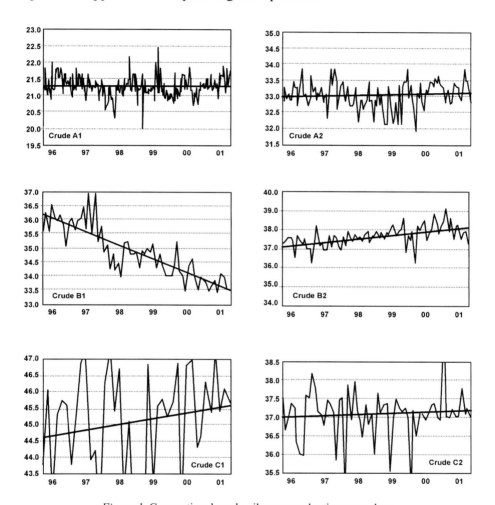

Figure 1. Conventional crude oil assay evaluation procedure

An increase in the quality of the received cargo does not assure a higher profitability, because LP models optimize crude mixtures according to the constraints defined, and perhaps the quality increase will be offset in operation to a specification that can not be reached with the planned yield/profit.

Some authors suggest an existing tradeoff between improving LP predictions with rigorous models and the cost associated to continuously update the crude assay libraries. A conventional crude oil assay evaluation is long and costly, and alternative methods are needed to take full advantage of the possibilities offered by these supply-chain solutions.

3.   NEW TECHNOLOGIES IN CRUDE OIL ASSAY EVALUATION

Crude oil evaluation methods have not significantly changed during the last decades. The crude oil is distilled in batch columns to determine the hydrocarbon boiling point distribution (ASTM D2892/5236), and the fractions obtained are analyzed to study the physical properties distribution through the crude oil. Although each company defines the required characterization for the cuts, Figure 2 shows the information commonly included in the assay. Some details of these conventional tests are described in Chapter 3.

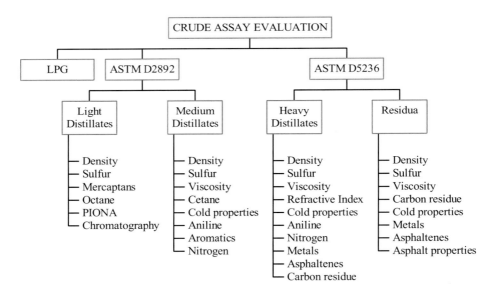

Figure 2. Conventional crude oil assay evaluation procedure

Once the laboratory work is finished, the preliminary assay obtained must be checked with computer assay synthesis models to assure consistency in the distribution of each property between the crude oil and the fractions.

The whole procedure is long and expensive (average 300 man-hours), and must be repeated each time a crude oil quality variation is suspected. Although there are no fixed rules about the recommended assay updating frequency (differences are illustrated in Figure 1), a maximum average assay age of 2-3 years has been proposed.[5] If the refinery has a limited crude basket this assay updating process is not a problem, but in most cases this maintenance implies a significant investment for the company.

Several references have suggested the need for alternative crude oil evaluation methods. These alternative methods (both analytical and mathematical) should allow keeping the assay information updated under a reasonable cost.

3.1   Analytical Methods

The new analytical developments appeared during the last decade are mostly focused on a faster determination of the distillation curve (TBP). HTSD is widely accepted as one of the best alternatives.[6-9] Originally developed as an extension of conventional GC simulated distillation (ASTM D2887/D5307), it provides better information on the crude oil back ends. The technique is currently under study to become an ASTM standard, overcoming the problems related with reproducibility of front and heavy end information.

Infrared spectroscopy[10,11] and NMR[12] are other alternatives proposed, more focused in on-line feed characterization and advanced control projects.

The main disadvantage of the previous methods is that they do not generate fractions to be characterized, and the updated information on product yields may be too limited for LP purposes, although complementary methods have been developed.[7]

GC-MS techniques have also been extensively used as alternative methods for assay evaluation. The method can be applied to the whole crude[13,14] or with a previous SFC fractionation.[15] The possibility to predict product properties relies on the availability of reliable property-composition correlations. Although there are commercial applications[16], the method is not standardized and problems have been reported on the accuracy of heavy fractions characterization and uncertainty in component identification and quantification due to mass overlapping.

Details of these alternative analytical methods for crude assay can be found in Chapter 3.

3.2   Chemometric Methods

Some recent references have demonstrated the power of statistical techniques applied to crude assay prediction.[17,18] The availability of large crude assay databases integrated with advanced statistical methods allows the prediction of updated crude oil assays or isolated physical properties.

Most of these technologies are based on neural networks models, and also provide an error estimate for the predictions, allowing the user to validate the updated crude oil update and quantify its benefits over the existing one. Usually a previous assay of the crude oil is required to improve accuracy in the predicted assay.

3.3   Other Alternatives

The acquisition of a commercial crude oil assay library with periodical updates is a common alternative, which could help refiners solve the planning problem appearing with crude oils type B (Figure 1). However, this approach is not effective with crude oils type C), and the cost is a factor to be considered.

Recently proposed web-based assay sharing systems provide similar benefits, but without the confidence on the availability of the required crude at the required time.

4.   CRUDE ASSAY PREDICTION TOOL (CAPT)

With the basis of the mentioned requirements and the shortage of commercially available tools, a new technology was developed with the objective to update/complete crude oil assays with limited experimental data using rigorous models. This new technology, registered under the name repCAPT (Repsol-YPF Crude Assay Prediction Tool) was intended to complement conventional evaluation procedures.[19]

The center of this technology is an expert system for crude assay generation. This expert system, which includes a set of mathematical algorithms based in powerful data-mining techniques and first-principle models, may generate a complete crude oil evaluation from variable levels of information available.

The methodology has been designed with flexible inputs, allowing different levels of information to be used in the predictions. It is possible to supply incomplete evaluations from different sources, with the purpose to fill the incomplete information in the crude oil assay according to the standard fractions and properties defined.

With this technique crude oil library quality control can be made at virtually no cost, selecting only specific crude oils to be evaluated in the laboratory and optimizing the information needed for LP planning.

4.1   Model Description

The generation of the assay takes places in two stages. The first one is the generation of preliminary TBP and physical properties distribution through predefined fractions, while the second is the analysis of these properties with first principle models to assure consistency in the assay.

The first stage of the expert system relies on the underlying relationship among thermophysical properties in crude oils. Figure 3 shows the weak correlation between API gravity and sulfur content for crude oils (conventional valuation method, and therefore available for all the cargos), arising from the natural distribution of sulfur compounds in light and heavy crude oil fractions.

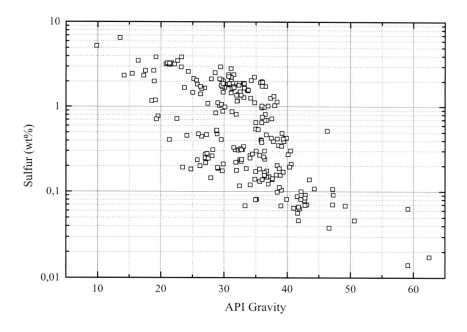

Figure 3. Correlation between crude oil API gravity and sulfur

However, with a third variable (kinematic viscosity at 40°C, Figure 4), the correlation among variables increases, indicating that they are not totally linearly independent. A further increase in this multivariate analysis also shows not only a significant degree of correlation between the bulk properties of a crude oil and the different fractions yields and properties, but even with any other underlying information. One illustrative example is included in Figure 5. A simple principal component analysis of a crude library using basic bulk properties (crude oil density, sulfur, viscosity and pour point) may identify additional information that was not explicit, in this case the geographical origin of the crude oil.

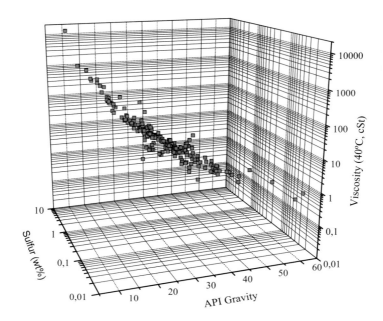

Figure 4. Correlation between crude oil API gravity, sulfur and kinematic viscosity

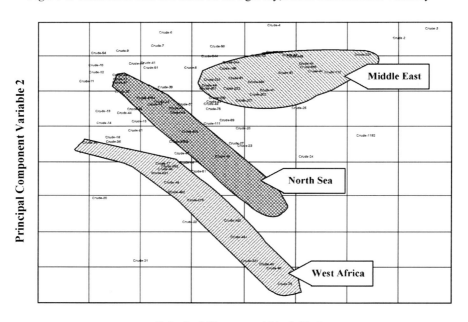

Principal Component Variable 1

Figure 5. Principal Component Analysis of crude oils bulk properties

From the operational point of view, only a minimum set of crude oil bulk properties will be available to generate the updated assay. This is the less favorable case for the application of the model, and minimum required properties must be carefully selected.

According to the generalized correlations proposed by Riazi and Daubert[20,21], based on two-parameter equations of state and the theory of intermolecular forces, a minimum of two physical properties are required to characterize non-polar petroleum fluids, one representing the molecular size and another one the molecular energy (or chemical structure). Although most of the published correlations are expressed in terms of boiling point and density, in the application to crude oils it is more appropriate to replace boiling point with kinematic viscosity as an input parameter, unless a reliable simplified distillation procedure is available. Crude oils and heavy petroleum fluids present some degree of polarity and a third bulk property is required to adequately represent the fluid (sulfur content, pour point, etc.).

Although crude oil density, sulfur and viscosity are the minimum typical bulk properties required by the model, sometimes optional laboratory analysis may be required to improve the accuracy of the predictions. The methodology is independent of the available information for the crude oil, and partial crude oil assays can be used. The output information is always a complete updated assay.

One of the expert system requirements is the availability of a homogeneous user-supplied reference crude assay library. The first versions of the technology included predefined correlations to generate preliminary properties. Due to the continuous incorporation of new and updated experimental information to the assay libraries, an on-line correlation development algorithm was implemented in later developments. This algorithm, based in crude assay data-mining through classification neural networks, selects the most appropriate information from the crude oil library to generate the best correlations suitable for the new crude oil. This new approach has greatly simplified the algorithm maintenance, and has allowed us to extend the original technology from crude oils to any intermediate product in the refinery.

The next stage is the validation of the updated assay with first principle models. Most of the petroleum fractions obtained from natural crude oils exhibit a thermodynamic behavior similar to hydrocarbon homologous series, and the consistency can be checked with asymptotic behavior correlations.[22-23]. Figure 6 illustrates the ABP behavior for MeABP and liquid molar volume in several crude oils. Similar consistent correlations can be applied for most of the PVT and transport properties of the crude oil.

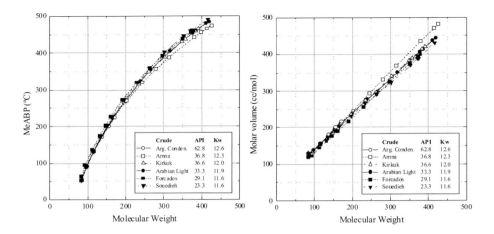

Figure 6. Asymptotic behavior correlations (ABC) for average boiling point and liquid molar volume of crude oil fractions

Property balances with bulk values are checked and consistency analysis of property values in the fractions is performed until the updated assay is validated. When the validation process is finished, the reliability of the updated assay, matching the quality of the received cargo, will be improved for LP optimization.

4.2   Potential Applications

The implementation of this technology may provide benefits in planning and operation areas compared with the conventional approach. Some of the benefits expected in the planning area include:

- Extended crude oil quality tracking system.
- Updated assays from existing crude oil libraries.
- Generation of preliminary full assays for new crude oils based on limited laboratory data. The availability of a previous assay is not required.
- Optimization of required laboratory assay updates.
- Conventional crude manager tools: historical crude oil library, crude oil analysis and comparison and mixtures, etc.

The availability of real tank characterization in the operation area allows a more accurate updated planning including final adjustments in crude blending.

Although the technology has been initially implemented in the refining area, additional applications could also be available in the upstream and trading areas.

The system is integrated with unified libraries, updated through the intranet, compiling all the crude oil assay information available. Instant access is provided to crude assay updates, and software interfaces assure integration with other supply chain tools available in planning and operation areas (Figure 7).

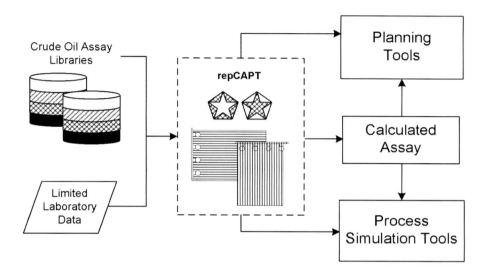

Figure 7. CAPT integration with planning and operation tools

4.3   Model Results

The following figures illustrate some representative results of the implementation of repCAPT technology. Figure 8 shows the satisfactory results obtained predicting pure crude oils. Two of the crude oils (i and iii) had not been previously analyzed, and there were no previous assays in the crude oil library. The other two (ii and iv) has a previous analysis in the library, and the updated predicted assays matched the quality of the new cargo better than the existing assays. Although all the predictions were considered satisfactory, the results present a slight higher accuracy when a previous assay was available.

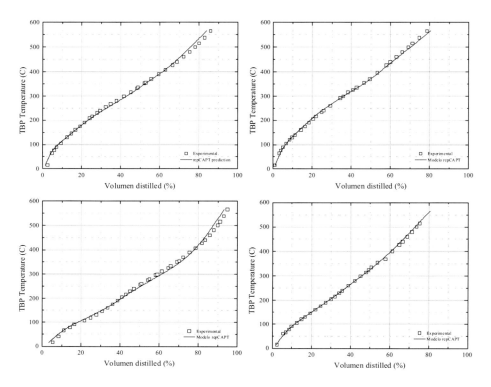

Figure 8. Comparison of predicted TBP for different crude oils

Similar results were also obtained when a real on-tank crude mixture was predicted.[19] The minimum laboratory analyses previously defined were used in all the predictions.

Physical properties are predicted with similar good accuracy, as shown in figure 6 for the crude oil fractions.

Figure 9 illustrates the deviations between experimental and predicted fraction properties. The differences in fraction yields (TBP) are well within the reproducibility of the method ASTM D2892. The rest of the properties of the fractions (specific gravity, sulfur and kinematic viscosity) also present differences close to the accuracy of their analytical techniques, if we consider the propagation of TBP reproducibility error to the physical properties.

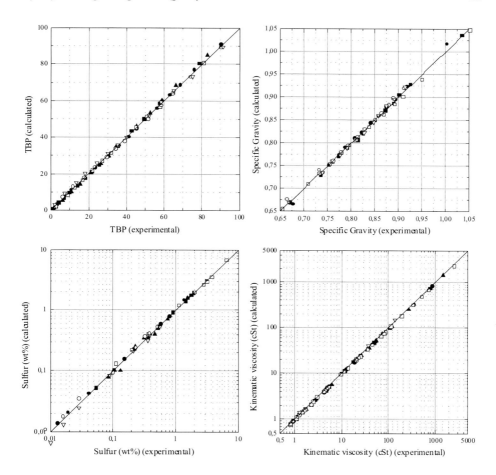

Figure 9. Comparison of experimental vs predicted properties for crude oil fractions (i – TBP, ii – specific gravity, iii – sulfur content, iv – kinematic viscosity)

The rest of the information included in a conventional assay can be predicted with similar quality, although in some cases additional laboratory measurements of crude properties (bulk metals, for example) are required to reach a higher accuracy in the reproduction of the physical properties distribution through the fractions.

5. CONCLUDING REMARKS

In this chapter we have reviewed a new methodology to generate complete and reliable crude oil assays from limited laboratory data. This technology, repCAPT, is an example of the synergies that can be obtained between IT technologies and rigorous chemical engineering models.

The technology continuously updates crude oil libraries with limited cost, allowing supply chain models to optimize refinery planning and operation with reduced uncertainties. Trading areas could also benefit from better crude oil valuations with the updated characterization.

Further research can be expected from the application of repCAPT to Advanced Control Projects. This promising alternative relies on available on-line analyzers for the bulk crude oil physical properties required by the prediction system, or their replacement/supplement in future developments by alternative on-line techniques (NIR, NMR).

Another potential application not covered in this chapter is the extension to the upstream area. Some preliminary results are available in adapting repCAPT to predict the quality of new reservoirs based on the information available for crude oils from the same geographical area. This could turn repCAPT in a valuable tool to generate preliminary quality data for new wells.

6. REFERENCES

1. Pavlovic, K.R., "Gravity and Sulfur-based Crude Valuations More Accurate Than Believed", *Oil & Gas J.,* **1999**, 97(47), 51-56.
2. Di Vita, V.B.; Trierwiler, L.D., "The New LP", *ERTC Computing 2003, Milan, Italy, June 23-25, 2003.*
3. Hartmann, J.C.M., "Crude Valuation For Crude Selection", *Petroleum Technology Quarterly* (www.eptq.com), March 2003
4. Ferritto, T., "Crude Oil Properties and their Effect on Linear Programming", *Crude Oil Quality Group Meeting, Denver, CO, February 2, 1999.*
5. COQG. "Panel Discussion On Crude Quality and Refining", *Crude Oil Quality Group Meeting, Houston, TX, September 27, 2001*
6. Villalanti, D; Maynard, J.; Raia, J.; Arias, A., "Yield Correlations Between Crude Assay Distillation and High Temperature Simulated Distillation (HTDS)", *AIChE Spring National Meeting, Houston, TX, March 9-13, 1997.*
7. Alexander, D., "How To Update An Old Assay", *Crude Oil Quality Group Meeting, New Orleans, LA, January 30, 2003.*
8. Thompson, T. "High Temperature Simulated Distillation", *Crude Oil Quality Group Meeting, Houston, TX, October 2, 2003.*
9. Song, C; Lai, W.C.; Reddy, K.M.; Wei, B. "Temperature-Programmed Retention Indices for GC and GC-MS of Hydrocarbon Fuels and Simulated Distillation GC of Heavy Oils" In *Analytical Advances for Hydrocarbon Research*, C.S. Hsu (Ed.), Kluwer Academic/Plenum Publishers: New York, 2003.
10. Hidajat, K; Chong, S.M., "Quality Characterization of Crude Oils by Partial Least Square Calibration of NIR Spectral profiles", *J. Near Infrared Spectrosc.,* **2000**, 8, 53-59.
11. Lambert, D.; Descales, B.; Llinas, R.; Espinosa, A.; Osta, S.; Sanchez, M.; Martens, A., "On-line Near Infrared Optimisation of Refining and Petrochemical Processes", *NIR-95 Conference, Montreal, August 6-11, 1995.*
12. Kane, L., "HPIn Control: Consider MRA", *Hydrocarbon Processing,* **2002**, 81(7), 17.

13. Hsu, C. S.; Drinkwater, D. "Gas Chromatography-Mass Spectrometry in the Petroleum Industry" In *Current Practice of Gas Chromatography-Mass Spectrometry*, W. M. A. Niessen (Ed.), Marcel Dekker: New York, 2001.

14. Roussis, S.G.; Fedora, J.W.; Fitzgerald, W.P.; Cameron, A.S.; Proulx, R. "Advanced Molecular Characterization by Mass Spectrometry: Applications for Petroleum and Petrochemicals" In *Analytical Advances for Hydrocarbon Research*, C.S. Hsu (Ed.), Kluwer Academic/Plenum Publishers: New York, 2003.

15. Yoshimoto, N; Kato, H., "Development of Advanced Evaluation Method for Crude Oil", *International Symposium on the Advances in Catalysis and Processes for Heavy Oil Conversion", San Francisco, CA, April 13-17, 1997*

16. Bunger, JW and Associates, Inc., "Z-Basic^{TM} Crude Oil Assay", www.jwba.com.

17. Morgan, N., "Crude Oil Intelligence", Re*view - The BP Technology Magazine*, **2001**, April, 20-21.

18. Davis, W., "The Assay Simulator – Update Assays Based On Monitoring Data"*, Crude Oil Quality Group Meeting, New Orleans, LA, January 30, 2003.*

19. Peña-Díez, J.L.; "Improved Refinery Planning and Operation Through Better Feedstock Characterization", *AspenWorld 2002, Washington,D.C.,October 27 – November 1, 2002.*

20. Riazi, M.R.; Daubert, T.E., "Simplify Property Predictions", *Hydrocarbon Processing*, **1980**, 59(3), 115-116.

21. Riazi, M.R.; Daubert, T.E., "Characterization Parameters for Petroleum Fractions", *Ind. Eng. Chem. Res.*, **1987**, 26(4), 755-759.

22. Riazi, M.R.; Al-Sahhaf, T., "Physical Properties of n-Alkanes and n-Alkyl Hydrocarbons: Application to Petroleum Mixtures", *Ind. Eng. Chem. Res.*, **1995**, 34(11), 4145-4148.

23. Marano, J.J.; Holder, G.D.; "General Equation for Correlating the Thermophysical Properties of n-Paraffins, n-Olefins, and Other Homologous Series. 2. Asymptotic Behavior Correlations for PVT Properties ", *Ind. Eng. Chem. Res.*, **1997**, 36(5), 1895-1907.

INDEX

Advanced Fluid Technology, 143-144
Advanced Process Control, 247-255
    Benefits, 250-253, 255
    Economics, 249-250
AGC-21, 102
Alkyl Benzenes, 126
Alkyl Naphthalenes, 128
Ammonia, 352-354
Aromatic Esters, 121-123
Aromatics, 150
Asphaltenes, 157-163, 209
    Chemical Structure, 159-160, 163
    Mesophase, 162, 172-173
    Pendant-Core Model, 162
    Precipitation, 175-178
    Properties, 157-159
    Thermal Chemistry, 160-163
Automatic Model Building, 196-198
Automation Infrastructure, 254
Autothermal Reforming (ATR), 330-331
β-Scission, 194-195
Bitumen,
    Composition, 150-154
    Processing, 149-181
CANMET Hydrocracking Process,
    162-163, 179
Catalysts,
    Amorphous Surface, 88
    Shape Selective, 88
Catalytic Dewaxing, 88
Chemometrics, 224
Clean Fuels,
    Models, 261-278
    Planning, 272-278
Coke Formation, 172-174, 178
Cold Cranking Simulator (CCS), 81
Corrosion of NMP Plants, 30
Crude Assay, 396-405
    Analytical Methods, 397
    Chemometric (Mathematical) Methods,
    397-398
    Conventional Evaluation Procedure,
    396-397
    Lube, 11-12
    Prediction Tool (CAPT), 398
    Quality Control, 394-396
CSTR, 230-231
Deasphalting, 4, 7, 174-175

Dehydrogenation, 192
Deprotonation, 192-193
Dibasic Esters, 107, 118-119
DISTACT Distillation, 176
Deoiling, 7, 59-63
Dewaxing (Solvent), 4, 6, 31-70, 90
    Cold Wash, 50-55
    Dewaxing Aids, 67-69
    DICHILL, 35-36, 43-45
    Filters, 45-50
        Hot Washing. 55-56
    Ketone, 6, 34-63
    Propane, 6, 63-67
        Asphaltene Contamination, 69
    Role of Solvent, 32-33
    Solvent Recovery, 57-58
Electronic Transfer Mechanism, 167
Engine Oils,
    Additives, 131-136, 138-143
    Detergents, 131-134
    Dispersants, 131-132, 134-136
        Mannich-type, 139
    Performance Chemistry, 137-138
Extraction-Hydroconversion, 99-101
FCC Models, 209-210, 262-266
Flue Gas Emission, 351-352
Formulated Lube Oils, 2
Fouling, 157, 171, 174, 178
    Resistance in Steam Reformers,
    299-300
    Tendency, 176
Gas-to-Liquids (GTL), 80, 83, 101-103
Hard Wax, 59
H-Oil Process, 163
Heavy Oil,
    Composition, 150-154
    Processing, 149-181
Heteroatom-Containing Compounds,
    154-157
High Temperature Shift Converter
    (HTSC), 341, 342
Hot Stage Micoscopy, 171-175
Hydride Shift, 194
Hydrocracker Model, 266-272
Hydrocracking,
    Combined with Hydrodewaxing, 97
    Heavy Paraffins, 187-203
    Kinetic Modeling, 187-203

Paraffin, 189
   Reaction Mechanism, 191
Hydrodesulfurization (HDS),
   Alternative Approaches, 155
Hydrofinishing, 4, 6-7
Hydrogen,
   Donors, 174, 179
   Economic of Production, 359-366
   Management, 371-392
   Network Optimization, 384-387
   Production and Supply, 323-367, 388
   Synthesis, 281-322
   Thermodynamics, 324-326
Hydrogenation, 192
Hydroisomerization, 89-91
Hydroprocessing, 4, 83-85
Hydrotreating on FCC Performance,
   274-278
Incompatibility, 178
Inhibition Reaction, 195
Ion Exchange Chromatography, 154, 172
IsoDewaxing, 89
Kinetic Modeler's Toolbox (KMT), 188,
   191
Kinetics Lumping, 187
   Bottom-Up Approach, 206-207, 211
   Mathematical Approaches, 208,
     220-224
   Partition-Based, 205-207, 209-219
   Structure-Oriented-Lumping (SOL),
     215-217
   Top-Down Approach, 206, 209-210
   Total, 205-207, 224-241
Linear Free Energy Relationship, 189-190
Low Temperature Viscosities, 107
Lube Business Outlook, 9
Lube Crude Assay, 11-12
Lube Feedstock (Crude) Selection, 9
Lubricant (Lube Oil) Basestocks, 1-5
   Composition, 4-5
   End Uses, 8-9
   Feedstock Selection, 9
   Group I, 2, 83
   Group II, 3, 83
   Group III, 3, 83
   Group IV, 3, 83
   Group V, 3, 83
   High Viscosity Index, 81
   Manufacturing, 1-70
     Hydroprocessing, 2
     Solvent Processing, 1-2
   Oxidative Stability, 92, 107
   Properties, 3-4
   Synthetic, 105-107

   Volatility, 81-82, 107
Lube Crude Assay, 11-12
Lube Oil Feedstocks, 4
   Selection, 9-11
Lube Plants, 8
Maltenes, 103, 209
MAXSAT, 98-99
Methanol, 352-354
Methyl Shift, 194
Micelles, 158, 160-161
MLDW, 188-189
Mobil 1, 108
Models,
   Hydrogen Plant-Wide Optimization,
     281-322
   Refinery-Wide Optimization, 257-278
   Steam Reformer, 282-283
   Validation, 304-311
Model-Preditive Control, 252-254
Modeling,
   Graph Theory, 188, 202
   Linear Free Energy Relationship,
     189-191, 217
   Mechanistic Kinetic for Heavy
     Paraffins, 187-203
   Monte Carlo, 188
   Petroleum Reaction Kinetics, 205-242
   Quantitative Structure-Reactivity
     Correlation (QSRC), 188-191, 198,
     207, 211, 217
MSDW, 89, 90
MSDW-2, 97
Multi-Grade Passenger Vehicle
   Lubricants, 80
   Additives, 80
   Evolution, 80
Nitrogen Compounds, 156
Olefin Copolymers (OCP), 138-141
Paraffins, 149
   Conversion, 88-91
Partial Oxidation (POX), 324-325,
   330-341
Peptization Ratio, 179
Performance Systems, 146
Petroleum,
   Compatibility, 175-180
   Composition, 149-150, 157
Phosphate Esters, 126
Pipestill Troubleshooting, 20-22
Plants,
   Carbon Monoxide, 332
   Hydrogen, 332
     Environmental Issues, 350-354
     Models for Optimization, 281-332

Modern, 342
Performance Improvements,
  357-359
Performance Monitoring, 354-357
Old Style, 341-342
HYCO, 332, 333
Conventional Solvent Lube, 4, 8
Synthesis Gas, 332, 333

Plug Flow Reactor, 229
Polyalkyleneglycol (PAG), 108-109,
  123-125
Polyalphaolefins (PAO), 107-118,
  139-141
Applications, 116-118
Composition, 111-112
Manufacturing Process, 112
Next Generation, 116
Properties, 111, 112-116
Synthesis, 110
Polyisobutylene (PIB), 125, 139-141
High Vinylidene, 139-141
PIB Based Dispersants, 134
Polymerization, 178
Polyol Esters, 107-109, 120-123
Pour Point, 107
Pressure Swing Adsorption (PSA) Unit,
  342
Process Models, 205
Propane Deasphaltening, 4
Protonated Cyclopropane (PCP)
  Isomerization, 194
Protonation, 192-193
Raffinate Hydroconversion (RHC), 85, 94,
  99
Refinery Planning, 393-394
Refinery-Wide Optimization, 259-261
Resid Thermal Cracking, 209
Ring Conversion, 85-88
Combined with Hydroisomerization-
  Hydrofinishing, 96-99
Ruthenium Ion Catalyzed Oxidization
  (RICO), 159
SARA Analysis, 150-151
Saturation, 91-95

Selective Hydrocracking, 88
Short Path Distillation (DISTACT), 176
Solvent Deasphaltening (see
  Deasphaltening)
Solvent Dewaxing (see Dewaxing)
Solvent Extraction, 4, 6, 22-31
Analytical Tests, 30-31
Characteristics of Good Solvent, 24-25
Process, 25-30
Solvent Lube Processes, 5-7
Steam Methane Reforming (SMR), 324,
  326-331, 333-350
Combined with Oxygen Secondary
  Reforming (SMR/O_2R), 330-336,
  333, 334
Design Parameters, 343-350
Steam Reformers, 282-283, 313-322
Catalyst Activity, 318-322
Catalyst Poisoning, 294-295
Coking, 292-294
Heat Balance, 295-296
Heat Loss, 302
Heat Transfer Rates, 295, 297-302
Oxo-Alcohol Synthesis Gas, 317
Pressure Drop, 302-303
Steam Reforming Kinetics, 283-291
Butane, 309-311, 315
Methane (SMR), 283-286, 324
Naphtha, 286-291, 305-308, 314
Structure-Oriented Lumping (SOL), 85
Succination Chemistry, 139-140
Suncor, 257-258
Supercritical Fluid Extraction (SFE), 152,
  171
Synthetic Lubricant Oils, 105-129
Thermal Diffusion, 87, 89-90
Total Acid Number (TAN), 133-134
Total Base Number (TBN), 133-134
Upgrading Chemistry, 163-171
Vacuum Distillation, 4, 6, 12-20
Visbreaking, 179
Viscosity Index (VI), 22-23, 80, 106
Wastewater, 354
Water Gas Shift (WGS), 325-326
Wax Deoiling (see Deoiling)